Springer Water

The book series Springer Water comprises a broad portfolio of multi- and interdisciplinary scientific books, aiming at researchers, students, and everyone interested in water-related science. The series includes peer-reviewed monographs, edited volumes, textbooks, and conference proceedings. Its volumes combine all kinds of water-related research areas, such as: the movement, distribution and quality of freshwater; water resources; the quality and pollution of water and its influence on health; the water industry including drinking water, wastewater, and desalination services and technologies; water history; as well as water management and the governmental, political, developmental, and ethical aspects of water.

Abdelkader Anouzla · Salah Souabi
Editors

A Review of Landfill Leachate

Characterization Leachate Environment Impacts and Sustainable Treatment Methods

Springer

Editors
Abdelkader Anouzla
Faculty of Science and Technology
University of Hassan II Casablanca
Rabat, Morocco

Salah Souabi
Faculty of Science and Technology
University of Hassan II Casablanca
Rabat, Morocco

ISSN 2364-6934 ISSN 2364-8198 (electronic)
Springer Water
ISBN 978-3-031-55512-1 ISBN 978-3-031-55513-8 (eBook)
https://doi.org/10.1007/978-3-031-55513-8

This Springer imprint is published by the registered company Springer Nature Switzerland AG
The registered company address is: Gewerbestrasse 11, 6330 Cham, Switzerland

Paper in this product is recyclable.

Preface

The amount of waste is constantly growing due to population growth and the evolution of socioeconomic activities. Burying this waste produces leachate, a contaminated effluent created during the decomposition of organic waste and containing harmful substances like heavy metals, polyphenols, volatile organic compounds, microplastic, and microorganisms. Pathogens: These effluents emit unpleasant odors associated with ammonium ions. These discharges significantly impact the environment.

The juice created when waste is fermented, called leachate, presents a challenge when creating and managing a landfill. As landfill leachate can infiltrate through soils and contaminate streams and groundwater if not adequately collected, treated, and disposed of, it has been identified as a possible cause of soil, groundwater, and surface water contamination. In addition, groundwater and surface water can be slowly degraded due to seepage of contaminants through soil or passage into waterways. In contrast to groundwater contamination, which results from leachate infiltration and diffusion in permeable or fissured subsurface, surface water pollution by leachate may occur through overflow and liquid flow in the water system, either abruptly or gradually. Landfill leachate, a complex liquid formed within landfills, represents a significant threat to environmental resources and ecosystems due to its composition of various pollutants. Numerous scientific studies have extensively documented the potential consequences of leachate and its management strategies on soil and water quality. Nevertheless, a comprehensive evaluation of the risks associated with landfill leachate necessitates a more holistic approach that considers the immediate impacts and the indirect repercussions that may not yet have fully manifested. To illustrate, an all-encompassing leachate impact assessment should encompass not only the environmental effects but also the resulting implications on human health, emissions of greenhouse gases, and the potential effects on water security for local communities. Despite the wealth of research on landfill leachate, the current focus remains predominantly on the direct impacts of leachate, with scant attention directed toward exploring the broader environmental footprint of leachate and its corresponding management practices. In alignment with the principles of sustainable development goals, a comprehensive understanding of the ecological footprint associated with leachate management is crucial to mitigating the environmental risks posed by leachate and advancing progress toward achieving sustainability objectives. Consequently, a thorough review of existing literature has been undertaken to evaluate the extent to which current assessment methodologies offer a holistic perspective on the crucial factors, influential variables, and environmental risks linked to landfill leachate. This review aims to shed light on the gaps in knowledge and highlight areas where further research is warranted to enhance our understanding and management of landfill leachate in a sustainable manner.

This book will be a ready reckoner of recent information regarding the impact and characterization of leachate landfills, leachate valorization, and heavy metals on a single platform. As a result, the master's and doctoral academics, researchers, and students will be able to comprehend the most recent developments in municipal solid waste landfill operations, supporting their research.

This book will inspire readers on how to deal with environmental pollution problems due to leachate contamination in freshwater and agricultural soils using various technologies.

Rabat, Morocco Abdelkader Anouzla

Contents

Chapter 1
Application of Electrical Resistivity Tomography in Landfill Leachate Detection Assessment

Muhammad Syamsul Imran Zaini and Muzamir Hasan

Abstract Over the years, the use of electrical resistance tomography (ERT) has expanded significantly in engineering applications owing to the competence of the programme involving project duration, cost and data limitation aspects. This investigations was particularly relevant for ERT to assess the level of pollution from waste dumps in Simpang Renggam, Johor, Malaysia. ERT conducted an in-depth investigation using ABEM Terrameter LS 2 devices and Schlumberger protocol configurations in the research area. Furthermore, seven (7) parameter parameters for characterization of leachate have been identified and evaluated for current leachate conditions such as chemical oxygen demand, biological oxygen demand, suspended solids, hydroelectric power, ammonia hydrocarbons, trium and biodegradability ratio. Furthermore, the study was able to determine the existence of chemicals in soils at a depth of 1.5–4.0 m, particularly at low resistivity of 1–10 m due to chemically apparent resistance. Based on the research conducted, the current characteristics of the Simpang Renggam gas station were 1 633.9 mg/L (chemical oxygen demand), 137.41 mg/L (biological oxygen demand), 359.8 mg/L (suspended solids), 7.61 (hydrogen output), 385.29 (hydrogen ammonia), 117.65 (turbidity), and 0.07 (biodegradability ratio), showing that other parameters beyond the value of the local standards set out in the Environmental Quality Leaching from the dump is thought to have been caused by system deficiency in controlling and managing waste, which have been intensified by the current excessive rainfall. In retrospect, the ERT results are workable for recognizing leachate and can therefore be used by the authorities to take instant action to end the ample water disruption in the studied region.

Keywords Pollutants · Electrical resistivity tomography · Landfill leachate

---

M. S. I. Zaini (✉) · M. Hasan
Faculty of Civil Engineering Technology, Universiti Malaysia Pahang Al-Sultan Abdullah, Lebuh Persiaran Tun Khalil Yaakob, 26300 Kuantan, Pahang, Malaysia
e-mail: syamsulimran94@gmail.com

A. Anouzla and S. Souabi (eds.), *A Review of Landfill Leachate*, Springer Water,
https://doi.org/10.1007/978-3-031-55513-8_1

1.1 Introduction

Over the past few years, the effect and occurrence of natural disasters and activities around the world has increased significantly and consistently [1, 2]. Inadequate disposal and long-term accumulation are ticking time bombs, resulting in a major ecological calamity. Mineral leakage into the atmosphere as a result of improper disposal can have serious environmental consequences and harm persons [1]. Rapid urbanisation [3] is anticipated to increase worldwide garbage creation by 73%, from 2.5 billion tonnes in 2016 to 3.5 billion tonnes in 2050. Internationally, MSW output averages 1.4 billion tonnes per year (1.25 kg per day) and is predicted to rise to almost 2.20 billion tonnes by 2025. India has a population of 1.35 billion people and is predicted to generate more than 55 tonnes of solid municipal garbage per year as the population grows by 5% [4].

According to Hamzah et al. [5] and Giang et al. [6], leachate is any liquid or waste contaminated by rainwater discharge, accumulated pollutants, and trash disposal that flows into the ground and nearby surroundings. Leachate (liquid waste) is produced when trash is deteriorated due to the amount of dissolved salt and increased conductivity [7–10]. The volume and content of garbage might vary greatly depending on the amount of waste and water, the type and age of the waste, and the meteorological conditions. The existing flow of garbage in earth conditions has been regarded as a significant risk [11]. According to a 2013 study by Feng et al. [10], the United States alone produced more than 62 million m^3 of liquids. In 2018, China generated around 24 million m^3 of leachate [12]. It is expected that the amount of waste produced by waste landfills worldwide would rise [12]. The effect of waste leaching on water is slow and gradual. Almost 70% of trash may be biodegradable, with 10% recycled as sewage [13]. A small parts of the leachates can be appropriately handled at the leachate rehabilitation clinic, while the remainder travels through the waste border, settles in the soil, and enters the water. Leachates have a variety of effects on the aquifers. The concentration of contaminants in trash disposal varies with time, precipitation, and kind of waste disposal [8, 9].

ERT is regarded as one of the greatest techniques for geological mapping of waste disposal, as contaminated waste contains a high concentration of soluble ions capable of transferring energy loads [11, 14]. The electrical properties of landfill leftovers are relatively low when compared to clean subsurface water [15, 16], and the variation [11] of the RRE can reveal alterations and pollutants [7, 17]. Finally, ERT delivers faster, simpler, and more comprehensive coverage in wide areas by gathering trustworthy and detailed data about the underlying [18, 19]. Several academics from around the world, notably in the domains of geology and geology, have provided useful analysis for basic profile categorization [20–26]. As a result, ERT started a legal investigation into the landfill in Simpang Renggam, Johor, Malaysia.

In essence, the purpose of this inquiry was to enable those engaged in ERT's optimistic forecast to adopt a forensic leachate analysis and landfill investigation strategy. Therefore, the hole is filling with the potential to use Schlumberger's network for the assessment of the leakage through ERT surveys and another choice strategy to modernize engineering applications and provide better solutions. There have been numerous research on the use of ERT for engineering properties, but few have examined its application as a technique for analysing landfill debris. The optimum mix of the ERT inspection and the leachate indiction test can demonstrate the ERT's performance in assessing the leachate that is both energy-efficient and cost-effective.

1.2 Preliminary Assessment

1.2.1 Study Area

The study site is located at the Simpang Renggam dump site in Simpang Renggam, Kuang, Johor, at 1° 53′ 41" N, 103° 22′ 35" E. The landfill, which is used daily by the garbage dump cover, is six hectares and collects up to 600 tons of household wet waste from three areas each day and has been operating since 13 years and is used to monitor the waste. According to Zailani et al. [27], the waste is dangerous and pollutes the nearby river, which is the provenience of freshwater for the Simpan Renggam population. The waste in landfills will overflow during the monsoon season, which will affect nearby areas, including palm oil plantations near landfills (Fig. 1.1).

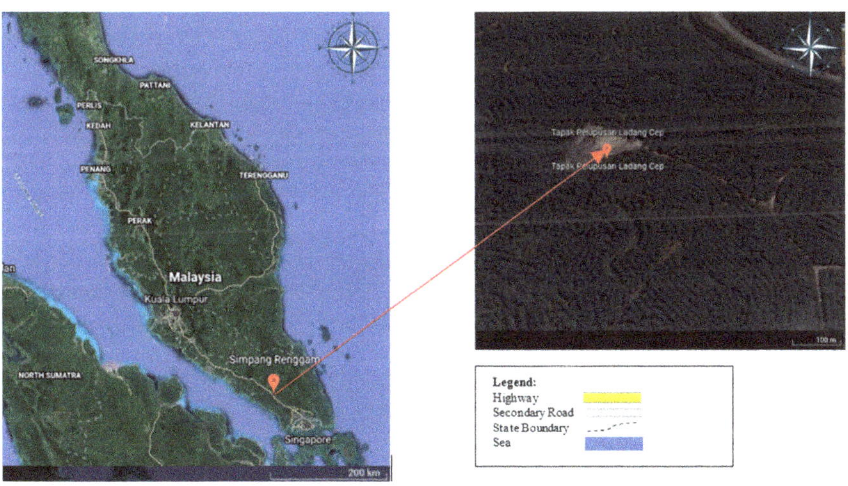

Fig. 1.1 Study location

1.2.2 *Electrical Resistivity Tomography (ERT)*

1.2.2.1 Basic Application

ERT technology is one of numerous electrical resistance technologies used in geophysical engineering to identify variations in electrical resistance in the subsurface [20–26]. This study used the ERT method because it improved the current flow in the media by the ability of liquid particles.

1.2.2.2 Equipment Description

ERT devices consist of three main parts: records, inducers, and sources [28]. The ERT battery is powered by 12 V DC batteries. As a current inducer, 61 steel electrodes were used and ABEM Terrameter LS 2 was used to evaluate and exhibit the resistance value. Finally, the primary data gathered in field studies were processed and explained through the RES2DINV program.

1.2.2.3 ERT Site Setup

The research is separated into desk research, field measurements, and data analysis. Desk research begins with the collection of important information and details about the study's area using maps and publications in order to gather correct data, such as geological location, and so on, on both global and local scales. The real measurements were then performed using ERT, and the original information from the site evaluations was assessed and analysed using the RES2DINV software. The ERT was carried out with the ABEM Terrameter LS 2 tools, and the ERI survey included two (2) distribution lines (the resistance lines RL1 and RL2) in a landfill site that is still in use, as shown in Fig. 1.2. Table 1.1 depicts electrode array and total electrical distance. Figure 1.3 depicts the site arrangement of the ERI. Furthermore, the Schlumberger array was implemented for data gathering because to the system's ability to generate extensive reasonably close resistance data.

1.2.2.4 ERT Data Collection

According to Abidin et al. [28], Schlumberger array can produce a clean image making them suitable for the detection of leaks. Throughout this approach, underground currents recorded are used to calculate resistance, and are used as important data for reverse modeling to construct a true resistance simulation of the undersea [24–26]. It is more precise when the vertical resistance of the undersoil 2D model moves along the line segment [14], and 2D measurements with the help of ERT

Fig. 1.2 RL1 and RL2 of ERT

Table 1.1 ERT field survey setup data

ERT cable	Array (m)	Takeout number	Distance (m)
RC1	5.0	20	100
RC2	2.5	20	50
RC3	2.5	20	50
RC4	5.0	20	100
Total ERT distance (m)			300

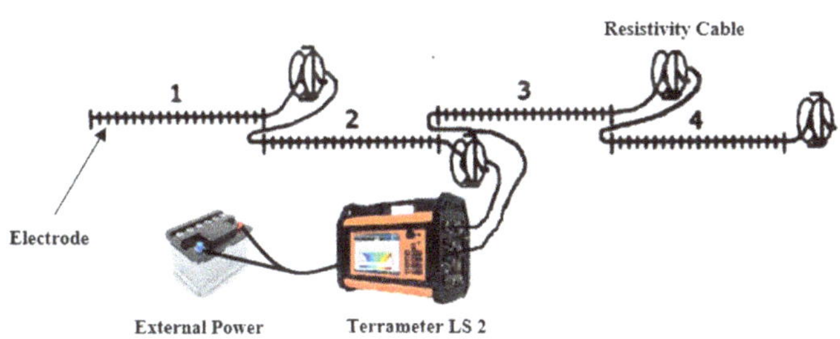

Fig. 1.3 ERT cables field survey

are obtained in a sequential sequence over a profile [20–26]. Furthermore, unprocessed data from the ERT site investigation was analysed utilizing commercially available RES2DINV programmes [22–26], resulting in an inversion model that ideally matched the actual subsurface structure. Subsurface 2D resistivity models were created from interface data acquired using RES2DINV code reversal, and RES2DINV may generate direct designs from resistivity input data using the least square technique with a linear constraint [28]. Figure 1.2 depicts the position of the entire resistance line at the Simpang Renggam landfill in Johor.

1.2.3 Basic Characterization of Soil

The soil sample was cumulated at leachate site, with samples taken to a depth of three metres. The laboratory tests used to characterise the characteristics of soil samples obtained at leachate sites includes mechanical sieve analyses to control sample particle size, pulse analysis to determine sample relative density, Atterberg limit testing to regulate soil plastics, liquids, and plasticity indexes, and standard loading tests to regulate soil moisture content (OMC) and maximum dry density (MDD).

The sieve analysis was conducted in line with BS 1377: Part 2: 1990. The size of the particles in the soil leaching samples was measured using this assay. Hydrometer testing and fine analysis were performed in line with BS 1377: Part 2:1990. This laboratory test is designed to discover the surface structure of a material using a 63-m sieve.

The relative density of the soil sample was calculated using a small pycnometer as prescribed in Part 2 of BS 1377 1990. The test used 10 g of dried samples in the oven, and passed through a 2 mm BS bag twice until the results differed by more than 0.03 Mg/m^3.

The consistency limit of the soil sample was quantified based on BS 1377: Part 2: 1990. A liquid limit (LL) is measured by using cone penetration methods for test specimens weighing at least 300 g passing through 425 micro meters, and a plastic limit (PL) is determined by measuring about 20 g of test collected from LL samples. The test was carried out to distinguish Silt from Clay.

1.2.4 Chemical Composition of Soil

The material composition of soil samples was determined utilizing the Bruker S8 Tiger X-ray Fluorescence (XRF) analyzer and fully met ASTM E1621-13 criteria. 10 g soil samples were obtained from the Simpan Rengam leach site, sealed in plastic bags, and delivered to the laboratory for testing. The 10 g soil sample was put in an XRF device, and the composition was determined. The existence of containments in soil specimens is determined by analysing its chemical makeup.

1.2.5 Sampling of Leachate

The sampling method used in this investigation was to collect a solvent sample from three distinct sites. A clean 30-L high-density polyethylene (HDPE) container was utilised to collect wastewater samples from the SRL site. To avoid extra chemical changes during the scenario, the collected samples are transferred directly to the wastewater laboratory and kept at ambient temperature in cold storage. The experimentation and analysis followed the American Public Health Association's (APHA 2012) guidelines [29]. The study looked at seven (7) parameters from the Simpang Rengam landfill: chemical oxygen demand (COD), biological oxygen demand (BOD), suspended solids (SS), hydrogen power (pH), ammonia hydrogen (NH_3-N), turbidity, and BOD5/COD ratio.

1.3 Results and Discussion

1.3.1 ERT Data

The RES2DINV tool is implemented to reverse two (2) resistivity data points acquired on the profile during site measurement, as shown in Fig. 1.2. The lines in the study area run from northeast to southwest for RL1 and southwest to northeast for RL2, and they encompass half of the waste side as well as parts of the adjoining palm plantation. Resistance values in both reverse models ranged from 0.5 to 130 Ωm, indicating areas of leachate with less than 10 Ωm or unpolluted areas with more than 10 Ωm, containing soil and freshwater from the top to the deepest depth of 35 m below the soil surface.

Resistance distribution is separated into three zones with strong apparent resistance between 45 and 110 Ωm, shwing buried garbage or waste piles at soil depths of 0–3.0 m. In the area far from landfills illustrated by RL2, the high resistance value at the top of the surface can be attributed to the sand surface above groundwater. Bai et al. [15] found that sandy and clay soils have lower apparent resistance at depths ranging from 0 to 5 m. When compared to groundwater resistance [11, 15], the electric constant of leaching is substantially smaller, and the poisons released into the atmosphere are nearly inactive at the disposal site. Most particles move through porous medium in three phases: molecular diffusion, diffusion, and mechanical dispersion [30]. Pollution penetration through the soil in landfills and flow water percolation through the soil contaminate groundwater [30], and landfill leaching is associated with an increase in electrolyte concentrations, resulting in a lower resistance value in the geological structures that contain them.

Figures 1.4 and 1.5 depict the reverse resistance models of RL 1 and RL 2. The dotted line in Figs. 1.4 and 1.5 represents the level of exploration for landfill leachate, which runs from 0.5 to 5.0 m. According to Park et al. [11], Maurya et al. [31], Moretto et al. [32], Chu et al. [33], and Ansari et al. [34], the apparent resistance at significant leaching levels is always less than 10 Ωm. Marya et al. [31] found that variations in landfill leachate ion intensity influence the low-resistance anomaly configuration. Additionally, Moretto et al. In 2017, [32], it was discovered that the ERT method can assess both low and high resistance ranges, with low resistance locations below 10 m suggesting the existence of leachate concentration zones. According to Ansari et al. [34], leachate can affect electrical conductivity, pH, and sulphur concentration. The researchers discovered that increasing pore water conductivity, pH, and sulphur

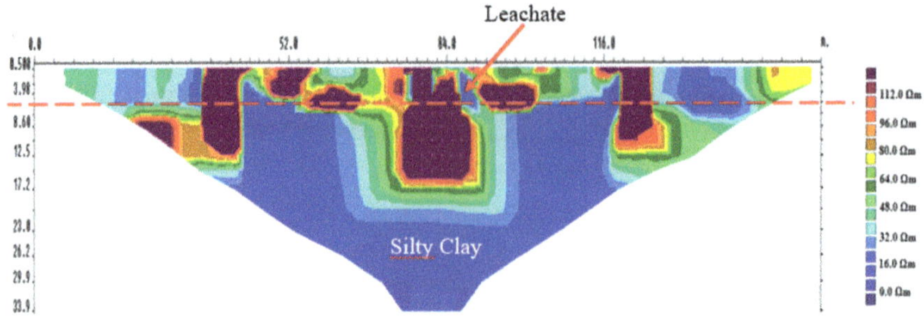

Fig. 1.4 The ERT inverse model for RL1

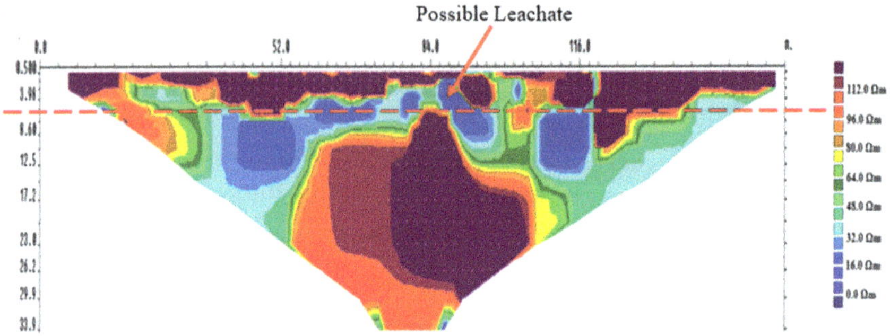

Fig. 1.5 The ERT inverse model for RL2

levels dramatically lowered sample electrical resistance. Furthermore, it helps to reduce soil sample resistance to a number of pollutants and heavy elements. Chu et al. [33], found that adding metals into soil pore fluid decreases resistance. They discovered that as water content increased, the resistivity of heavy metal-polluted soils reduced.

Coherently, severe oxidation leachate area was hardly evident in reverse modelling performed in both lines in clay sections with approximately similar electrical resistances. Resistance data from RES2DINV indicates that RL1 profiles in effective landfills have resistance values ranging from 2.0 to 9.5 Ωm, and are oriented towards the southwest boundary of oil palm cultivation. Leachate was identified at 1.5–4.0 m depth and 85.0–90.0 m distance.

A line of resistance mapping (RL2) was conducted out outside the oil palm deforestation location. RL2 resistance values ranged from 1.5 to 9.5 Ωm, indicating the presence of leachate at the site's northeast perimeter. The leak was located at 1.5–5.0 m depth and 88.0–90.0 m distant. RL2 was conveyed via the sewage tanks, where waste was dumped. Figure 1.5 shows the leachate detected in a reverse simulation of RL2 resistance.

1.3.2 *Physical Properties of Soil*

Soil samples from the Simpang Renggam Leap Site as well as other Simpang Renggam localities unrelated to the Leap Site were tested in the laboratory for physical attributes. The primary physical parameters include soil categorization, which influences various elements including as specific gravity, Atherton limitations, and particle size analysis. Table 1.2 summarises laboratory tests on the soil parameters of the Simpang Renggam wastewater site as well as Simpang Renggam's natural soil properties.

1.3.2.1 Specific Gravity

Table 1.2 reveals that the average specific gravity of soil samples obtained from the leachate site is 2.73. Because of the variability in mineralology, the specific gravity of tropical soil residuals can range from very low to very high. As shown in Table 1.2, leachate has no significant effect on the special gravity value of Simpang Renggam soil, as the difference in margin values between leachate and natural soil is tiny, with a difference of 0.03.

Table 1.2 Physical properties of soil

Parameter	Simpang Renggam leachate soil	Simpang Renggam natural soil
Natural moisture content (%)	17.95	17.86
Gravel (%)	4.2	3.9
Sand (%)	27.9	27.4
Silt (%)	45.1	44.7
Clay (%)	22.8	22.5
[a]LL (%)	70.0	38.0
[b]PL (%)	31.0	30.0
[c]PI (%)	39.0	37.0
[d]SG (G_s)	2.73	2.70
[e]V.R. (e)	0.69	0.64
Porousness (n)	0.41	0.39
[f]B.D. (g/cm^3)	1.99	1.97
[g]D.D. (g/cm^3)	1.67	1.65

[a] Liquid Limit
[b] Plastic Limit
[c] Plasticity Index
[d] Specific Gravity
[e] Void Ratio
[f] Bulk Density
[g] Dry Density

1.3.2.2 Consistency Limits

Table 1.2 shows that the cosistency limits for soil samples obtained at the sewage plant were 70%, the plastic limits were 31.0%, and the plastic index (PI = LL − PL) was 39.0%. The British soil classification system (BSCS) (MHS) classifies soil as a high-plastic SILT based on field testing. As shown in Table 1.2, leachate has no significant impact on the Atterberg boundary value of Simpang Renggam soil, as the differences in liquid boundary values, plastic boundary values, and plasticity indexes between leachate soil and natural soil are small, with boundary values of 0.02%, 0.01%, and 0.01%, respectively.

1.3.2.3 Particle Size Distribution

Figure 1.6 shows a particle size distribution at the leachate site. The British soil classification system (BSCS) classifies remaining soils as very high plastic soils (MVS) of sandy soils. 100 g soil samples had grain sizes of 4.2% larger than 2 mm (gravel), 27.9% larger than 2 mm and 0.063 mm (sand), 45.1% smaller than 0.063 mm and 0.002 mm (silt), and 22.8% smaller than 0.002 mm (clay). Figure 1.6 shows the average distribution of particle sizes from several soil generating sources gathered in the research area.

According to Table 1.2, leachate did not have a significant impact on the value of soil particle size at the site because the margins of soil samples containing gravel, sand, silt, and clay are small, with margin values of 0.3%, 0.5%, 0.4%, and 0.3%, respectively. However, the sewage is harsher than the sewage.

1.3.3 Chemical Analysis of Soil

Landfill emissions are known to impact soil because they contain toxic chemicals such as metals, chromium, arsenic, lead, copper, zinc, and salinity [1, 36, 37]. According to Salleh et al. [36], Malaysia is one of the countries that has already taken the necessary steps by implementing the Second Schedule to the Environmental Quality Act 1974. The 2009 Regulation on Environmental Quality (Containment of Pollution by Solid Waste Transfer Stations and Waste Collections) was enacted in accordance with the authority provided by articles 21, 24, and 51. Table 1.3 differentiates the chemical composition of the dump site to uncontaminated natural soil, while Table 1.4 compares Malaysian landfill discharge authorised conditions to current conditions.

On the basis of Table 1.3, hazardous metals can be found in both soils. Table 1.3 obviously highlights that the chemical contents of iron, copper, and zinc in landfill soils in Simpang Renggam increased from 126.4 g/kg, 0.108 g/kg, and 0.267 g/kg [35, 38] to 143.6 g/kg, 0.204 g/kg, and 1.256 g/kg, respectively. Furthermore, three (3) dangerous elements (Cr, Ar, and Pb) were discovered in the waste soil at

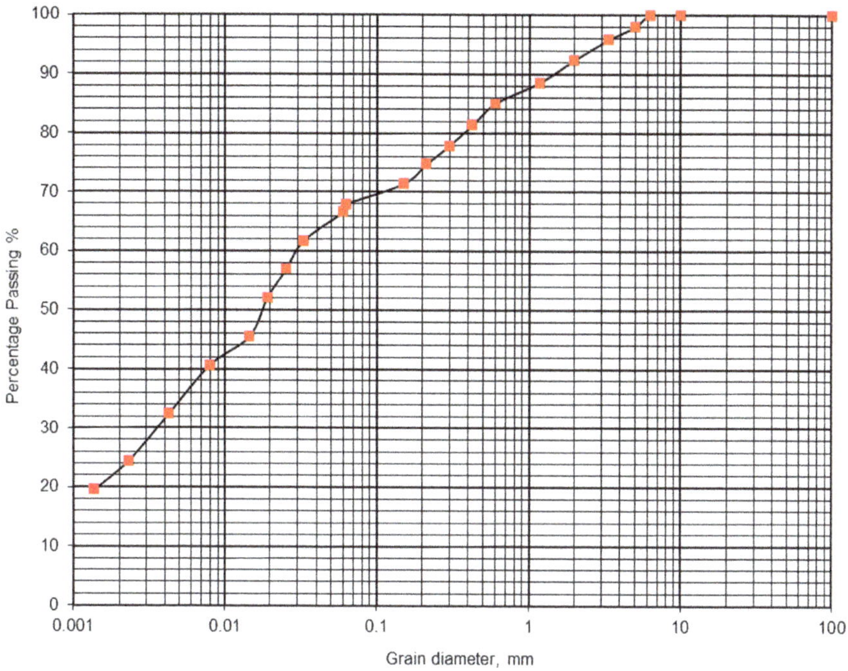

Fig. 1.6 Particle size analysis of natural soil

Table 1.3 Natural soil and leachate soil compositions

Toxic metals	Natural soil (g/kg)	Leachate soil (g/kg)
Iron (Fe)	126.4	143.6
Chromium (Cr)	0	0.236
Arsenic (Ar)	0	0.395
Lead (Pb)	0	0.059
Copper (Cu)	0.108	0.204
Zinc (Zn)	0.267	1.256

Table 1.4 Wastewater leachate compositions as compared to the EQA regulations

Toxic metals	Wastewater leachate (mg/L)	EQA regulations [39]
Iron (Fe)	8.457	5.00
Chromium (Cr)	0.453	0.20
Arsenic (Ar)	0.054	0.05
Lead (Pb)	0.110	0.10

quantities of 0.236, 0.395, and 0.059 g/kg, indicating that the current waste landfill soil is contaminated.

The study by Kadir et al. [40] exclusively analyses hazardous chemicals for Fe and Cr. According to Kadir et al. [40], Fe and Cr were found at concentrations of 16.10 and 0.623 mg/L. According to the study, the amount of heavy metals decreased over the year. Although heavy metal concentrations in leachate specimens are decreasing, they remain over Malaysia's leachate limits, especially for iron (Fe) and total chromium (Cr), which are 8.46 mg/L and 0.45 mg/L, respectively. Toxic substances accumulate in biological systems and become a substantial source of danger [1]. According to the collation, most heavy metal concentrations do not reach the high-concentration discharge standard. Because the gathered samples are unprocessed, most variables should not fall below a tolerable threshold [38]. To ensure the success of rehabilitation, sewage must be controlled according to its type and substance.

1.3.4 Leachate Analysis

Previous studies have found disparities in dumping circumstances between landfills [27, 36, 41, 42]. Table 1.5 displays the cleaning characteristics utilized in this study. As seen in Fig. 1.5, leachate contains a high quantity of NH_3-N and COD. The COD and BOD5 values are 1633 and 137.41 mg/L, respectively. Table 1.5 displays the biodegradability ratio (BOD5/COD) of fresh leaching effluent samples, which is 0.07. The investigation found that the COD and BOD5 values indicate that the leachate is successfully stable. The amounts of COD (3000 mg/L) and NH_3-N (> 400 mg/L) in stable aqueducts are substantial, although the proportion of BOD5/COD is low [27, 36, 42]. The pH of the leakage ranges from 7.50 to 9.00, with a mean of 8.25, and has been increasing over time [27, 42]. Previous research suggested that the pH of stabilised leaching was greater than 7.5. [27, 36, 40, 42].

Table 1.5 Leachate characterization

Parameters	Units	Result	EQA [39]
COD	mg/L	1633	400
BOD5	mg/L	137.41	20
SS	mg/L	359.8	50
pH	N/A	7.61	6.0–9.0
NH_3-N	mg/L	385.39	5
Turbidity	NTU	117.65	–
BOD5/COD	N/A	0.07	–

1.3.4.1 Analysis of Chemical Oxygen Demand (COD)

COD measurements, which show the aquifer's organic strength, diminish in intensity as the waste ages. Our study's COD data were from the landfill, where the average was 1633 mg/L. Previous research by Daud et al. [43] discovered that COD levels ranged from 13.166 to 13.500 mg/L, with an average of 13.333 mg/L. According to Fatihah [44], the highest COD values range from 9839 to 1568, with an average of 25,519. Other authors' COD values (such as Salleh et al. [36], Detho et al. [42], with COD values of 1993, 233.4, and 1712, respectively, indicate negligible changes, as COD levels began to fall in 2018. Data reveal that COD levels declined between 2015 and 2017 due to landfill age (> 10 years). The COD content of the leachate surpassed the 400 mg/L emission limit before it was let out into the surrounding. This is due to the properties of the leachate that were altered over time as the landfill has aged [36, 42].

Figure 1.7 depicts the inaccuracy of the data on COD concentrations in wastewater. The experimental values in the figures reveal that COD levels in sewage effluent changed dramatically between 2013 and 2015, as standard errors do not overlap and are large in magnitude. Furthermore, there were no significant differences in the COD level of wastewater discharge between 2018 and 2021, owing to overlapping standard errors and large error margins.

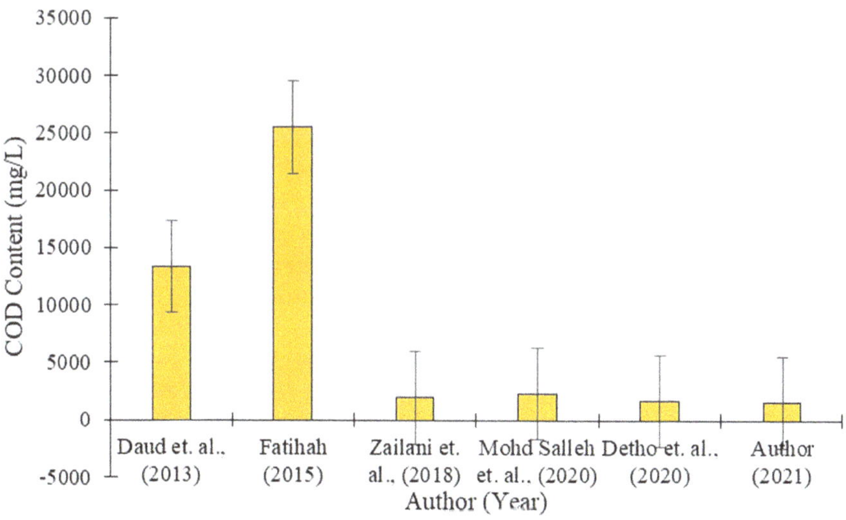

Fig. 1.7 COD value from 2013 to 2021

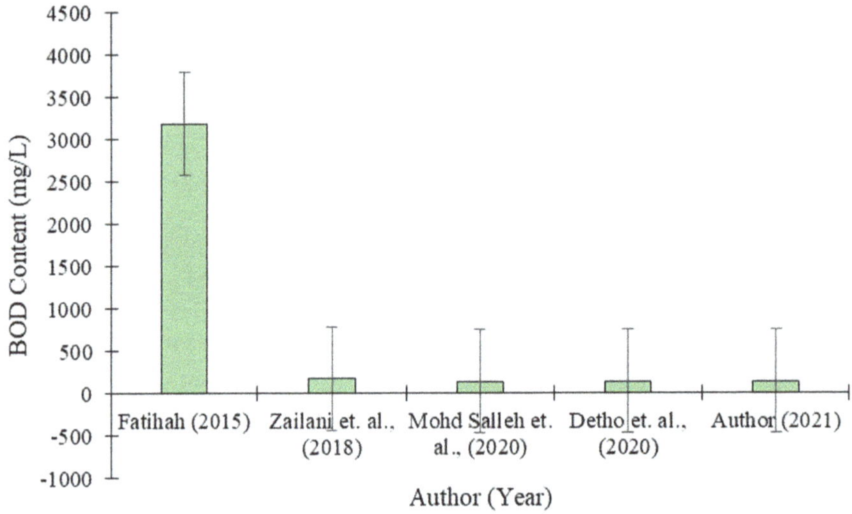

Fig. 1.8 BOD value from 2013 to 2021

1.3.4.2 Analysis of Biological Oxygen Demand (BOD)

In contrast to prior studies on the demand for biological oxygen, the new study's result was 137.41 mg/L (BOD). Fatihah [44] found that the average BOD was 3183 mg/L. According to Salleh et al. [36], mature waste landfills have BOD values ranging from 100 to 200 mg/L. BOD values are calculated using SRLS and provided in the range of 100–200 mg/L. Between 2015 and the following years, BOD levels varied dramatically, reaching 170, 139.57, 138.66, and 137.41 mg/L in 2017, 2018, 2020, and 2021. The garbage landfill was identified as mature based on the data. Compared to EQA [39], the majority of the data published by various sources between 2015 and 2021 above 50 mg/L, indicating the need for improved remediation to reduce BOD concentration. There is a considerable link between COD and BOD5 readings, which could serve as early indicators of wastewater contamination levels [36].

Figure 1.8 depicts the data inaccuracy of BOD concentrations in water effluents. The error bar in this figure illustrates that there are significant changes in BOD contents in wastewater between 2015 and 2018, as standard errors do not overlap much and are high in size. Furthermore, the BOD concentration of leaching wastewater does not differ much between 2018 and 2021, owing to strongly overlapping standard errors and large error margins.

1.3.4.3 Analysis of Suspended Solid (SS)

According to a 2013 study by Daud et al. [43], the SS level starting from 270 to 1200 mg/L, with a mean of 735 mg/L. On the other hand, Fatihah [44] reported a

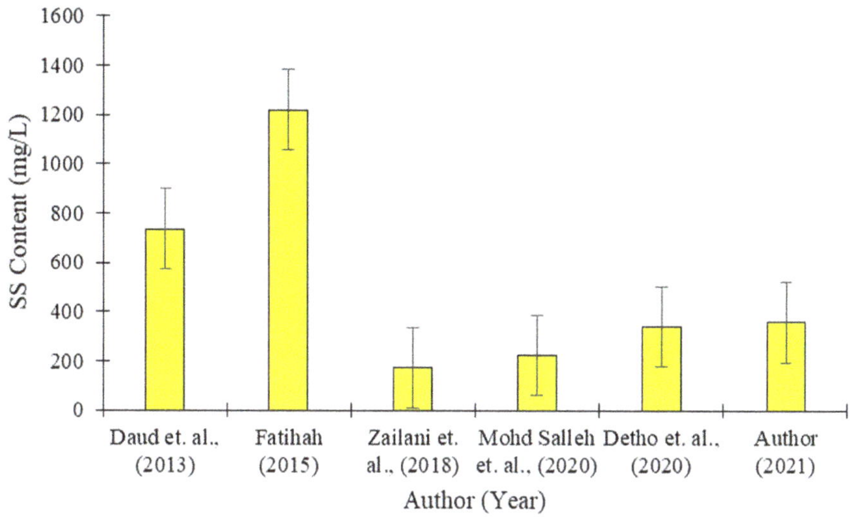

Fig. 1.9 SS value from 2013 to 2021

maximum SS of 1200–1240 mg/L and a total average of 1220 mg/L. The SS levels measured at the Wastewater Treatment Plant varied with an average of 359.8 mg/L. As demonstrated in Fig. 1.9, the SS value reduced substantially between 2015 and 2018, with values of 1220 mg/L and 173 mg/L. The academic community originally announced that the value of the SS will steadily increase between 2018 and 2021. In contrast to EQA [39], SS readings exceeded the 50 mg/L limit. In leachate investigations, SS components are one of the variables linked to biological coloid particles [36, 42].

Figure 1.9 depicts data inconsistencies in the SS content of wastewater from the levee. Data about the diversity of data sets obtained by various researchers between 2013 and 2021 may be found in SS content data. The experimental data on the graph show a considerable change in the concentration of SS in wastewater between 2013 and 2015, as standard errors do not overlap and are large in magnitude. Furthermore, the SS concentration of leachate water remains constant between 2018 and 2021, with considerable standard error overlap and a huge error margin.

1.3.4.4 Analysis of Power of Hydrogen (pH)

The previous landfill can help predict the types of waste produced. The change from a shorter to a longer period of waste decomposition involves two distinct phases: acidity and fermentation [27, 36, 42]. The pH of the dump site is 7.61. The previous conclusions were reached in 2013, when pH levels ranged from 8.31 to 8.47 [43]. The solvent's pH exceeds 7.5 and is categorised as ancient [36]. The pH of the leached residue remains in the alkali zone based on sample properties analysis (Table 1.5). The

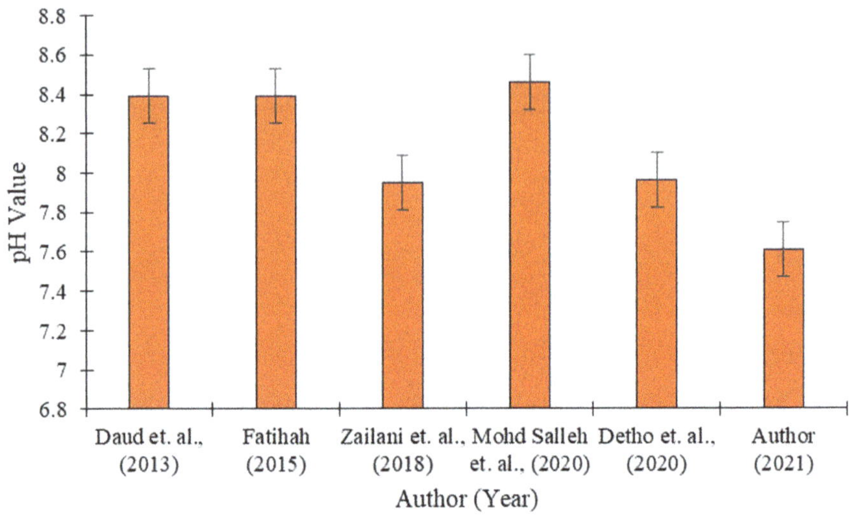

Fig. 1.10 The pH value from 2013 to 2021

pH of the discharged leachate waste is a crucial indicator of the waste's biochemical process. Compared to the EQA [39], the pH levels measured were deemed acceptable at 6.0–9.0. Coherent to that, leaks can be emancipated without requiring additional pH correction.

The data regarding the pH values of wastewater sewage are erroneous, as illustrated in Fig. 1.10. The pH values reported by many scientists between 2013 and 2021 vary. The experimental figures provided in the graph reveal that there was no significant difference in pH in discharge wastewater solutions from 2013 to 2015, as the margins of error were excessively overlapping and large. Furthermore, the standard error was not considerably overlapped and was considerable, indicating a substantial shift in pH values of the liquid outflow between 2015 and 2018, as well as 2018 and 2020.

1.3.4.5 Analysis of Ammonia Hydrogen (NH_3-N)

The greater acid concentration of leachate makes it very harmful to life. The ammonium indicator showed little change, indicating that the level had decreased over the year. The mean result at the landfill was 385.39 mg/L. Daud et al. [43] discovered that ammonia concentrations ranged from 200 to 1000 mmol/L, with an evaluated value of 1712.5 mmol/L. It has no adverse effects on the anaerobic process. According to a previous report filed by researchers [36], the amount of nitrogen in ammonia increased in 2013 from 755 to 2670 mg/L, and bioremediation is not suggested at the site. The absence of organic molecules causes a decrease in acid nitrogen content [42]. In contrast to EQA [39], the resulting ammonium concentrations should exceed the

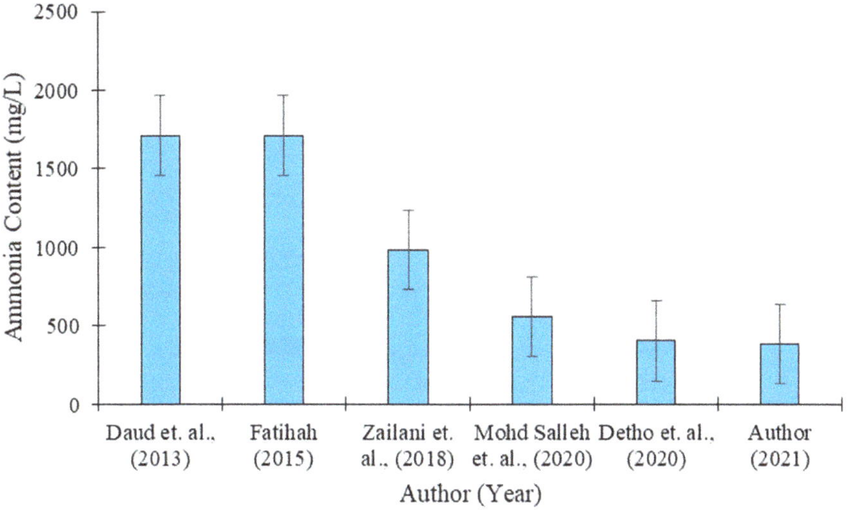

Fig. 1.11 Ammonia value from 2013 to 2021

allowed limits (5 mg/L). To fulfil designated discharge restrictions, ammonia nitrogen can be decreased by a variety of means, including physical and chemical treatment. The present global average ammonium concentration is 385.39 mg/L, which should be decreased to < 5 mg/L in order to fulfil production targets.

Figure 1.11 depicts data inconsistencies for ammonia concentrations in effluent wastewater. The evaluated values of the data show that there were no significant changes in the proportion of ammonia in wastewater from 2013 to 2015 or 2018 to 2021, as the error margin was wide. Furthermore, the standard error does not significantly overlap, and the error margins are large, showing that the concentration of ammonia in wastewater leachate varied greatly between 2015 and 2018.

1.3.4.6 Analysis of Turbidity

The detected turbidity was 2250 NTU, according to prior data gathered in 2015. Turbidity measurements taken at SRL's site in 2018 dropped to 2064 NTU. Turbidity values remain constant between 2018 and 2021, ranging from 117 to 142 NTU. The ancient nature of landfills and the stability of leachate may cause variations in the upsurge of turbidity levels.

Figure 1.12 depicts the unreliability of data on turbidity concentrations in sewage effluent. The evaluated values in the graph show that the turbidity % of wastewater leakage varies greatly between 2015 and 2018, since the standard error is not significantly overlapped and is large in size. Furthermore, standard errors overlap significantly and error margins are large, indicating that there is no difference in the

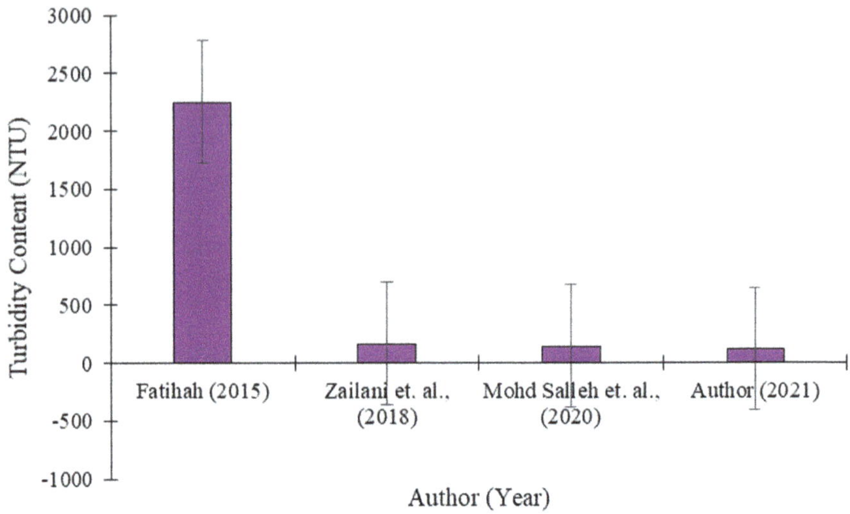

Fig. 1.12 Turbidity value from 2015 to 2021

turbidity content of sewer effluent between 2018 and 2021. Between 2018 and 2021, turbidity content discrepancies are diminishing.

1.3.4.7 Analysis of Biodegradability Ratio (BOD5/COD)

The biological degradation ratio (BOD5)/COD was utilised to determine landfill age [27, 36, 42]. The permitted range of 0.07 varies depending on the biodegradability features of the landfill. Leach is categorised as an older (stabilised) leach effluent with a BOD5/COD value of 0.1, according to the data. Coherently, as landfills age, their BOD5/COD ratios fall [42, 44, 45]. The biodegradable ratio is often referred as the most effective approach for evaluating SRLS biological and chemical degradation. The rise in COD value compared to BOD5 shows that more compounds may be oxidised chemically rather than physiologically. A low BOD5 result shows that the waste produced has a low biodegradability value and is incompatible with physiological treatment procedures [36, 46]. The mean value of BOD5 and BOD5/COD was 137.41 mg/L and 0.076, respectively.

Figure 1.13 illustrates the inaccuracy of data on the biodegradability of wastewater purification debris. The experimental value in the figure shows that there is a definite difference in the biodegradability ratio of wastewater dewatering from 2015 to 2018, as the standard error is not very overlapping and huge. Nevertheless, there is no significant contrast in the ratio of biodegradability to wastewater leaching between 2018 and 2021, as the error bars overlap and the error margins are large. Nevertheless, from 2018 onwards, the margins between the biodegradability ratios got lower than in 2015.

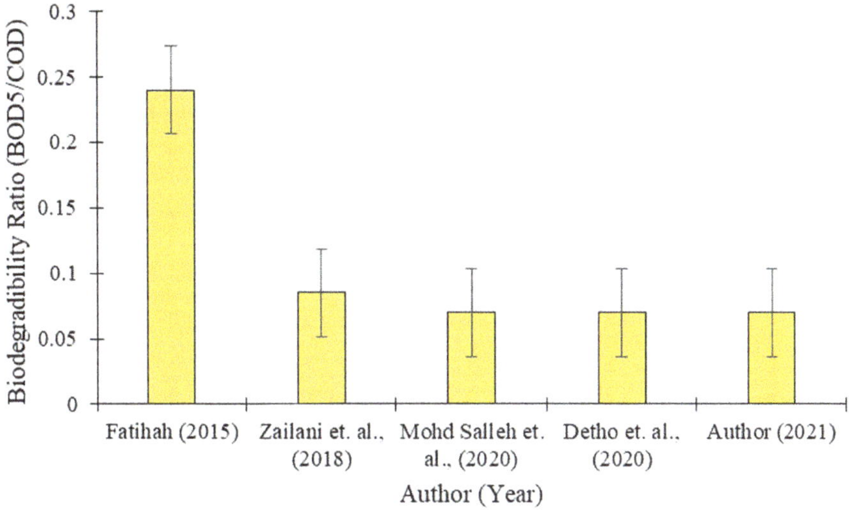

Fig. 1.13 Biodegradability ratio value from 2015 to 2021

According to Figs. 1.7, 1.8, 1.9, 1.10, 1.11, 1.12 and 1.13, the course of changes in parameter concentrations (COD, COD, SS, pH, NH3-N, turbidity, and COD5/COD) from 2013 to 2021 is likely to decrease as landfill sites age by more than ten years, which has a direct impact on the change in waste properties. As the present concentration of parameters exceeds the local standard, leachate may flow and contaminate neighbouring soil and groundwater. It is widely assumed that leachate overflows into rivers, lakes, and groundwater degrade water quality and harm natural systems.

1.4 Conclusions

The 2D ERT was out in Simpang Renggam in Johor revealed the existence and transit of wastewater from landfills into soils and rivers. Leachate affects oil palm cultivation in the neighbouring southern landfills. At the SRL site, leachate penetration ranged from 1.5 to 4.0 m vertically under the soil base and 80 m horizontally. Reverse resistance models (RL1 and RL2) indicate that locations with resistance values of 1–10 Ωm are the most contaminated. Finally, by assessing the effectiveness of leakage under thc soil, previous researchers have demonstrated that the ERT technology is cost-effective and environmentally friendly owing to the clear image of the visualization obtained from the RES2DINV software, as well as one of the most effective methods for leakage assessment. Based on the findings, it is concluded that leachate has little impact on the physical charactersitics of soil in the landfill due to the special gravity value of the consistency limit and the particle size of the leachate soil specimen compared to the landfill's natural soil. In soil chemical content analysis, the contents of fer, chromium, ar, pb, cu, and zinc in contaminated soils are higher than in natural soils, and the contents of fer, chromium, ar, and pb in contaminated specimens are also higher than the values set out in the EQA Regulations, demonstrating that heavy metal concentrations can accumulate in biological systems and become substantial hazardous substances. In terms of wastewater discharge characteristics, measures such as COD, BOD, SS, NH3-N, turbidity, and BOD5/COD are decreasing in value from 2015 to 2021, however they still exceed the EQA [39] standard.

Acknowledgements The study leading to these conclusions received financing from the Universiti Malaysia Pahang Al-Sultan Abdullah (UMPSA) under Grant Agreement Numbers RDU202701, RDU223309, and PGRS2003185, and from Hokoku Engineering Co. Ltd. under Grant Agreement Number UIC201503.

References

1. Khanna R, Konyukhov YV, Ikram-ul-haq M, Burmistrov I, Cayumil R, Belov VA, Rogachev SO, Leybo DV, Mukherjee PS (2021) An innovative route for valorising iron and aluminium oxide rich industrial wastes: recovery of multiple metals. J Environ Manage. https://doi.org/10.1016/j.jenvman.2021.113035
2. Zaini MSI, Hasan M (2023) Effects of industrial and agricultural recycled waste enhanced with lime utilisation in stabilising kaolinitic soil. Int J Sustain Constr Eng Technol. https://doi.org/10.30880/ijscet.2023.14.04.025
3. Zaini MSI, Hasan M, Masri KA (2023) Stabilization of Kaolinitic soil using crushed tile column. Mag Civ Eng. https://doi.org/10.34910/MCE.123.4
4. Tyagi VK, Kapoor A, Arora P, Banu JR, Das S, Pipesh S, Kazmi AA (2021) Mechanical-biological treatment of municipal solid waste: case study of 100 TPD Goa plant, India. J Environ Manage. https://doi.org/10.1016/j.jenvman.2021.112741
5. Hamzah U, Jeeva M, Mohd Ali NA (2014) Electrical resistivity techniques and chemical analysis in the study of leachate migration at Sungai SEDU landfill. Asian J Appl Sci. https://doi.org/10.3923/ajaps.2014.518.535
6. Giang NV, Kochanek K, Vu NT, Duan NB (2018) Landfill leachate assessment by hydrological and geophysical data: case study NamSon, Hanoi, Vietnam. J Mater Cycles Waste Manag. https://doi.org/10.1007/s10163-018-0732-7
7. Zaini MSI, Hasan M (2023) Effectiveness of silica fume eggshell ash and lime use on the properties of kaolinitic clay. Int J Eng Technol. https://doi.org/10.46604/ijeti.2023.11936
8. Zaini MSI, Hasan M, Zolkepli MF (2023) Influence of *Alstonia angustiloba* tree water uptake on slope stability: a case study at the unsaturated slope, Pahang, Malaysia. Bull Geol Soc. https://doi.org/10.7186/bgsm75202305
9. Zaini MSI, Hasan M (2023) Effect of optimum utilization of silica fume and lime on the stabilization of problematic soils. Int J Integr Eng. https://penerbit.uthm.edu.my/ojs/index.php/ijie/article/view/9573
10. Feng D, Song C, Mo W (2021) Environmental, human health, and economic implications of landfill leachate treatment for per- and polyfluoroalkyl substance removal. J Environ Manage. https://doi.org/10.1016/j.jenvman.2021.112558
11. Park S, Yi MJ, Kim JH, Shin SW (2016) Electrical resistivity imaging (ERI) monitoring for groundwater contamination in an uncontrolled landfill, South Korea. J Appl Geophys. https://doi.org/10.1016/j.jappgeo.2016.07.004
12. Zaini MSI, Hasan M, Jusoh WNBW (2023) Utilization of bottom ash waste as a granular column to enhance the lateral load capacity of soft kaolin clay soil. Environ Sci Pollut Res

13. Zolkepli MF, Ishak MF, Yunus MYM, Zaini MSI, Wahap MS, Yasin AM, Sidik MH, Hezmi MA (2021) Application of unmanned aerial vehicle (UAV) for slope mapping at Pahang Matriculation College, Malaysia. Phys Chem Earth
14. Ishak MF, Zolkepli MF, Yunus MYM, Ali N, Kassim A, Zaini MSI (2021) The effect of tree water uptake on suction distribution in tropical residual soil slope. Phys Chem Earth
15. Bai L, Huo Z, Zeng Z, Liu H, Tan J, Wang T (2021) Groundwater flow monitoring using time-lapse electrical resistivity and self potential data. J Appl Geophys. https://doi.org/10.1016/j.jappgeo.2021.104411
16. Awang H, Salmanfarsi AF, Zaini MSI, Mohamad Yazid MAF, Ali MI (2021) Investigation of groundwater table under rock slope by using electrical resistivity imaging at Sri Jaya, Pahang, Malaysia. IOP Conf Ser Earth Environ Sci
17. Ishak MF, Zolkepli MF, Yunus MYM, Ali N, Kassim A, Zaini MSI (2021) Verification of tree induced suction with numerical model. Phys Chem Earth
18. Goh JR, Ishak MF, Zaini MSI, Zolkepli MF (2020) Stability analysis and improvement evaluation on residual soil slope: building cracked and slope failure. IOP Conf Ser Mater Sci Eng
19. Zaini MSI, Hasan M, Yie LS, Masri KA, Jaya RP, Hyodo M, Winter MJ (2022) The effect of utilizing silica fume and eggshell ash on the geotechnical properties of soft kaolin clay. J Teknol
20. Ishak MF, Zaini MSI, Zolkepli M, Wahap MS, Sidek JIJ, Yasin AM, Zolkepli MN, Mohmad Sidik MH, Mohd Arof KZ, Abu Talib Z (2020) Granite exploration by using electrical resistivity imaging (ERI): a case study in Johor. Int J Integr Eng
21. Crawford MM, Bryson LS, Woolery EW, Wang Z (2018) Using 2-D electrical resistivity imaging for joint geophysical and geotechnical characterization of shallow landslides. J Appl Geophys. https://doi.org/10.1016/j.jappgeo.2018.06.009
22. Zaini MSI, Ishak MF, Zolkepli MF (2020) Monitoring soil slope of tropical residual soil by using tree water uptake method. IOP Conf Ser Mater Sci Eng
23. Zaini MSI, Ishak MF, Zolkepli MF (2020) Forensic assessment on landfills leachate through electrical resistivity imaging at Simpang Renggam in Johor, Malaysia. IOP Conf Ser Mater Sci Eng
24. Zolkepli MF, Mohamad Rozar N, Ishak MF, Sidik MH, Ibrahim NAS, Zaini MSI (2021) Slope mapping using unmanned aerial vehicle (UAV). Turk J Comput Math Educ
25. Yue LJ, Ishak MF, Zaini MSI, Zolkepli MF (2019) Rainfall induced residual soil slope instability: building cracked and slope failure. IOP Conf Ser Mater Sci Eng
26. Zolkepli MF, Ishak MF, Zaini MSI (2019) Slope stability analysis using modified Fellenius's and Bishop's method. IOP Conf Ser Mater Sci Eng
27. Zaini MSI, Hasan M, Zolkepli MF (2022) Urban landfills investigation for leachate assessment using electrical resistivity imaging in Johor, Malaysia. Environ Chall
28. Hasan M, Zaini MSI, Zulkafli AS, Wahab A, Hokabi AA, Masri KA, Jaya RP, Hyodo M, Winter MJ, Sholichin M, Haribowo R, Zulkafli MS (2021) Geotechnical properties of bauxite: a case study in Bukit Goh, Kuantan, Malaysia. IOP Conf Ser Earth Environ Sci
29. APHA (2012) Standard methods for the examination of water and wastewater. American Public Health Association, Washington DC
30. Hasan M, Zaini MSI, Hashim NAA, Wahab AA, Masri KA, Jaya RP, Hyodo M, Winter MJ, Sholichin M, Haribowo R (2021) Stabilization of kaolin clay soil reinforced with single encapsulated 20 mm diameter bottom ash column. IOP Conf Ser Earth Environ Sci
31. Maurya PK, Rønde VK, Fiandaca G, Balbarini N, Auken E, Bjerg PL, Christiansen AV (2017) Detailed landfll leachate plume mapping using 2D and 3D electrical resistivity tomography-with correlation to ionic strength measured in screens. J Appl Geophys 138:1–8
32. Moretto RL, Siqueira Neto AC, Elis VR, Miguel MG (2017) Detection of leachate pockets in experimental cell of municipal solid waste with aid of geophysics. Margherita di Pula, Proceedings Sardinia
33. Chu Y, Liu S, Wang F, Cai G, Bian H (2017) Estimation of heavy metalcontaminated soils' mechanical characteristics using electrical resistivity. Environ Sci Pollut Res 24(15):13561–13575

34. Ansari S, Norooz R, Zeynolabedin A, Foadi H, Ghiassi R (2021) Experimental investigation of correlations between electrical resistivity, moisture content and voltage values for leachate contaminated clayey sand. J Appl Geophys. https://doi.org/10.1016/j.jappgeo.2021.104391
35. Hasan M, Zaini MSI, Yie LS, Masri KA, Putra Jaya R, Hyodo M, Winter MJ (2021) Effect of optimum utilization of silica fume and eggshell ash to the engineering properties of expansive soil. J Mater Res Technol. https://doi.org/10.1016/j.jmrt.2021.07.023
36. Mohd-Salleh SNA, Shaylinda MZN, Othman N, Azizan MO, Yashni G, Afnizan WMW (2020) Sustainability analysis on landfilling and evaluation of characteristics in landfill leachate: a case study. IOP Conf Ser Mater Sci Eng
37. Li R, Zhang B, Wang Y, Zhao Y, Li F (2019) Leaching potential of stabilized fly ash from the incineration of municipal solid waste with a new polymer. J Environ Manage. https://doi.org/10.1016/j.jenvman.2018.11.036
38. Aja D, Okolo CC, Nwite NJ, Njoku C (2021) Environmental risk assessment in selected dumpsites in Abakaliki metropolis, Ebonyi state, southeastern Nigeria. Environ Challenges. https://doi.org/10.1016/j.envc.2021.100143
39. Government of Malaysia (1974) Environmental quality act 1974. Environ Act Qual
40. Kadir AA, Mustafa M, Bakri A, Sandu AV, Noor NM, Lisanah A, Latif A, Hussin K (2014) Usage of palm shell activated carbon to treat landfill leachate. Int J Conserv Sci
41. Ishak MF, Koay BK, Zaini MSI, Zolkepli MF (2018) Investigation and monitoring of groundwater level: building crack near to IIUM Kuantan. Int J Eng Technol Sci
42. Detho AA, Daud Z, Samo SR, Khan Z, Memon DA, Awang H (2020) Physicochemical characteristics of landfill leachate from Simpang Renggam landfill site, Johor, Malaysia. Quaid-e-Awam Univ Res J Eng Sci Technol
43. Daud Z, Muhamad Hanafi NF, Awang H (2013) Optimization of COD and colour removal from landfill leachate by Electro-Fenton method. Aus J Basic App Sci
44. Fatihah MH (2015) Landfill leachate treatment by combination of Electron-Fenton and sequencing bathc reactor method master thesis. Universiti Tun Hussein Onn Malaysia, Johor
45. Ishak MF, Zaini MSI (2018) Physical analysis work for slope stability at Shah Alam, Selangor. J Phys Conf Ser
46. Hasan M, Zaini MSI, Hong NAW, Wahab AA, Masri KA, Jaya RP, Hyodo M, Winter MJ, Sholichin M, Haribowo R (2021) Sustainable ground improvement method using encapsulated polypropylene (PP) column reinforcement. IOP Conf Ser Earth Environ Sci

Chapter 2
Biological Test Used in the Assessment of Cytotoxicity and Genotoxicity of Olusosun Landfill: The Largest and Unregulated Landfill in Nigeria

Chibuisi Gideon Alimba

Abstract Indiscriminate disposal of solid wastes into the environment generates landfill gases and leachates. Leachate, a mixture of deleterious substances including metals, inorganic and organic compounds, particulate matters, radioactive elements, and microorganisms, contaminates soils and underground and surface water resources increasing risk to human and wildlife health. This study is a compendium of studies on the cytotoxic and genotoxic assessment of Olusosun landfill (the largest unregulated landfill in African). Leachates emitted from Olusosun solid waste landfill induced cytotoxicity and DNA damage via direct and indirect mechanisms. Different biomarkers of genotoxicity were used to evaluate the cytotoxic and genotoxic potentials of OSL leachates. Biomarkers of chromosome aberration, micronucleus assay, comet assay, sperm head abnormality test and toxicogenomic signatures, were assessed in somatic and germline cells of exposed organisms. The mechanisms of DNA damage include generation of free radicals and oxidative stress, inhibition of DNA repair process, alkylation on DNA molecules that may lead to chromosome breakage, down-regulation of transcriptional genes related to apoptotic expressions, and increased DNA fragmentation in cells and tissue. The damaged DNA can increase genome instability and genetic related diseases. Heavy metals, physico-chemicals and organic substances present in the leachates are known mutagens, genotoxin and carcinogens. It is suggested that proper landfill management be instituted in various countries to avert exposure to landfill emissions.

Keywords Olusosun landfill · Genotoxicity biomarkers · Xenobiotics in leachates · Mechanisms of leachate induced genotoxicity · Health impacts · Biomarkers of DNA damage · Cytotoxicity · Genotoxicity · Municipal landfill leachates

C. G. Alimba (✉)
Cell Biology and Genetics, Department of Zoology, University of Ibadan, Oyo, Nigeria
e-mail: chivoptera@yahoo.com

A. Anouzla and S. Souabi (eds.), *A Review of Landfill Leachate*, Springer Water,
https://doi.org/10.1007/978-3-031-55513-8_2

2.1 Introduction

There is a close link between the healthiness of an environment and human and wildlife well-being. A healthy environment, associated with cleanliness and relatively free of pollutants and pathogens, will possibly enhance long life in good health of wildlife and humans. While polluted environment, may pose significant health risk to humans and other biota in the ecosystems. Studies are increasingly reporting that anthropogenic activities including industrial, agricultural, medical, municipality and educational activities generate wastes (which contain numerous chemical, physical and microbial agents) that end up in the environment [12, 44]. Reduction in waste generation during anthropogenic activities is almost inevitable, since waste generation is closely linked with economic development and human population growth [24]. Therefore, appropriate and effective waste management methods could protect the environment and human health from the numerous chemicals, radiations, microorganisms and deleterious physical agents usually contained and emitted from the large stream of solid wastes daily generated during anthropogenic events. Unprecedented increase in solid waste generation, though a global issue, is a major problem that requires urgent attention in most developing countries. Due to the waste management methods in these countries are mostly inappropriate and involve direct disposal of solid wastes and effluents directly into the environment. Adequate and proper solid waste management methods including recovery from source of generation, re-use, treatment and eventual disposal of the residual wastes into engineered landfill, are usually lacking in most developing nations. Standard waste management methods are expected to reduce contamination of the environmental metrics and hence, risk to human and wildlife health.

Inappropriate and indiscriminate disposal of the increasingly generated solid wastes into the environment, is a major cause of ecocatastrophe (environmental degradation arising from chemical and physical xenobiotic discharged into the environmental media, water, soil and air, which will eventually lead to mortality and morbidity of biota). Globally, landfilling is the most utilised solid waste management option. It accounts for the management of about 95% of the total solid waste collection [20]. Most developing countries with high population size generate huge quantity of solid wastes on daily basis. For instance, Nigeria (in Africa) with an estimated population size of 180 million and India (in Asia) with 1.21 billion [52], respectively generate daily per capital solid waste of not less than 0.44–0.66 kg [41] and 0.12–2.1 kg [14]. These were respectively estimated as 25 million and 68.8 million tons per year [8]. Virtually all volume of the generated solid wastes, in these countries, end-up in landfills. Unfortunately, most of these landfills are unequipped and not engineered. They lack landfill liners, leachate collection systems and landfill caps to prevent the emission of leachates and landfill gases into the environment [7, 12, 14]. Landfills in developing countries are sometimes open dumpsites that are usually located in cities and around residential quarters [7, 12]. In some instances, solid wastes are illegally disposed into water bodies and around river banks, in gullies excavated by erosions and human activities like quarry, in gutters and channels constructed to prevent flooding. Also, burning of solid wastes on major roads and public places is an eyesore in these countries [12]. Leachates and landfill gases released through these unsanitary methods of solid waste management, contaminate surrounding waters, soil and air with polycyclic aromatic hydrocarbons (PAHs) and polychlorinated biphenyl (PCB) congeners [15, 34], heavy metals [34], particulate matters and radioactive substances [39] and microorganisms and pathogens [43].

Olusosun landfill (OSL) in Lagos, Nigeria, is a mega dumpsite seating on 100 acres of land and receives about 10,000 tons of solid wastes generated from hospitals, industries, electronic and domestic wastes etc. on daily basis [22]. This dumpsite, without landfill liners and leachate collection system, is surrounded by communities, with many houses built within its circumference and is the largest landfill in Africa [33]. Human activities within and around Olusosun landfill including sorting and reclamation of recyclable materials by scavengers (Persons that sort wastes to reclaim recyclable materials. However, they usually do this job without wearing safety costumes and adhering to necessary precautionary measures) and residing within 500 m of the landfill circumference, are increasingly exposed to chemicals and microorganisms via consumption of leachate polluted underground and surface water, inhalation of particulate matter. They are also exposed to chemicals in landfill gases via inhalation. Also, commonly observed on the landfills are wildlife activities such as birds and rodents scavenging for remains of food materials in the landfill [4, 8, 12]. This wildlife is exposed to hazardous substances and microorganisms via drinking of leachate as source of water on the landfill and inhalation of particulate matters and landfill gases from the landfill [4, 5, 12, 26].

Increasing incidence of cancers and other genetic related diseases among residents exposed to chemicals leaching out of landfill, along with the incidence of low birth weight among residents exposed to hazardous chemicals from landfills in countries where there are epidemiological data, have increased awareness about the adverse human health effects associated with emissions (leachates and gases) from unsanitary landfills [37]. Hence, this required that adequate studies be conducted to increase the understanding of the extent to which emissions from landfills can cause toxicity. Chemical analysis showed that Olusosun landfill leachates contain heavy metals and organic xenobiotic substances that are classified as mutagens, genotoxin, teratogens and carcinogens even at trace concentrations [8, 15, 37, 39]. These chemicals are capable of interacting with the DNA molecule to induce alterations in its structure and function, and may lead to concomitant adverse effect on biological integrity (alteration in the functions of the DNA molecule) [4, 5, 8].

The objective of this chapter is to describe the different genotoxic biomarkers utilized in assessing the effects of DNA damage induced by emissions from Olusosun landfill and discuss the implications on health.

2.2 Leachate Production at Olusosun Dumpsite

Olusosun dumpsite, an open dumpsite, uncontrolled disposal sites or unsanitary landfill, receives over 40% of all forms of solid wastes generated in Lagos State (The commercial nerve-centre of Nigeria with the largest concentration of industrial and commercial establishments), Nigeria. Co-disposal of medical wastes, industrial solid wastes, radioactive wastes, agricultural wastes and domestic wastes occur in the dumpsite [12]. The dumpsite is located in public place surrounded by residential quarters and in wetland with seasonally high-water tables. It is not engineered with landfill gas and leachate collection system, and lack landfill liners and caps [12, 38]. During precipitation, surface water runoff, groundwater intrusion due to high water table and moisture content present in the wastes, solid wastes in the dumpsite undergo combination of physical, chemical and microbial processes [17]. These processes transform chemicals locked up in the solid wastes (numerous water-soluble and insoluble compounds) into leachates which contain mixture of toxic metals, organic and inorganic hazardous contaminants, microbial population and particulate matter [8, 38, 39, 43]. The potential for dumpsites or unlined sanitary landfills to release large quantity of hazardous and otherwise deleterious chemicals to nearby groundwater, soil and air, via leachate and landfill gas, underscores its importance in solid waste management. Chemicals and microorganisms present in landfill leachates and gases are potential carcinogens, mutagens, teratogens and genotoxin that readily contaminate the environment and increase threat to public health [37].

The significant adverse effects of leachates from landfills and municipal dumpsites on groundwater quality had been a recognizable problem since the 1950's. During this period, municipal solid waste (MSW) leachate contaminated groundwater was confirmed to be unfit for domestic water supply purposes. This culminated into a global recognition in the 1970's that landfill leachate contained a variety of highly hazardous chemicals that represent a significant public health threat to those who consume water contaminated by leachate [16].

The rate of leachate generation from a landfill depends on a number of variables, which include waste characteristics, moisture content (including the volume of liquid in disposed wastes), temperature, pH and availability of microbes [20]. When water flows through the wastes, it transports a wide variety of chemicals, particulate matter and microbes from one extreme of the landfill to another. The volume of leachate produced strongly correlates with the volume of water in or around the landfill, while the content of the leachate correlates with landfill size and the quantity and composition of solid wastes disposed in the landfill [51]. Larger landfills are known to generate higher volumes of leachate than smaller landfills. An estimated 100 acre (40.47 hectares) of landfill in the northern United States can produce 57 million gallons (215.78 million litres) of leachates every year [51]. Considering that Olusosun contains virtually all the variables that enhance leachate formation, it then can be estimated that the annual leachate generation from the 42 hectares (103.78 acres) Olusosun dumpsite that receives approximately 3.65 million tons of wastes (annually) is 59.1546 million gallons (223.92 million litres) of leachate. This value is enormous and its impact on the environment and health of biota including residents around the dumpsite is far reaching. Although no estimated value of leachate generated from Olusosun landfill had been previously reported to the best of my knowledge, however, that Olusosun landfill is situated in a wetland, an area where groundwater is not far beneath the landfill's base, suggests that groundwater can also rise from below the landfill in addition to precipitation, providing additional liquid to mix with solid wastes and produce higher volume of leachate [12, 20, 54]. The quantity of leachate generation from a landfill, though is site-specific, also greatly depends on water availability, weather conditions including temperature, landfill surface area, stage of waste decomposition in the landfill, waste composition whether co-disposal of industrial, municipal and agricultural wastes [20].

Hazardous substances characterized in the leachates generated from Olusosun include particulate matters, pathogenic microbes, heavy metals, inorganic and organic compounds capable of eliciting grievous health consequences. Reports have shown that groundwater from communities around the landfill contained higher concentrations of harmful chemicals and microbes than permissible limits for both national and international environmental regulatory organizations [38, 39, 43]. Humans and other biota are exposed to these toxic substances in Olusosun landfill leachates [12, 26]. Research output on cytotoxicity and genotoxicity of Olusosun landfill leachates is enormous and overwhelming. This chapter is a compendium of articles that reported on biomarkers of cytotoxicity and genotoxicity of Olusosun landfill leachates. It also describes the mechanisms of DNA damage in eukaryotic system, and the implication of DNA damage on human and other biota.

2.3 Genetic Assays in the Assessment of DNA Damaging Potentials of Olusosun Landfill Leachates

Genetic assays are test batteries used to detect substances capable of inducing DNA damage through various mechanisms. Damage to DNA may take the forms of gene mutations (point mutations), chromosomal aberration, recombination and numerical chromosome changes, micronuclei and abnormal nuclear formation, damage to DNA repair mechanisms and cell cycle interactions. These alterations are essential since it can predispose harbouring organisms to express heritable genetic disorders and genotoxic hazards including cancer and other syndromes related to genomic instability. Therefore, various batteries of genetic biomarkers have been used to assess the genotoxic potentials of many environmental xenobiotics. This section aims to describe the various applications of genetic biomarkers of effects in the assessment of cytotoxicity and genotoxicity of Olusosun landfill leachate.

2.3.1 *Chromosome Aberration Assay in the Cytotoxicity and Genotoxic Assessment of Olusosun Landfill Leachates*

Chromosomal aberration (CA) assay was first applied in *Tradescantia* to study the mechanisms of structural and numerical chromosome damage induced by radiations in plant system [32]. Afterwards it became popular in the genotoxic assessment of chemical, physical and biological agents using plant and animal models, in in vivo, in vitro and in situ test systems. It is a direct cytogenetic assay that detect chromosome damage based on direct observation of changes in structures and number of the sets of organismal chromosomes. Chromosome aberration as cytogenetic biomarker has been applied to confirm that constituents of OSL is capable of damaging the structure of chromosomes in both animal and plant models. According to details in Table 2.1, Alimba et al. [7] confirmed that exposure to leachates from OSL induced various aberrant chromosomes including chromosomal breaks, ring, gap and acentric chromosomes, in bone marrow cells of exposed Wistar rats. Ogunyemi et al. [40] also confirmed that leachates from OSL induced chromosome aberrations in the meristematic cells of the root tips of *Allium cepa* (plant). The tested leachates further induced mitotic inhibition (cytotoxicity) by affecting the mitotic indices of both the Wistar rat and *Allium cepa* [7, 40]. The authors posited that the various organic and inorganic constituents observed in the leachates which were higher than permissible limits set by the national and international environmental monitoring bodies. May be responsible for the observed chromosome damage by acting individually or in combinations and also via different mechanisms.

Authors from some other parts of the world similarly observed that landfill leachates are capable of causing chromosome aberration in various model organisms. For instance, Sang and Li [46] reported that leachates collected from Xinguo municipal solid waste landfill increased chromosome aberration (gap, break, ring and triradial CAs) and decrease mitotic index of exposed male and female Kunming mice. Sang and Li [45] further observed that the same leachate caused acute toxicity by inhibiting root growth, cytotoxicity by reducing cell division (mitotic index), and chromosomal aberrations (genotoxicity) in *Vicia faba*. Shekha et al. [47] exposed male mice to Erbil municipal landfill leachates and observed polyploidy, centromeric break and gap, chromatid break, gap and fragment, ring chromosome and pulverization chromosome aberrations in the bone marrow cells. It is evident that xenobiotics present in emissions from municipal landfill leachates are deleterious to organisms' genetic materials.

Table 2.1 Biomarkers used to assess Olusosun landfill leachates (OSL) induction of cytotoxicity and genotoxicity in different in vivo and in vitro models

Model organisms	Exposure methods	Biomarker/end points	Findings/observations	References
Bone marrow cells from Wistar rats	1, 2.5, 5, 10 and 25% of raw and simulated leachates from solid waste, was intraperitoneally injected for 48 h	Bone marrow chromosome aberrations (CA) assay was used to assess chromosome damage induced by the tested leachates	Significant increase in CA was observed. Chromosome breakage, acentric chromosome and ring chromosome were observed in exposed rats	Alimba et al. [7]
Epididymal sperm cells from Boar	1, 2, 4 and 8% of raw leachate from OSL was added to sperm suspension in culture medium in vitro for 3 h	Abnormal sperm morphology assay was used to examine damage done on sperm cell DNA via observing different shapes of the sperm head and tail	Significant increase in abnormal sperm morphology was observed compared to the control	Adedara et al. [2]
Epididymis from Wistar rats	Wistar rats were allowed to drink 12.5 and 25% of raw leachate from OSL for 28 days. Another group was left for another 28 days to recovery after 28 days exposure	Abnormal sperm morphology assay and Sperm count analysis were performed on the smeared sperm cells	Increase in abnormal sperm morphology, decrease in daily sperm production and sperm count. There was recovery from damage induced by OSL on the examined parameters after 28 days	Adedara et al. [1]
Peripheral blood and bone marrow from *Coturnix coturnix japonica* (Japanese quail)	Japanese quail was orally gavage 10, 25 and 50% raw leachate for 14 days	Peripheral erythrocytes and bone marrow cells were examined for micronucleated erythrocytes and nuclear abnormalities in the exposed Japanese quail	There was significant increase in peripheral erythrocytes with micronucleus and abnormal nucleus morphology in exposed Japanese quail. Also, there was significant increase in bone marrow cells with micronucleus in the exposed Japanese quail	Alimba and Bakare [5]

(continued)

Table 2.1 (continued)

Model organisms	Exposure methods	Biomarker/end points	Findings/observations	References
Hepatocarcinoma (HepG2), lymphoma (Jurkat) and Osteosarcoma (HOS) cells (in vitro)	Cells were treated with 0, 6.25, 12.5, 25, 50, 75 and 100% of OSL added to the cell culture medium	Cytotoxic (cell viability) effect of OSL was determined using MTT assay	With increasing concentration of OSL, there was significant decrease in viability of the exposed cells	Alimba et al. [8]
Clarias gariepinus (African catfish)	Fish were exposed by immersion to 5, 10, 15, 20 & 25% of OSL/dechlorinated tap water for 6 and 14 days	Peripheral erythrocytes were examined for micronucleated erythrocytes and nuclear abnormalities in exposed fish compared to the control fish	There was significant increase in peripheral erythrocytes with micronucleus and abnormal nucleus morphology in exposed fish	Alimba and Bakare [5]
Hepatocarcinoma (HepG2), lymphoma (Jurkat) and Osteosarcoma (HOS) cells (in vitro DNA damage using comet assay)	Cells were treated 0.2xLC_{50}, 0.5xLC_{50} and LC_{50} values obtained for each cell (for which at least 80% viability was obtained during the MTT assay) for 24 h and DNA damage evaluated using alkaline comet assay	OSL exposed cells and the control were examined for DNA damage using Comet assay	There was concentration dependent increase in tail length, percentage tail moment and Olive tail moment of OSL treated cells compared to the untreated cells	Alimba et al. [8]
Bone marrow cells from Wistar rats (*Rattus norvegicus*)	Rats were orally gavage 1, 2.5, 5, 10 and 25% raw leachate for 30 days	Bone marrow cells from OSL treated rats were investigated for micronucleated polychromatic (PCE) erythrocytes and normochromatic (NCE) erythrocytes using the micronucleus test	There was significant increase in micronucleated PCE and NCE in bone marrow cells of OSL treated rats. Also, there was significant decrease in PCE/NCE ratio (cytotoxicity index)	Alimba and Bakare [5]

(continued)

Table 2.1 (continued)

Model organisms	Exposure methods	Biomarker/end points	Findings/observations	References
Onions (*Allium cepa*)	Plant roots were grown in leachate solutions; 0.05, 0.1, 0.5, 1.0, 5, 10, 25, 50 and 100% for 72 h	Root tips were collected and examined for EC_{50}, mitotic index and frequency of chromosome aberrations	The leachate caused decrease in percentage root growth inhibition. Also, there was decrease in mitotic index and significant increase in formation of chromosome with various aberrations were observed in OSL exposed Onion roots compared to the controls	Ogunyemi et al. [40]
Male mice (*Mus musculus*)	Mice were exposed to 0.5 mL of 5, 10, 25, and 50% of OSL (v/v, Leachate/distilled water) for 5 consecutive days	Cauda epididymis of the exposed and control mice was examined for abnormal sperm morphology	The tested leachate significantly increased teratozoospermia with various abnormal sperm cells including amorphous sperm head, sperms with two heads and two tails, sperms without hooks, sperm with banana head, sperms with cytoplasmic droplets, sperms with abnormal hooks, no tail, pin head, wrong position of tail attachment and folded tails	Ademola et al. [3]
Male and female mice (*Mus musculus*)	Mice were exposed to 0.5 mL of 5, 10, 25, and 50% of OSL (v/v, leachate/distilled water) for 5 consecutive days. The exposed male mice were used to mate the untreated nulliparous females at 30 days post-exposure, until vaginal plugs were observed in the females	The pregnant females were then sacrificed on day 14 post-mating and the uterine contents were examined for number of implants per female, resorptions, and live and dead fetuses	There were high dead implants in females mated by OSL treated males. Also, there was high resorptions in females mated by OSL exposed males. Although, this was not statistically significant compared with the negative control. This suggests the tendencies of OSL to induce dominant lethal mutation. Also, low fertility was observed in the treated mice	Ademola et al. [3]

(continued)

Table 2.1 (continued)

Model organisms	Exposure methods	Biomarker/end points	Findings/observations	References
Wistar rats (*Rattus norvegicus*)	Male rats were placed in OSL at three different locations for 4, 8, 12, 16, 20 and 24 weeks, to assess the impact of exposure to ambient air and underground water (via drinking) at the landfill	Bone marrow micronucleus assay was used to assess the DNA damaging effects and cytotoxicity of exposure to chemicals in the ambient air and toxic metals in consumed underground water	There was significant increase in micronucleated polychromatic and normochromatic erythrocytes induction in rats exposed to OSL emissions. Also, exposure to OSL contaminants induced cytotoxicity via reduction in PCE/NCE ratios in the exposed rats	Gbadebo et al. [26]
Lymphoma (Jurkat) cells (in vitro DNA damage)	Cells were cultured in RPMI 1640 media containing 4.41, 11.02 and 22.04% of OSL for 24 h	DNA fragmentation assay was used to assess OSL ability to induce genomic DNA damage in the exposed Jurkat cells. Apoptotic cells were analyzed by evaluating nuclear morphology of the treated cells using Hoechst 33,258 and Propidium Iodide (PI) staining. qPCR analysis was used to determine alteration in apoptotic genes	Agarose gel electrophoresis of DNA isolated from OSL treated Jurkat cells were characterized by fragmented ladders of damaged genomic DNA. Nucleus condensation and cell shrinkage of exposed Jurkat cells were the observed morphological alterations	Alimba et al. [10]

2.3.2 *Application of Abnormal Sperm Morphological Assay in the Germ-Line Cytotoxicity and Genotoxicity Assessment of Olusosun Municipal Landfill Leachates*

Abnormal sperm morphology test is the assessment of abnormalities in sperm head shape and or tail in mammalian and non-mammalian species resulting from exposure to physical, chemical and biological genotoxic agents. This genotoxicity assay is a reliable germline genotoxic test capable of detecting germ cell mutagens that can induce spermatogenic dysfunctions and possible heritable mutations. Since it can detect abnormal alterations in the stages of mitotic and meiotic processes occurring during spermatogenesis [56, 57]. A review of earlier studies that applied abnormal sperm shape test to screen mutagens, genotoxins and carcinogens, showed that the mouse sperm morphology test was highly sensitive to germ-cell mutagens. The authors observed that 100% of the reviewed studies showed that the screened mutagens were correctly identified as positive mutagens by the sperm morphology test [57]. They therefore concluded that the mouse sperm morphology test has the potential to identify chemicals or agents that can induce spermatogenic dysfunction and heritable mutations [57]. Increase in the frequency of abnormal sperm morphology induced by endogenous and xenobiotic physical, chemical and biological agents have been associated with interference with the integrity of spermatogonia chromosomes, deletion of the Y chromosome long arm, and or alteration in genes responsible for stabilizing the fidelity of DNA during spermatogenesis [49, 50]. In most cases, sperms with abnormal morphologies rarely fertilize oocytes. Peradventure fertilization is successful, the resultant embryo may experience perinatal pre-natal death, or if survived until birth terms may express congenital malformations with varying degrees of severities [50]. Table 2.1 showed some reviewed articles that utilized sperm abnormality assays to assess OSL leachates for genotoxic effects. Adedara et al. [2] conducted an in vitro experiment, where sperm cells obtained from Boar was exposed to leachates from OSL for 3 h and observed abnormal deformities in the shape of the Boar's sperm heads and tails. The authors proceeded to investigate if the observation from the in vitro study is reproduceable in in vivo by exposing Wistar rats to the same leachate via free drinking for 28 days. Sperm analysis from the caudal epididymis of the exposed rats further confirmed abnormality in the shape of the sperm head and tail induced by OSL leachates in rats [1]. In another study, Ademola et al. [3] exposed mice to OSL leachate and observed various abnormal sperm morphology in the exposed mice compared to the control. The curiosity whether the observed abnormal sperm cells in landfill leachate exposed rodents, is capable of fertilizing receptive oocytes and produce viable embryo was further examined by Ademola et al. [3]. The authors observed that male mice which were exposed to OSL leachate and confirmed to harbour abnormal sperm cells in the caudal epididymis had low fertility when mated to nulliparous female. Also, successful mating produced females with resorption and death foetuses.

Other authors have shown that leachates from landfills are capable of eliciting abnormal sperm morphology. For instance, Shekha et al. [47] exposed male mice to

Erbil municipal landfill leachates and observed abnormal shapes of the sperm head and tail in the caudal epididymis. Also, Alimba et al. [4] confirmed that emissions (leachates and gases) from Abule Egba landfill induce germ-line genotoxicity in free ranging wild black rats (*Rattus rattus* Linnaeus, 1758). The authors caught free ranging rats from Abule Egba unsanitary landfill, Lagos, which were potentially exposed in in situ to xenobiotics in leachates and gases emitted from the landfill. They examined the rats and observed significantly high abnormal sperm cells than unexposed rats.

2.3.3 Application of Micronucleus Assay in the Cytotoxicity and Genotoxic Assessment of Olusosun Landfill Emissions

Micronucleus (MN) test is the most frequently used cytogenetic assays to assess structural chromosome aberrations induced by clastogens and aneugens. It is currently the most suitable assay for assessing the potentials for environmental xenobiotics to induce genotoxicity. Protocol has been suitably developed to assess MN in invertebrate and vertebrate cells that are actively dividing. Also, the assay finds applicability in laboratory and field-based studies. It can also be used following in vivo, in situ and in vitro exposures. Micronuclei are small nuclear bodies that contain either chromosome fragment or a whole chromosome. MN are formed and can be accessed in virtually all kind of cells, from the somatic and germinal linages of plant and animal origins. According to Table 2.1, OSL leachates have shown positive significance in causing structural chromosome breakage via MN and abnormal nuclear formation in peripheral blood erythrocytes of *Clarias gariepinus* (African catfish) and *Coturnix coturnix japonica* (Japanese quail), and in bone marrow cells of *Coturnix coturnix japonica* (Japanese quail) and *Rattus norvegicus* (Wistar rats) [5]. Also, the OSL leachate reduced polychromatic erythrocytes (PCE) and normachromatic erythrocyte (NCE) ratio in bone marrow of Wistar rats and increase cells with two nuclei (cytotoxicity biomarkers) [5]. Wistar rat exposed to ambient air and underground water (via drinking) at OSL, expressed high MN-PCE and MN-NCE and decreased PCE/NCE ratio [26]. Similar findings of high frequency of MN and other nuclear abnormalities including karyolitic and karyorrhectic cells were observed in women occupationally exposed to emissions from Cateura landfill [19].

2.3.4 Application of Comet Assay in the Genotoxic Evaluation of Olusosun Landfill Leachates

The comet assay or single cell gel electrophoresis assay was first developed as comet assay based on fluorescence measurement of single cells embedded in agarose on a glass slide after DNA unwinding in weak alkali and stained with acridine orange. It became popular after it was used to evaluate the DNA damaging potentials of radiation in exposed mammalian cells [42]. It was subsequently modified to the alkaline version of the assay, simplifying the technique to measure even low levels of DNA damage in cells [48]. Following these developments in the discovery of the assay, it has been described as a versatile, simple, sensitive and rapid tool for detecting primary DNA damage even at the early stage of exposure and is widely applied in in situ, in vitro, and in vivo studies. The approach which consists of the analysis of electrophoretic migration of DNA from the nucleoid after lysing a cell in a thin layer of agarose, requires a small quantity of the sample. It also has an advantage of being conducted in both proliferating and non-proliferating cell populations from any tissue or organ from virtually all experimental models [18]. It has been well adapted to detect and measure different types of DNA damage (including single and double strand breaks and alkaline labile sites, oxidative DNA damage, DNA cross-links, DNA adducts, apoptosis, and necrosis), induced in cells from somatic and germ tissues irrespective of their ability to actively proliferate [18]. The comet assay was employed to evaluate the potentials for OSL leachate to induce DNA damage in three human cells, hepatocarcinoma (HepG2), lymphoma (Jurkat) and Osteosarcoma (HOS) cells in an in vitro study [8]. The evaluated leachate also caused reduced cell viability in a concentration dependent pattern in the exposed HepG2, Jurkat and HOS cells [8]. Landfill leachate obtained from Lokva Vidotto sanitary landfill, Croatia, significantly increased DNA damage in human lymphocytes by altering virtually all comet assay parameters [23].

2.3.5 Application of Toxicogenomic and Molecular Signatures in the Genotoxicity Assessment of Olusosun Landfill Leachates

Molecular assays that employed toxicogenomic and proteomic tools applicable to genetic toxicology, by using information derived from changes in the expressions of genes and proteins or formation of some metabolites, that strongly correlate with the genotoxicity potentials of landfill emissions, are increasingly be used to delineate the possible mechanisms of leachate induced genotoxicity. According to Alimba et al. [10], Jurkat cells exposed to OSL leachates expressed DNA fragmentation, an indication of OSL leachate ability to induce genomic DNA damage. The leachate further induced alterations in genes regulating apoptosis in the exposed cells as detected using quantitative polymerase chain reaction (qPCR). This was confirmed by alterations in nuclear morphology of the treated cells as detected by using Hoechst 33,258 and Propidium Iodide (PI) staining. Also, Bhanpur khanti landfill leachate, significantly increased 8-Oxo-2′-deoxyguanosine (8-oxo-dG) accumulation, level of nuclear factor erythroid 2–related factor 2 (Nrf-2) and alteration in the mitochondrial genome, but decreased the mitochondrial membrane potential in human peripheral blood lymphocytes [28]. The authors further observed that the leachate increased DNA fragmentation, release nucleosomes, increased level of polyadenosine diphosphate ADP-ribose polymerase (PARP), and activity of caspase-3 in the exposed cells [28]. Using some of these toxicogenomic and molecular tools enhanced the deduction of the molecular mechanisms of landfill leachate induced cytotoxicity and genotoxicity (DNA damage). It can be concluded from the reports that exposure to leachates distorted mitochondrial redox homeostasis (machinery) and exhibited strong ability to induce pro-oxidant state inside the cell [10, 11, 28].

2.4 Possible Mechanisms of Action of Olusosun Municipal Landfill Induction of Cytotoxicity and Genotoxicity

Chemicals, radiations, microbes and particulate matters described as mutagenic, clastogenic and carcinogenic substances to human and wildlife health, are present in Olusosun landfill [8, 38, 39, 43] Also, see Table 2.2 for the physico-chemical properties and heavy metals, and Table 2.3 for the organic compounds, analysed in leachates from Olusosun landfill, Lagos, Nigeria. These substances are capable of causing cytotoxicity and genotoxicity on biological material. Although, the mechanisms for cytotoxicity and DNA damage induction is not fully understood. However, there is evidence that some of the compounds in OSL leachates can inhibit DNA repair processes. For instance, some metals present in OSL can exert unique mechanism(s) of DNA damage repair inhibition [29, 35].

The most commonly posited mechanism of leachate induced genotoxicity is generation of free radicals and oxidative stress induction in exposed biological model. Free radicals when generated may bind to purines and pyrimidines of the nucleic acids to cause base substitution and breakage of DNA double strand and eventually induce mutation [6, 9, 11, 28]. These radicals are also capable of distorting proteins that interact with DNA to maintain its structure and functions [25]. Furthermore, many of the organic compounds in OSL (see Table 2.3, [8]), including benzene, naphthalene, persistent organic pollutants, dioxins, polychlorinated biphenyls are alkylating agents with electrophilic properties. This property attracts, with high affinity, the nucleophilic centres of DNA molecule [58]. They alkylate with the nitrogen and oxygen rings of the DNA to form phosphodiester bond, which weakens the DNA backbone, modifying the N-glycosylic bond that leads to the formation of depurination or depyrimidination of the DNA [25].

Table 2.2 Physico-chemical parameters analyzed in Olusosun and Aba-Eku landfill leachates [5, 6]

Parameters*	OSL (raw)	OSL (simulated)	NESREA[a]	USEPA[b]
Colour	Dark brown	Brown	–	–
pH	8.1	7.6	6.0–9.0	6.5–8.5
Nitrate	72.3	43.2	10	10
Ammonia	122.10	32.8	10	0.02
BOD#	594	306	50	–
COD**	487	398	90	–
Phosphate	215.70	106.13	2.0	–
Chloride	1099	1012	250	250
Sulphate	218.12	122.51	250	250
Hardness	1200	809	150	0–75
Alkalinity	623	456	–	20
TS##	4100.30	2200.60	–	–
Copper	3.86	1.02	0.5	1.3
Iron	4.71	2.03	–	0.3
Lead	2.00	1.18	0.05	0.015
Cadmium	2.20	0.65	0.2	0.05
Manganese	3.10	1.04	0.2	0.05
Arsenic	2.60	1.20	–	0.01
Nickel	2.51	2.12	0.05	–
Chromium	2.43	1.80	0.05	0.1

*All values are in mg/L except pH
**COD—Chemical oxygen demand
#BOD—Biochemical oxygen demand
##TS—Total solid
[a]National Environmental Standards and Regulation Enforcement Agency [36] (Nigeria) maximum permissible limits for wastewater
[b] United State Environmental Protection Agency (2006) (www.epa.gov/safewater/mcl.html). OSL (raw)—Olusosun landfill raw leachate; OSL (simulated)—Olusosun landfill simulated soil leachate

2.5 Health Implication of Olusosun Landfill Leachate Induced DNA Damage on Humans and Wildlife

Damage to chromosome is a major heritable genetic problem which has been associated with cancer predisposition, pathophysiological disturbances and many other genetic related syndromes in wildlife and lower eukaryotes. Although, there is no report on genetic damage induced by emissions from Olusosun landfill among exposed humans, however, rats placed in different locations on the landfill for 4 to 24 weeks expressed dose related genetic damage [26]. Also, exposure to the leachate in laboratory controlled experiments has unequivocally been confirmed to

Table 2.3 Identified organic compounds at their various peak retention times in Olusosun simulated soil leachate (OSL) sample [10]

Peak retention time (min)	Identified organic compounds
2.03	Methylene chloride
2.64	[1,4]Dioxino[2,3-B]-1,4-hexahydro-2,3-dimethyl dioxin
6.32	1-[N-dodecyl] aziridine
6.81	N-methyl-N-nitroso-1-octanamine
8.26	1-isopropoxy-2,2,3-trimethylaziridine(sin)
8.81	3,3′-Iminobispropylamine
9.66	1,2-Ethanediamine, N-(2-aminoethyl)
10.91	1,3-bis-t-butylperoxy-phthalate
11.80	N-2-dimethyl-N-nitroso-1-propanamine
12.25	2-methyl-3-propyl oxaziridine,
12.97	4-Ethylamino-N-butylamine
13.29	Azacyclodecan-5-ol
15.07	1,6-anhydro-2,4-dideoxy-beta-D-ribo-hexopyranose
16.02	2-trifluoroacetylamino propenoic acid
19.36	Cyanoacetylurea
20.32	Methanol, chloro-Acetate
20.97	Gamma-guanidinobutyric acid
21.38	Methyl iminodiacetic acid
21.56	2,2′-Iminobis[N-Hydroxyl ethane imidamide]
21.84	2,3,4,5,6,7-hexahydro-3,6-dihexyl-10,11-diphenyl-bis[1,3] oxazino[6,5F]
22.03	2-Formyl-9-[Beta-D-ribofuranosyl] hypoxanthine
23.46	2-Decenoic acid
26.08	Pterin-6-carboxylic acid

induce DNA damage in both in vivo and in vitro test systems (Table 2.1). Women working within Cateura landfill in Asunción, Paraguay as waste pickers and are exposed to emissions from the landfill expressed significant increase in frequency of micronuclei and other nuclear abnormalities such as karyolytic and karyorrhectic cells in their peripheral erythrocytes. Also, there was significant high percentage of damaged DNA in the tail of comets in the analysed peripheral blood lymphocytes [19]. De Felice et al. [21] observed that emissions from illegal waste sites increased oxidative stress, shortened telomere length and lowered telomerase activity in peripheral blood mononuclear cells from healthy pregnant women in Campania, Italy. The altered biomarkers are determinants of cell senescence and aging-related meiotic dysfunction in women. Also, the presence of short telomeres is associated with some complex inherited disorders including dyskeratosis congenita, Alzheimer's disease, neoplastic transformation, and fragile X-associated tremor/ataxia syndrome [13, 31].

DNA damage has been implicated with mild to severe genetic and developmental disorders both in humans and wildlife. This corroborated evidence from epidemiological studies wherein exposure to emissions from landfills was associated with risk of congenital anomalies including Down syndrome, chromosomal congenital anomalies, foetal and neonatal deaths and developmental anomalies [4, 11, 27, 30, 55] among human and wildlife populations residing within 2–3 km circumference of landfill facilities. Figure 2.1 shows schematic presentation of xenobiotics in Olusosun landfill leachate generate reactive oxygen species (ROS) or acting directly on exposed organisms to induce genotoxicity. Which can be detected by chromosome aberration, micronucleus assay, Comet assay, sperm head abnormality test and toxicogenomic assays. The induced cytotoxic and genotoxic effects may lead to deleterious health impacts on individuals and populations occupationally and environmentally exposed to the emissions from landfills.

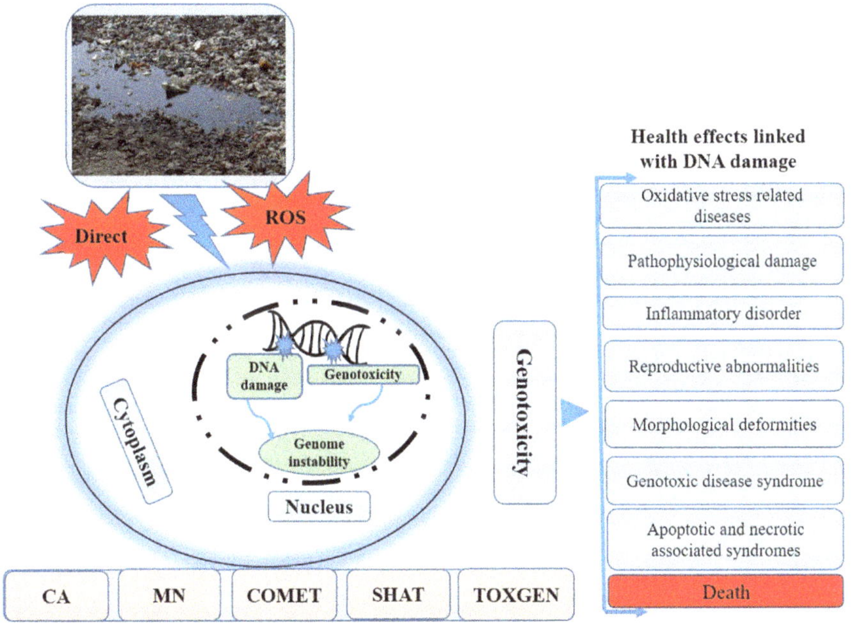

Fig. 2.1 Schematic presentation of xenobiotics in Olusosun landfill leachate generate reactive oxygen species (ROS) or acting directly on exposed organisms to induce genotoxicity as detected by chromosome aberration (CA), micronucleus assay (MN), Comet assay, sperm head abnormality test (SHAT) and toxicogenomic assays (TOXGEN). Induced genotoxic effects may lead to deleterious health impacts on populations

2.6 Conclusion

The potentials for emissions from solid waste landfills to induce cytotoxicity and DNA damage is increasing being investigated due to emissions from landfill are ubiquitous in the environment and human and wildlife are daily exposed. The reports herein showed that emissions from OSL is capable of increasing DNA damage via direct and indirect mechanisms. Different biomarkers of genotoxicity have been used to evaluate the genotoxic potentials of OSL leachates. Biomarkers of chromosome aberration, micronucleus assay, comet assay, sperm head abnormality test and toxicogenomic signature, were assessed in somatic and germline cells of exposed organisms. The mechanisms of DNA damage include generation of free radicals and oxidative stress, inhibition of DNA repair process, alkylation on DNA molecules that may lead to chromosome breakage, down-regulation of transcriptional genes related to apoptotic expressions, and increased DNA fragmentation in cells and tissue. The DNA damaging potentials of xenobiotics in OSL leachates may increase genome instability, the hallmark of genetic related diseases (Fig. 2.1). It is recommended that the ability of landfill leachates to induce genome instability related genetic disorder in human including cancer, should be adequately investigated.

References

1. Adedara IA, Lawal TA, Adesina AA, Oyebiyi OO, Ebokaiwe AP, Farombi EO (2014) Sperm functional parameters and erythrocytes oxidant–antioxidant imbalance during municipal landfill leachate treatment withdrawal in rats. Environ Toxicol Pharmacol 37:460–467
2. Adedara IA, Oyebiyi OO, Lawal TA, Adesina AA, Farombi EO (2013) Involvement of oxidative stress in municipal landfill leachate-induced toxicity in boar sperm. Environ Toxicol Pharmacol 36:972–978
3. Ademola OJ, Alimba CG, Bakare AA (2020) Reproductive toxicity assessment of Olusosun municipal landfill leachate in *Mus musculus* using abnormal sperm morphology and dominant lethal mutation assays. Environ Anal Health Toxicol 35(2) (Article ID: e2020010, 10 pages). https://doi.org/10.5620/eaht.e2020010
4. Alimba CG, Adewumi OO, Binuyo OM, Odeigah PGC (2021) Wild black rats (*Rattus rattus* Linnaeus, 1758) as zoomonitor of genotoxicity and systemic toxicity induced by hazardous emissions from Abule Egba unsanitary landfill, Lagos, Nigeria. Environ Sci Pollut Res Int 28(9):10603–10621
5. Alimba CG, Bakare AA (2016) In vivo micronucleus test in the assessment of cytogenotoxicity of landfill leachates in three animal models from various ecological habitats. Ecotoxicology 25:310–319
6. Alimba CG, Bakare AA, Aina OO (2012) Liver and kidney dysfunction in Wistar rats exposed to municipal landfill leachate. Resour Environ 2(4):150–163
7. Alimba CG, Bakare AA, Latunji CA (2006) Municipal landfill leachates induced chromosome aberration in rat bone marrow cells. Afr J Biotechnol 5(22):2053–2057
8. Alimba CG, Gandhi D, Sivanesan S, Bhanarkar MD, Naoghare PK, Bakare AA, Krishnamurthi K (2016) Chemical characterization of simulated landfill soil leachates from Nigeria and India and their cytotoxicity and DNA damage inductions on three human cell lines. Chemosphere 164:469–479

9. Alimba CG, Onajole AI, Bakare AA (2015) Brain dysfunctions in Wistar rats exposed to municipal landfill leachates. Beni-Suef Univ J Basic Appl Sci 4:284–290
10. Alimba CG, Rudrashetti AP, Sivanesan S, Krishnamurthi K (2022) Landfill soil leachates from Nigeria and India induced DNA damage and alterations in genes associated with apoptosis in Jurkat cell. Environ Sci Pollut Res 29(4):5256–5268
11. Alimba CG, Sivanesan S, Krishnamurthi K (2022) Mitochondrial dysfunctions elicited by solid waste leachates provide insights into mechanisms of leachates induced cell death and pathophysiological disorders. Chemosphere 307:136085. https://doi.org/10.1016/j.chemosphere.2022.136085
12. Alimba CG (2013) DNA and systemic damage induced by landfill leachates and health impacts of human exposure to landfills in Lagos and Ibadan, Nigeria. Ph.D thesis. Department of Zoology, University of Ibadan, Nigeria, pp 1–258
13. Alter BP, Rosenberg PS, Giri N, Baerlocher GM, Lansdorp PM, Savage SA (2012) Telomere length is associated with disease severity and declines with age in dyskeratosis congenita. Haematologica 97(3):353–359
14. Annepu RK (2012) Sustainable Solid Waste Management in India. MSc thesis. Department of Earth and Environmental Engineering, Fu Foundation School of Engineering and Applied Science, Columbia University in the City of New York, pp 1–189
15. Aralu CC, Okoye PC, Akpomie KG, Chukwuemeka-Okorie HO, Abugu HO (2022) Polycyclic aromatic hydrocarbons in soil situated around solid waste dumpsite in Awka, Nigeria. Toxin Rev 42(1):122–131. https://doi.org/10.1080/15569543.2021.2022700
16. Boyle WC, Ham RK (1974) Biological treatability of landfill leachate. J Water Pollut Control Feder 46:860–872
17. Christensen TH, Kjeldsen P, Bjerg PL, Jensen DL, Christensen JB, Baun A, Albrechtsen HJ, Heron C (2001) Biogeochemistry of landfill leachate plumes. Appl Geochem 16:659–718
18. Dhawan A, Bajpayee M, Parmar D (2009) Comet assay: a reliable tool for the assessment of DNA damage in different models. Cell Biol Toxicol 25:5–32
19. de Diana DF, Abreu JS, Serafini DC, Ortız JF, Samaniego MJ, Aranda AC, Zamorano-Ponce E (2018) Increased genetic damage found in waste picker women in a landfill in Paraguay measured by comet assay and the micronucleus test. Mutat Res 836:19–23
20. El-Fadel M, Findikakis AN, Leckie JO (1997) Environmental impacts of solid waste landfilling. J Environ Manag 50:1–25
21. De Felice B, Nappi C, Zizolfi B, Guida M, Di Spiezio Sardo A, Bifulco G, Guida M (2012) Telomere shortening in women resident close to waste landfill sites. Gene 500:101–106
22. Freeman A (2012) Seven of the largest landfills in the world. Takepart.com https://en.wikipedia.org/wiki/Olusosun_landfill. Accessed 2 Jan 2022
23. Gajski G, Orescanin V, Vera G-V (2012) Chemical composition and genotoxicity assessment of sanitary landfill leachate from Rovinj, Croatia. Ecotoxicol Environ Saf 78:253–259
24. Gardiner R, Hajek P (2020) Municipal waste generation, R&D intensity, and economic growth nexus—a case of EU regions. Waste Manage 114:124–135
25. Gates KS (2009) An overview of chemical processes that damage cellular DNA: spontaneous hydrolysis, alkylation, and reactions with radicals. Chem Res Toxicol 22(11):1747–1760
26. Gbadebo AM, Alabi OA, Alimba CG, Bakare AA (2022) Metal bioaccumulation, cytogenetic and clinico-biochemical alterations in *Rattus norvegicus* exposed *in situ* to a municipal solid waste Landfll in Lagos, Nigeria. Biol Trace Element Res. https://doi.org/10.1007/s12011-021-02758-3
27. Gilbreath S, Kass PH (2006) Fetal and neonatal deaths and congenital anomalies associated with open dumpsites in Alaska Native villages. Int J Circumpolar Health 65(2):133–147
28. Gupta P, Bhargava A, Kumari R, Lodhi L, Tiwari R, Gupta PK, Bunkar N, Samarth R, Mishra PK (2019) Impairment of mitochondrial nuclear cross talk in lymphocytes exposed to landfill leachate. Environ Health Insights 13:1–11
29. Hartwig A, Asmuss M, Ehleben I, Herzer U, Kostelac D, Pelzer A, Schwerdtle T, Bürkle A (2002) Interference by toxic metal ions with DNA repair processes and cell cycle control: molecular mechanisms. Environ Health Perspect 110(suppl 5):797–799

30. Jarup L, Morris S, Richardson S, Briggs D, Cobley N, de Hoogh C et al (2007) Down syndrome in births near landfill sites. Prenat Diag 27(12):1191–1196
31. Jenkins EC, Tassone F, Ye L, Gu H, Xi M, Velinov M, Brown WT, Hagermen RJ, Hagerman PJ (2008) Reduced telomere length in older men with premutation alleles of the fragile X mental retardation 1 gene. Am J Med Genet A 146A(12):1543–1546
32. Lea DE, Catcheside DG (1942) The mechanism of the induction by radiation of chromosome aberrations in *Tradescantia*. J Genet 44:216–245
33. Mills C (2018) 15 of the world's largest landfills with photos and statistics. Updated on June 6, 2018 (cited 2019 Aug 12). https://owlcation.com/stem/15-of-the-Worlds-Largest-Landfills
34. Mohammed AA, Iniaghe PO, Abu TO, Bello MO, Abdulkadir MD (2020) Source analysis of heavy metals and polycyclic aromatic hydrocarbons from a popular dumpsite, Lagos State, Nigeria. J Turk Chem Soc 7(2):489–504. https://doi.org/10.18596/jotcsa.687322
35. Morales ME, Derbes RS, Ade CM, Ortego JC, Stark J, Deininger PL, Roy-Engel AM (2016) Heavy metal exposure influences double strand break DNA repair outcomes. PLoS ONE 11(3):e0151367. https://doi.org/10.1371/journal.pone.0151367
36. National Environmental Standards and Regulation Enforcement Agency (Federal Republic of Nigeria Official Gazette) (2011) National Environmental (sanitation and waste control). Published by Federal Government of Nigeria printer, Abuja FGP 112/102009/L000 (OL54) 60(96):1057–1102
37. Ncube F, Ncube EJ, Voyi K (2017) A systematic critical review of epidemiological studies on public health concerns of municipal solid waste handling. Perspect Public Health 137(2):102–108
38. Ogundiran OO, Afolabi TA (2008) Assessment of the physico-chemical parameters and heavy metals' toxicity of leachates from municipal solid waste open dumpsite. Int J Environ Sci Technol 5(2):243–250
39. Ogungbemi KI, Adedokun MB, Ibitoye AZ, Oyebola OO, Owoade RL (2023) Estimation of radiological impact of the activities of Olusosun Dump Site on workers and dwellers of Olusosun, in Lagos Southwest Nigeria. J Radiat Res 64(1):53–62. https://doi.org/10.1093/jrr/rrac067
40. Ogunyemi AK, Samuel TA, Amund OO, Ilori MO (2018) Genotoxicity evaluation of landfill solid waste leachates from a waste dumpsite in Lagos State, Nigeria using the *Allium cepa* Assay. Plant Animal Res J 1(3):68–76
41. Ogwueleka TC (2009) Municipal solid waste characteristics and management in Nigeria. Iran J Environ Health Sci Eng 6:173–180
42. Ostling D, Johanson KJ (1984) Microelectrophoretic study of radiation-induced DNA damages in individual mammalian cells. Biochem Biophys Res Commun 123(1):291–298
43. Oziegbe O, Oluduro AO, Oziegbe EJ, Ahuekwe EF, Olorunsola SJ (2021) Assessment of heavy metal bioremediation potential of bacterial isolates from landfill soils. Saudi J Biol Sci 28:3948–3956
44. Pimienta-Serrano EV, Pacheco-Bustos C (2022) Perspectives on the environmental impact of anthropogenic activities and the generation of solid waste in beaches of the Colombian Caribbean. Environ Eng 24(2):1–12. https://doi.org/10.25100/iyc.v24i2.11365
45. Sang N, Li G (2004) Genotoxicity of municipal landfill leachate on root tips of *Vicia faba*. Mutat Res 560:159–165
46. Sang N, Li G (2005) Chromosomal aberrations induced in mouse bone marrow cells by municipal landfill leachate. Environ Toxicol Pharmacol 20:219–224
47. Shekha YA, Al-Attar MS, Saleem MA, Toma JJ, Abdulla SM (2017) Effect of landfill leachates extract of Erbil city on abnormal sperm morphology and chromosomal aberrations in male albino mice. ZANCO J Pure Appl Sci 29(6):18–27
48. Singh NP, Mccoy MT, Tice RR, Schneider EI (1988) A simple technique for quantitation of low levels of DNA damage in Individual cells. Exp Cell Res 175:184–191
49. Styrna J, Kilarski W, Krzanowska H (2003) Influence of the CBA genetic background on sperm morphology and fertilization efficiency in mice with a partial Y chromosome deletion. Reproduction 126:579–588

50. Toure A, Szot M, Mahadevaiah SK, Rattigan A, Ojarikre OA, Burgoyne PS (2004) A new deletion of the mouse Y chromosome long arm associated with the loss of *Ssty* expression, abnormal sperm development and sterility. Genetics 166:901–912
51. USEPA (1998) U.S. Environmental protection agency. Characterization of municipal solid waste in the United States: 1997 Update. EPA/530-R-98-007
52. United Nation (2013) Population Division, Department of Economic and Social Affairs of the United Nations Secretariat. World Population Prospects: the 2012 Revision (New York)
53. United States Environmental Protection Agency. National recommended water quality criteria 2006
54. United States Environmental Protection Agency (USEPA) (1997) Greenhouse gas emissions from the management of some selected materials in municipal solid wastes. Document prepared for Office of Solid Waste and Office of Policy, Planning and Evaluation, U.S Environmental Protection Agency by ICF incorporated EPA Contract No. 68-W6-0029 Work Assignment 0-06. https://nepis.epa.gov/Exe/ZyNET.exe/1000118X.TXT?
55. Vrijheid M, Dolk H, Armstrong B, Boschi G, Busby A, Jorgensen T, Pointer P (2002) Hazard potential ranking of hazardous waste landfill sites and risk of congenital anomalies. Occup Environ Med 59(11):768–776
56. Wyrobek AJ, Bruce WR (1975) Chemical induction of sperm abnormalities in mice (25 chemicals/mutagens/teratogens/carcinogens). Proc Natl Acad Sci USA 72:4425–4429
57. Wyrobek AJ, Gordon LA, Burkhart JG, Francis MW, Kapp RW Jr, Letz G, Malling HV, Topham JC, Whorton MD (1983) An evaluation of the mouse sperm morphology test and other sperm tests in nonhuman mammals: a report of the U.S. environmental protection agency Gene-Tox program. Mutation Res 115(1):1–72
58. Xu G, Zhao J, Yu H, Wang C, Huang Y, Zhao Q, Zhou X, Li C, Liu M (2022) Structural insights into the mechanism of high-affinity binding of ochratoxin A by a DNA aptamer. J Am Chem Soc 144(17):7731–7740

Chapter 3
Landfill Leachate Characteristics

Yasmin Cherni, Sarra Hechmi, Samira Melki, Mohamed Ali Wahab, and Ismail Trabelsi

Abstract Landfilling is an engineered alternative for the safe disposal of municipal solid waste (MSW) and hazardous waste. To dispose of solid waste, most countries use the initial phase of landfilling in the form of open dumping. Landfill leachate, gas emissions, and odor control are presented as issues of designing and operating a landfill. In fact, leachate is considered highly contaminated wastewater because it contains dissolved organic compounds, inorganic macro compounds and heavy metals. Therefore, leachate characterization is a crucial step towards preventing pollution and evaluating the performance leachate treatment alternatives. Furthermore, it is important to discuss some challenges in landfill leachate management and treatment. Accordingly, this chapter will introduce and summarize different topics related to landfill leachate characterization, including physical/chemical characterization and leachate generation and composition factors. Finally, future trends in landfill Leachate Management that could improve leachate quality are further discussed.

Keywords Landfill leachate · Waste management · Organic matter · Heavy metals · Biological characteristics

3.1 Introduction

Even after carrying out the three R's rule (reduce, recover, recycle), landfilling remains one of the most important components of a solid waste management strategy. In most cases, not all waste can be recovered and recycled. Most developing countries dispose of their waste in one of two ways: open dumping or regulated dumping [17]. In general, these countries lack the financial resources to implement costly

Y. Cherni (✉) · S. Hechmi · S. Melki · M. A. Wahab · I. Trabelsi
Laboratory of Wastewater Treatment and Valorization, Water Research and Technologies Center, CERTE, Carthage University, Technopark Tourist Route of Soliman Nabeul, B.P. 273, 8020 Süleyman, Tunisia
e-mail: yasmin.cherni@gmail.com

A. Anouzla and S. Souabi (eds.), *A Review of Landfill Leachate*, Springer Water,
https://doi.org/10.1007/978-3-031-55513-8_3

treatment systems such as waste-to-energy technology and materials recovery facilities. Typically, MSW in landfills is complexly degraded through chemical, physical, and biological processes. Rainwater percolating through waste results in the leaching of degraded matter [21]. The primary product of the MSW degradation process is landfill leachate [19]. In light of the information gathered from the literature, leachate is defined as the liquid effluent produced by rainwater percolating through solid waste disposed of in a landfill, as well as the moisture present in the waste and the degradation products of residues [2, 23]. Evapotranspiration, surface runoff, precipitation, groundwater infiltration, and the degree of compaction within landfill are the main factors that affect the amount of leachate [12]. The composition of leachate is primarily determined by the age of the landfill, the composition of solid waste, climatic conditions, temperature, hydrological conditions, and landfill operation processes [16]. The produced leachate is often highly toxic because the waste can contain a variety of chemicals, organic and inorganic compounds in large quantities, and these compounds also decompose. The massive quantities of MSW generated contain toxic and recalcitrant substances [5]. Thus, its control and management has become critical in recent years [1]. Waste segregation and recycling are two principal methods of reducing pollution in a landfill. In fact, microplastics are found among the hazardous materials containing in landfill leachates in many landfill leachates [18]. This has been developed over a long period of time, and the major types are polyethylene and polypropylene, which have become more prevalent over the years due to its simplicity of use [11]. Moreover, toxic trace metals such as mercury, lead, arsenic, as well as xenobiotic compounds such as halogenated organic compounds, have long been an issue of concern [25]. To control the pollution that can occur to groundwater, soil, and air, proper and timely observing of the landfill and its leachate is required [25]. Knowledge in improving the treatment for this landfill leachate management control is still an ongoing process that has increased the interest of experimental researchers and scientists, all over the world [12]. Several leachate treatment alternatives have been developed, including biological, physical, and chemical processes, as well as their combinations [8]. Further enhancements, especially in terms of reducing expenses optimisation, are part of the ongoing efforts. The main issue with landfill leachate is the quantity and degree of variability. To select the performant and efficient leachate treatment alternative, the leachate must be characterized and its risk assessed.

3.2 Importance of Understanding Landfill Leachate Characteristics

3.2.1 *Landfill Leachate Formation*

Leachate is formed in landfills through chemical and biological mechanisms that occur in four stages: (1) the aerobic stage; (2) the anaerobic acidogenic stage; (3) the methanogenic stage; and (4) the stabilising stage [4]. During the first aerobic stage, the oxygen available in the freshly buried waste empty spaces is rapidly depleted, resulting in the formation of CO_2. Thus, in this condition, several studies proved that the temperature of the waste rises. Leachate is produced in this phase as a result of moisture loss during the compacting process and precipitation short-circuiting via the deposited waste [13]. Once oxygen sources are reduced enough for fermentation to occur, waste enters the anaerobic phase, also known as the second step, and turns anaerobic. This phase is characterised by the hydrolytic, fermentative, and acetogenic bacteria, which causes a build-up of carboxylic acid and a decrease in pH [24]. When oxygen sources are reduced, fermentation processes can occur, and the anaerobic phase, or second step, begins. In this phase, hydrolytic, fermentative, and acetogenic bacteria predominate, resulting in carboxylic acid buildup and a pH drop. The BOD_5/COD ratio has been estimated to be over 0.4 or 0.7 in the acid phase. Due to its high chemical reactivity and acidic pH, acid stage leachate enhances the dissolution of many chemicals. A quantifiable amount of methane is produced, and this marks the beginning of the third phase which is the first stage of methanogenesis [9]. This stage is primarily related to when the pH of the waste reaches a sufficiently neutralised level to permit the germination of methanogenic bacteria, at least in a limited capacity. In addition, the bacteria that produce methane convert acids created during the acidic stage into CH_4 and CO_2, increasing the rate at which methane is produced. The rate of methane production will peak during the stable methanogenic phase and then start to slow as the amount of carboxylic acids decreases. The rate at which cellulose and hemicellulose break down during the acid phase is used to gauge the rate of CH_4 generation. As pH levels rise, the BOD/COD ratio frequently drops to as low as 0.1 during the stabilising phase because carboxylic acid is digested as quickly as it is produced [9]. The different stages of leachate formation are summarized in Fig. 3.1.

3.2.2 *Physical and Chemical Characteristics of Landfill Leachate*

The properties of leachate are definitely affected by the refuse components and moistures, as well as their sources. Rainfall is a dominant parameter that influences the leachate characteristics of landfill leachate. Rainwater enters the landfill and transports soluble pollutants from the solid phase to the liquid phase. At the same

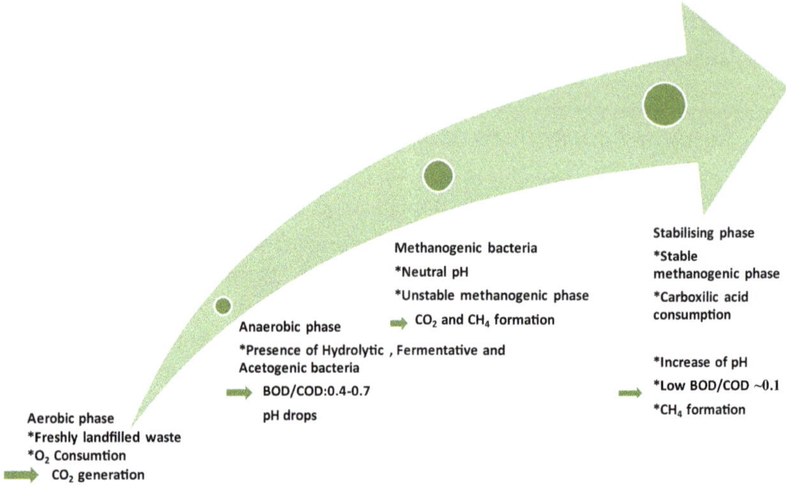

Fig. 3.1 Synoptic diagram of leachate formation process

time, organic waste is decomposed into soluble organic waste (such as volatile fatty acids), which penetrate the leachate via the action of microorganisms [3]. Leachate contains not only a high concentration of organic pollutants, but also a wide range of organic and inorganic, toxic and harmful compounds. It is worth mentioning that if industrial waste is mixed with refuse in a municipal solid waste landfill, the composition of leachate will be more complex due to pollutant dissolution in the industrial wastes [5].

The characteristics of landfill leachate are summarized in Table 3.1. Several reports proved significant differences in the components of the leachate. Furthermore, There are three types of leachates: young, intermediate, and old. Young leachate (less than 5 years old) has high concentrations of BOD and COD, moderate levels of ammoniacal nitrogen, a high BOD_5/COD ratio (> 0.6), and a pH of approximately 4. Furthermore, humic and fulvic acids, as well as NH_3-N, are produced at this stage as a result of anaerobic decomposition [5], indicating low biodegradability due to the release of high concentrations of ammonia, nitrogen, and recalcitrant macromolecular organic molecules. Furthermore, some organic matter interacts with heavy metals, indicating that both of these substances are major issues in landfill leachate treatment [10].

- **PH level**

The variation in the concentration of H^+ ions in landfill leachate depends on the number of years the landfill has been functional and the amount of volatile acid produced when methanogenic Bacteria can be found. The pH value in the initial stages of landfill is between 6 and 7, indicating a weak acidity. After the landfill stabilisation stage, the pH is between 7 and 8 [6].

Table 3.1 Typical compositions of landfill leachate

Parameter	Young (< 5 years)	Intermediate (5–10 years)	Stabilised (> 10 years)
PH	< 6.5	6.5–7.5	> 7.5
COD (mg/L)	> 10,000	4000–10,000	< 4000
BOD_5/COD	0.5–1.0	0.1–0.5	< 0.1
TOC/COD	< 0.3	0.3–0.5	> 0.5
Kjeldahl nitrogen (g/L)	0.1–0.2	NA[a]	NA[a]
Trace metals (mg/L)	Low to medium	Low	Low
Biodegradability	Important	Medium	Low

[a]NA is not available

- **Turbidity, Color and TDS**

Color is a common contaminant in landfill leachate. Water may change from yellow to dark brown as a result of the breakdown of organic molecules such as humic acid. According to literature, leachate has a high chroma and appears brown, dark brown, or black, with a strong odour of corruption. The color and turbidity of LFL are caused by chemicals and particles. Leachate has a high total dissolved solids concentration. The maximum value (1000 mg/L) is usually reached between 6 and 15 months. At the same time, it has high concentrations of Na^+, K^+, Cl^-, SO_4^{2-}, and other inorganic salts, which will gradually decrease until the landfill reaches its final stability. TDS demonstrate the integrative effect of specific cations and anions on wastewater, such as Ca^{2+}, Cl^-, Mg^{2+}, Na^+, K^+, and bicarbonates. Furthermore, total dissolved solids can be formed from small amounts of dissolved organic matter, and it has been proposed that higher EC (electrical conductivity) and TDS (total dissolved solid) can be used to identify dissolved organic and inorganic compounds in leachate [14].

- **Organic compounds**

Aromatic hydrocarbons, alkanes and alkenes, organic acids, lipids, alcohols and phenols, aldehydes and ketones, and amide are among the organic compounds found in leachate, which can be toxic or nontoxic. The suspected carcinogen and carcinogens, as well as blacklist priority pollutants, may be present in these organic matters. In the same vein, Dissolved organic matter (DOM) presents approximately 80% of TOC (total organic compound) in landfill leachate, with the remainder consisting primarily of refractory humic chemicals and VFA (volatile fatty acid). Typically, BOD_5 and COD are two measurements of dissolved organics. In landfill leachate, BOD and COD are both associated with organic and inorganic pollution, and the BOD/COD ratio has been applied to determine the age of landfill and organic contaminants containing in leachate. In general, BOD/COD ratio is approximately 0.7 in young leachate. If the ratio is between 0.1 and 0.5, the leachate will be in the intermediate stage. The BOD/COD ratio of stabilised or old leachate is less than 0.1 [14].

- **Inorganic macro compounds**

According to the literature, Cl^-, SO_4^{2-} and NH_3 are among the anions and cations found in landfill leachate. Nitrogen is a major pollutant in leachate because it can remain in the atmosphere for an extended period of time and stable even in anaerobic conditions. On the other hand, several researchers affirmed that geological and biological processes presents moderately effect on the concentration of Cl^-. Consequently, it is worthy to study the hydro-geological pathways. Concerning the nitrogen, for the most part, the concentration change as a result of protein breakdown, hydrolysis, and microorganism biodecomposition of organic compounds, resulting in nitrogen (NH_4-N) build-up.

Middle mature leachate is characterised by highly concentrated ammonia-nitrogen. In fact, this compound is hydrolyzed and fermented which increases the difficulty of ammonia nitrogen removal treatment. Currently, the majority of landfill sites use anaerobic technology, increasing pollutant concentrations even during the methane production. When the landfill reaches the stable stage, a high concentration of NH_3-N (1000 mg/L) can be obtained. During a 160-day leachate study, the NH_3-N concentration raised from 200 to 1500 mg/L. The chloride content was consistently high in all of the leachates. The Cl^- concentration validates a wide range of TDS values. According to several studies, leachate from landfills may contain a higher concentration of chlorides due to the dissolution of alloys and salts found in industrial waste, which is also disposed of in MSW landfills. Regardless of the sources, leachate concentrations are always much lower, particularly dissolved phosphate concentrations. For instance, in parts of North American landfills, the ratio of BOD_5/total phosphorus (TP) in the leachate exceeds 300, far below the 100:1 required for microbial growth. $Ca_5OH(PO_4)_3$ is found to be the most common form of dissolved phosphate in leachate. Calcium ion and total alkalinity concentrations in leachate can reach 7200 and 2500 mg/L, respectively, while TP concentration is only 125 mg/L. It indicates that calcium ion and alkalinity influence the content of soluble phosphate in leachate, following in phosphorus inadequacy in the biological treatment. It is evident that the phosphate inadequacy phenomenon persists in the anaerobic-aerobic process when the mixing ratio of leachate and sewage volume is 4:6 to 4:5. Furthermore, sulphate concentrations were found to be high in leachate samples, probably due to the presence of various inorganic waste products [12].

- **Xenobiotic organic compounds (XOCs)**

Because of the large differences in landfill and waste composition, as well as landfill time, the concentration range of XOCs varies. Aromatic compounds (benzene, toluene, ethylbenzene, and xylene) and halogenated hydrocarbons (tetrachloroethylene and trichloroethylene) are the most common XOCs. The herbicide phenoxy acid has also been found in the leachate [12].

- **Heavy metals concentration**

Heavy metals are one of the most hazardous substances found in landfill leachate. Because prior separation of non-hazardous and hazardous wastes is typical in most developing countries, higher concentrations of various heavy metals are found in landfill leachates [22], and the elimination of heavy metal is a difficult process. Heavy metal pollution from various sources may leach into landfill leachate. Heavy metals can be found in MSW in the form of discarded batteries, light bulbs, old motor lubricants, plastic, and paint chips, among other things. Tannery waste, electroplating waste, and used rechargeable and home batteries are the primary sources of Cd, Cr, Mn, and Ni in landfills [25]. Metal ions, particularly heavy metal ions, should be given special attention among the various pollutants in leachate due to their environmental toxicity and influence on biological treatment processes. The leachate contains a variety of heavy metal ions, and the insoluble metals in the waste are converted into soluble metal ions and then dissolved in the leachate via physical and chemical reactions [25].

3.2.3 Biological Characteristics of Landfill Leachate

Nitrite bacteria, nitrifying bacteria, denitrifying bacteria, desulfobacter, thiobacillus denitrificans, iron bacteria, sulphate reducing bacteria, methanogenic bacteria, pathogenic bacteria, and pathogenic microorganisms are all present in leachate and play an important role in its degradation. Furthermore, the high concentrations of heavy metal elements and ammonia nitrogen create an imbalance of microbial nutrient elements, which inhibits microorganism growth and regeneration. Corynebacterium and streptococcus are the most common bacteria isolated from leachate. Achromobacter, granular bacteria, aerobic bacteria, clostridium, micrococcus, Morax's bacteria, pseudomonas aeruginosa, Neisseria bacteria, Serratia, and *Staphylococcus* aureus are some other common bacteria.

3.3 Challenges in Leachate Management and Treatment

The effective management and treatment of landfill leachate present numerous challenges due to the complex and dynamic nature of leachate composition and the environmental risks associated with its improper handling. Addressing these challenges is crucial to mitigate negative impacts on both human health and the environment.

3.3.1 Composition Variability and Complexity of Landfill Leachate

Landfill leachate composition can vary widely depending on factors such as waste types, landfill age, and climatic conditions. This variability makes it challenging to develop standardized treatment approaches that can effectively address the diverse range of contaminants present. Treatment processes must be adaptable to handle changes in leachate characteristics over time. Landfill leachate is often characterized by high levels of biochemical oxygen demand (BOD) and chemical oxygen demand (COD), indicating the presence of organic pollutants. The complexity and toxicity of these organic compounds can hinder biological treatment processes and require advanced treatment techniques, often leading to higher operational and maintenance costs. Leachate can contain persistent organic pollutants (POPs) and heavy metals, many of which are resistant to degradation and pose long-term environmental risks. Effective removal of these pollutants necessitates specialized treatment technologies, which might not be readily available in all regions or might come at a significant expense.

3.3.2 Operational and Resource Challenges in Leachate Management and Treatment

In many regions, especially in developing countries, the lack of proper infrastructure and resources presents a significant challenge in managing and treating landfill leachate. Constructing and maintaining advanced treatment facilities require substantial investments in terms of both capital and skilled personnel. Several leachate treatment processes, such as advanced oxidation processes or membrane filtration, demand substantial energy inputs.

This can lead to an increased carbon footprint and potentially offset the environmental benefits of leachate treatment. Finding sustainable treatment approaches that minimize energy consumption is a critical challenge. Landfill sites vary in scale from small local landfills to large regional facilities. Developing cost-effective treatment solutions for both small-scale and large-scale operations is a challenge. Smaller landfills might lack the resources to implement advanced treatment technologies, while larger landfills require scalable solutions to manage the significant volume of leachate generated.

3.3.3 Regulatory and Environmental Considerations in Leachate Management

Stringent environmental regulations demand that landfill leachate treatment meet certain quality standards before discharge. Achieving compliance can be complex due to the diverse contaminants present in leachate. Continuous monitoring and adjustment of treatment processes are necessary to ensure adherence to these regulations. Some treatment alternatives can generate secondary waste streams, such as sludges or concentrates, that require proper management themselves. These waste streams can pose disposal challenges and raise questions about the overall sustainability of the treatment process. Even after leachate treatment, the potential for residual contaminants to impact groundwater and surface water remains. Long-term monitoring and maintenance are essential to ensure that treated leachate does not pose a threat to the environment or public health. The issue of long-term liability for potential pollution can be challenging to address. Thus, addressing these challenges requires a multidisciplinary approach, involving researchers, engineers, policymakers, and stakeholders. Collaboration and innovation are key to developing sustainable and effective solutions for landfill leachate management and treatment.

3.3.4 Recommendations to Consider in the Context of Landfill Leachate Management

In the context of landfill leachate management, a set of comprehensive recommendations can guide the development of effective strategies. Regular and thorough monitoring of leachate composition is imperative to grasp its evolving characteristics and trends over time. This data serves as a foundation to shape suitable treatment approaches and to promptly address any deviations that demand immediate attention.

Waste minimization and source separation strategies should be prominently emphasized to curtail leachate generation. Implementing rigorous waste management practices at the source significantly reduces both the volume and toxicity of leachate. Simultaneously, investing in research and development of advanced treatment technologies is crucial to tackle the intricate composition of leachate. Exploring emerging technologies, including advanced oxidation processes and membrane filtration, can enhance treatment efficiency and efficacy.

Acknowledging the diverse composition of leachate among landfill sites is vital. Rather than relying on uniform solutions, tailoring treatment methodologies according to specific site conditions is recommended. Collaboration amongst governmental bodies, researchers, engineers, and stakeholders within the waste management sector can foster holistic solutions. This collaborative effort addresses regulatory compliance, environmental concerns, and public health aspects effectively.

Raising public awareness about responsible waste disposal's direct impact on leachate generation is essential. Encouraging community participation in recycling programs and waste separation initiatives can bring positive change. Recognizing leachate management as an ongoing process is imperative. Establishing protocols for continuous monitoring, even after landfill closure, ensures treated leachate remains environmentally benign over time.

Exploring avenues to extract resources and energy from leachate aligns with circular economy principles. Utilizing treated leachate for energy generation, material recovery, or agricultural purposes where applicable can maximize its utility. To ensure adherence to quality standards, stringent regulations related to leachate management need reinforcement. Robust enforcement mechanisms should be implemented to deter non-compliance and uphold environmental integrity.

Education remains a cornerstone. Providing training programs for landfill operators, waste management personnel, and engineers involved in leachate management enhances their skills, ultimately elevating the overall effectiveness of treatment processes. Supporting ongoing research and knowledge sharing in the field bolsters the collective understanding of landfill leachate management. Collaboration with academic institutions and industry experts accelerates the continuous improvement of treatment methods and strategies.

By adopting these recommendations, stakeholders can contribute to effective landfill leachate management, minimize environmental impacts, and work towards a more sustainable and cleaner environment.

3.4 Future Trends in Landfill Leachate Management

As environmental concerns continue to grow and technology advances, the field of landfill leachate management is undergoing significant changes. Future trends are moving towards more efficient and sustainable approaches, driven by the need to minimize environmental impacts and maximize resource recovery.

3.4.1 Advancements in Leachate Treatment Technologies

Advanced oxidation processes, such as photocatalysis and ozonation, are gaining attention for their ability to effectively degrade persistent organic pollutants and contaminants resistant to conventional treatment methods [3, 15]. These processes applied reactive species to break down complex organic compounds into simpler, less harmful substances. Membrane filtration techniques, including nanofiltration and reverse osmosis, are becoming more efficient in removing contaminants from leachate [7]. Continued improvements in membrane materials and designs are enhancing the effectiveness and energy efficiency of these processes. Researchers are working on enhancing the performance of biological treatment processes by optimizing microbial consortium and conditions. This includes bioaugmentation, which involves introducing specialized microorganisms to enhance degradation efficiency, and designing biofilm reactors for improved treatment outcomes. Innovative approaches aim to convert the organic content in leachate into energy through anaerobic digestion, microbial fuel cells, and other technologies [8]. These methods not only treat leachate but also harness its energy potential, contributing to a more sustainable waste management cycle.

3.4.2 Sustainable Approaches to Leachate Management

By promoting source separation of waste at the source, the amount of waste entering landfills can be reduced. This not only lessens leachate generation but also minimizes the potential contaminants in the leachate. Public awareness campaigns and waste separation initiatives play a vital role in this approach.

Incorporating vegetation and engineered natural systems into landfill sites can aid in leachate treatment. Phytoremediation, the use of plants to remove pollutants, and green infrastructure like constructed wetlands can contribute to leachate treatment while enhancing ecological value. Small-scale, decentralized treatment systems offer advantages for smaller landfills or regions with limited resources [20]. These systems can be tailored to local conditions and reduce the need for extensive infrastructure development. Embracing the principles of the circular economy, landfills are being viewed as potential sources of valuable resources. Recovery of materials, energy, and water from leachate can contribute to reducing the environmental footprint of landfills. Future trends involve integrating landfill leachate management into broader waste management strategies. This includes considering leachate treatment alongside waste-to-energy, recycling, and composting initiatives to create synergies and optimize resource utilization.

These future trends reflect the evolving understanding of the environmental impacts of landfill leachate and the need for sustainable waste management practices. Advances in technology, coupled with a shift towards holistic and resource-efficient approaches, hold the promise of reducing the environmental footprint of landfills and protecting our ecosystems for generations to come.

3.5 Conclusion

In conclusion, landfill leachate is a complex and dynamic liquid waste generated during the decomposition of organic materials within landfills. Its composition is influenced by a multitude of factors, including waste types, landfill age, and environmental conditions. Key characteristics of landfill leachate include varying physical appearances, odors, and temperatures, along with chemical components like pH, organic pollutants, heavy metals, and nutrients. The presence of pathogens and a diverse microbial community also contribute to its unique nature. Effective management of landfill leachate is of paramount importance to ensure environmental protection and human health. The challenges associated with leachate management and treatment are multifaceted, ranging from the variability in leachate composition to the toxicity of its contaminants. These challenges demand innovative and sustainable approaches to mitigate the potential risks posed by untreated or improperly managed leachate. As we move forward, advancements in leachate treatment technologies, such as advanced oxidation processes, membrane filtration, and biological enhancements, hold promise for more efficient and comprehensive treatment. Embracing sustainable approaches, such as source separation, phytoremediation, and circular economy concepts, will further contribute to minimizing the environmental impact of leachate. The significance of effective leachate management extends beyond the boundaries of landfill sites. Untreated leachate has the potential to contaminate groundwater, surface water, and soil, leading to far-reaching ecological and public health consequences. By implementing robust treatment strategies and integrating leachate management into broader waste management practices, we can safeguard our environment for current and future generations.

In essence, the responsible management and treatment of landfill leachate represent an essential aspect of sustainable waste management. As our understanding of leachate characteristics and treatment options continues to evolve, our collective efforts to address the challenges associated with leachate will contribute to a cleaner and healthier environment.

References

1. De Almeida R, Porto RF, Quintaes BR, Bila DM, Lavagnolo MC, Campos JC (2023) A review on membrane concentrate management from landfill leachate treatment plants: the relevance of resource recovery to close the leachate treatment loop. Waste Manage Res 41(2):264–284
2. Anand N, Palani SG (2022) A comprehensive investigation of toxicity and pollution potential of municipal solid waste landfill leachate. Sci Total Environ 838:155891
3. Anjum M, Anees M, Qadeer S, Khalid A, Kumar R, Barakat MA (2023) A recent progress in the leachate pretreatment methods coupled with anaerobic digestion for enhanced biogas production: feasibility, trends, and techno-economic evaluation. Int J Mol Sci 24(1):763
4. Ateş H, Argun ME (2023) Fate of phthalate esters in landfill leachate under subcritical and supercritical conditions and determination of transformation products. Waste Manage 155:292–301
5. Bai F, Liu S, Zhang Y, Ma J (2023) Effective and mechanistic insights into reverse osmosis concentrate of landfill leachate treatment using coagulation-catalytic ozonation-bioaugmentation-based AnMBR. Chem Eng J 463:142430
6. De Carluccio M, Sabatino R, Eckert EM, Di Cesare A, Corno G, Rizzo L (2023) Co-treatment of landfill leachate with urban wastewater by chemical, physical and biological processes: Fenton oxidation preserves autochthonous bacterial community in the activated sludge process. Chemosphere 313:137578
7. Cherni Y, Elleuch L, Messaoud M, Kasmi M, Chatti A, Trabelsi I (2021) Recent technologies for leachate treatment: a review. Euro-Mediterr J Environ Integr 6:1–20
8. Dhamsaniya M, Sojitra D, Modi H, Shabiimam MA, Kandya A (2023) A review of the techniques for treating the landfill leachate. Mater Today Proc 77:358–364

9. Ghosh A, Kumar S, Das J (2023) Impact of leachate and landfill gas on the ecosystem and health: Research trends and the way forward towards sustainability. J Environ Manage 336:117708
10. Guzmán-Fierro V, Salamanca D, Arriagada C, Espinoza C, Campos V, Gallardo JJ, Roeckel M (2023) Efficient removal of nitrogen and organic matter strategy from landfill leachate under high seasonal substrate variations. Environ Technol Innov 32:103284
11. Kabir MS, Wang H, Luster-Teasley S, Zhang L, Zhao R (2023) Microplastics in landfill leachate: sources, detection, occurrence, and removal. Environ Sci Ecotechnol 100256
12. Kundu A, Reddy CV, Singh RK, Kalamdhad AS (2023) Critical review with science mapping on the latest pre-treatment technologies of landfill leachate. J Environ Manage 336:117727
13. Nalladiyil A, Prakash P, Babu GS (2023) Garbage enzyme-mediated treatment of landfill leachate: a sustainable approach. Bioresour Technol 129361
14. Obiri-Nyarko F, Duah AA, Karikari AY, Tagoe R (2023) Characterization of leachate, groundwater quality analysis, and evaluation of hydrogeochemical processes at the Kpone engineered landfill site, Ghana. Sustain Water Resour Manage 9(1):15
15. Samy M, Alalm MG, Khalil MN, Ezeldean E, El-Dissouky A, Nasr M, Tawfik A (2023) Treatment of hazardous landfill leachate containing 1, 4 dioxane by biochar-based photocatalysts in a solar photo-oxidation reactor. J Environ Manage 332:117402
16. Sangeetha A, Gandhimathi R, Nidheesh PV (2023) Treatment of stabilized landfill leachate using pyrite-activated persulfate oxidation process. Process Saf Environ Prot 171:413–422
17. Siddiqua A, Hahladakis JN, Al-Attiya WAK (2022) An overview of the environmental pollution and health effects associated with waste landfilling and open dumping. Environ Sci Pollut Res 29(39):58514–58536
18. Singh S, Malyan SK, Maithani C, Kashyap S, Tyagi VK, Singh R, Malhotra S, Sharma M, Kumar A, Panday BK, Pandey RP (2023) Microplastics in landfill leachate: occurrence, health concerns, and removal strategies. J Environ Manage 342:118220
19. Sliusar N, Filkin T, Huber-Humer M, Ritzkowski M (2022) Drone technology in municipal solid waste management and landfilling: a comprehensive review. Waste Manage 139:1–16
20. Wdowczyk A, Szymańska-Pulikowska A (2020) Differences in the composition of leachate from active and non-operational municipal waste landfills in Poland. Water 12(11):3129
21. Yaashikaa PR, Kumar PS, Nhung TC, Hemavathy RV, Jawahar MJ, Neshaanthini JP, Rangasamy G (2022) A review on landfill system for municipal solid wastes: insight into leachate, gas emissions, environmental and economic analysis. Chemosphere 136627
22. Yang X, Liu L, Wang Y, Lu T, Wang Z, Qiu G (2023) Sustainable and reagent-free cathodic precipitation for high-efficiency removal of heavy metals from soil leachate. Environ Pollut 320:121002
23. Yuan Y, Liu J, Gao B, Sillanpää M (2022) Landfill leachate treatment in-depth by bio-chemical strategy: microbial activation and catalytic ozonation mechanism. Chem Eng J 444:136464
24. Zhang D, Cheng L, Zhang S, Zheng J (2023) Denitrification performance and microbial community analysis of sulfur autotrophic denitrification filter for low-temperature treatment of landfill leachate. J Environ Chem Eng 11(2):109314
25. Zheng L, Lin H, Dong Y, Li B, Lu Y (2023) A promising approach for simultaneous removal of ammonia and multiple heavy metals from landfill leachate by carbonate precipitating bacterium. J Hazard Mater 456:131662

Chapter 4
Mechanism of Leachate Formation Pollutant and Category

Mai M. Badr

Abstract Solid waste management is becoming more and more difficult. Finding a solution that is both affordable and environmentally friendly is a challenge for many nations. When water undergoes filtering downward through a landfill, it collects dissolved components from the decaying waste and creates leachate. The leachate may be either safe or extremely harmful depending on the properties of the landfill and the waste it houses. An initial aerobic phase, an anaerobic acid phase, an initial methanogenic phase, and a stable methanogenic phase are the minimum stages of decomposition that landfills go through. Leachates may contain considerable concentrations of heavy metals, inorganic salts, ammonia, halogenated hydrocarbons suspended solids, chemical oxygen demand (COD), biochemical oxygen demand (BOD5), and organic pollutants in huge quantities. Also abundant in it are phenol, nitrogen, and phosphorus. Due to its propensity to seep into soils and subsoils and seriously pollute receiving waters, landfill leachate could be a significant cause of surface and groundwater contamination if it is not properly handled and disposed of. Therefore, throughout the past few decades, pollutant reduction from leachate treatment has gained substantial attention.

Keywords Pollutants · Young leachate · Old leachate

4.1 Introduction

The most popular method for handling the dumping of solid waste around the world is landfilling. Other than that, the best low-cost, economical, and environmentally responsible disposal methods are for industrial and municipal garbage [9, 11].

Leachate, a liquid that is produced when water or another liquid comes into contact with solid waste, is one of the biggest pollution issues brought on by municipal solid

M. M. Badr (✉)
Environmental Chemistry and Biology, Department of Environmental Health, High Institute of Public Health, Alexandria University, 165 El-Horreya Avenue, El-Ibrahimia, Alexandria 21561, Egypt
e-mail: mai.badr@alexu.edu.eg

A. Anouzla and S. Souabi (eds.), *A Review of Landfill Leachate*, Springer Water,
https://doi.org/10.1007/978-3-031-55513-8_4

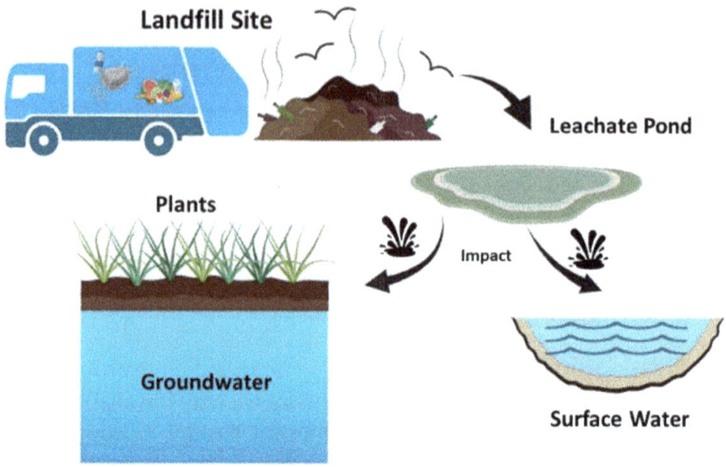

Fig. 4.1 Landfill leachate and its impacts on surface water and groundwater

waste landfills. Many dissolved and suspended elements are present. Upon closure of the municipal solid waste dump, the process of producing contaminated leachate may continue for 30 to 50 years. The problem of landfill leachate is getting worse because of population growth and industry development. Leachate, which migrates from the landfill and taints nearby lands and water, has the potential to contaminate ground and surface water and endanger human health. It is highly challenging and expensive to clean up the contaminated water once the leachate has entered the water bodies [16].

Heavy metals, organic chlorinated substances, and organic and inorganic materials are often present in higher concentrations in leachate effluent [5]. Landfill leachate has the potential to pollute receiving waterways severely if it is not properly treated and disposed of. It could seep into soils and subsoils and contaminate surface and groundwater [18] (see Fig. 4.1).

4.2 Leachate Formation Mechanism

In landfills, food waste, green trash, and other wastes that easily break down and appear in the leachate after a few months of dumping are examples of biodegradable waste. Most inorganic chemicals are easily soluble, and the released ions typically manifest quickly. The process of gently releasing heavy metals from the solid waste into the leachate can take up to many years [19].

The physical, chemical, and biological processes as well as moisture from precipitation all contribute to the degradation of trash in landfills. The solid phase, liquid phase (leachate), and gas phase are all present in the landfill. The liquid phase is chemically exceedingly complex and is characterized by the presence of several types of dissolved organic molecules, inorganic compounds, and heavy metals. The gas phase is composed of carbon dioxide (CO_2) and methane (CH_4). Leachate is the term for this liquid, which gathers at the bottom of the dump and slowly seeps into the soil to contaminate the aquifer underneath it and nearby surface water bodies. The governing factors that determine the pace of production and characteristics of the leachate are the solid waste content, particle size, degree of compaction, hydrology, age of the landfill, moisture, temperature, and oxygen that is accessible [17].

Through the subsequent stages of aerobic, acetogenic, methanogenic, and stabilization, the production of leachate also varies greatly. In a landfill, the trash decomposition process goes through several stages. The aerobic breakdown of organic materials characterizes the first phase, which is often brief. Anaerobic deterioration continues after the oxygen is gone. The two main fermentation phases of the anaerobic degradation process are the acidogenic phase, which produces young, biodegradable leachate, and the methanogenic phase, which produces old, stabilized leachate [6].

4.3 Pollutant Groups in Landfill Leachate

The nature of the deposited solid wastes, soil characteristics, passing time, temperature, moisture, available oxygen, rainfall patterns, the age of the landfill, landfill engineering and operational factors, pH, landfill chemical and biological activities, and others all affect the leachate composition at different sites and under different environmental conditions [1, 6, 7, 10, 20, 24]. Leachate is a liquid that is largely created when rainwater percolates through an open landfill or the cap of a finished site [4, 12, 19, 23].

Pollutants can be grouped into four classes. Dissolve organic matter, which accumulates during the acid phase of waste stabilization and is measured as Chemical oxygen demand (COD), total organic carbon (TOC), and volatile fatty acids. The second class consists of Inorganic macro-components like calcium, magnesium, sodium, potassium, ammonium, iron, manganese, chloride, sulfate, and hydrogen carbon. Heavy metals like chromium, cadmium, lead, iron, nickel, zinc, and copper represent the third class. Finally, Organic molecules, which are found in relatively low amounts, are derived from common or industrial chemicals [2].

In general, the biological degradation of organic molecules and the precipitation of soluble components like heavy metals cause leachate strength to decrease over time. The organic compounds degrade more quickly than the inorganic compounds as leachate production ages. As a result, as a landfill gets older, the ratio of total volatile solids to total fixed solids declines [21]. Low BOD5/COD ratio readings, most frequently around 0.1, are indicative of stabilized leachates. Leachate's dark color is mostly caused by humic material. An increase in pH is the result of a decrease in VFAs. For stabilized leachate, a typical pH value is close to 8 [13]. Due to the solubility of many metal ions decreasing with rising pH, the concentration of metal ions is often low. Lead and humic acids combine to generate very stable compounds.

The reduction of sulfate to sulfide during the methanogenic phase and the impact of pH changes on metal ions both contribute to an increase in the precipitation of metal ions [14].

4.4 Leachate Categorization

The performance of the landfill that produces leachate is impacted by the garbage age in a complicated combination of physical, chemical, and biological processes [15]. Leachate can be categorized and distinguished as either young (unstabilized) or old (stabilized) leachate based on its properties [3].

According to the age of the landfill and the rate of decomposition of landfill leachate wastewater, a change in the biodegradable ratio in leachate wastewater indicates the quality of different types of leachate wastewater. In general, the biodegradable ratios of young (less than 1-year-old), intermediate (between 1 and 5 years), and old (stabilized) leachate (more than 5 years) were reported as being, respectively, 0.5–1.0, 0.1–0.55, and 0.1 [3].

Large volumes of easily biodegradable organic materials are present in young leachate from the early acidogenic phase. Anaerobic fermentation of the complex organic molecules produces mostly soluble organic acids, including free volatile fatty acids (VFAs), amino acids, various low molecular weight chemicals, and gases including H_2 and CO_2 [14]. Low pH (less than 5) is caused by the presence of VFAs, which can account for up to 95% of the TOC. High BOD5/COD ratios of 0.5–0.7 suggest significant levels of biodegradable organic matter [8, 13]. The COD ranges from 3000 to 60,000 mg/l [2]. Due to the lower pH and the metals' increased solubility during this phase, there are relatively large quantities of Fe, Mn, Ni, and Zn [14].

Old landfill leachate wastewater is characterized as having a low biodegradable ratio value (0.1), making it challenging to treat with biological methods. Leachates from old landfills often have a pH between 7.0 and 7.6 and are either neutral in nature or mildly alkaline in nature. A reduced concentration of VFAs will be present in old leachate during the methanogenic phase. This is because, during this second fermentation stage, they are transformed into the gaseous end products methane and carbon dioxide. The amount of VFAs and other easily degradable organic compounds in the leachate declines, and refractory molecules like humic and fulvic acid-like substances start to dominate the organic matter in the leachate [8].

[22] claim that the extremely ancient landfill produces leachates with an alkaline pH range of 8.0–8.5. However, landfills with leachates that range in pH from 3.5 to 6.5 are likely to have created leachates during the early stages of waste decomposition. The leachate's acidic composition suggests that carbonate ions or carboxylic acids are present as well [3].

4.5 Conclusion

In conclusion, landfills go through at least four stages of decomposition: (a) an initial aerobic phase, (b) an anaerobic acid phase, (c) an initial methanogenic phase, and (d) a stable methanogenic phase. Chemical oxygen demand (COD) and total organic carbon (TOC), as well as ammonia nitrogen, heavy metals like copper, zinc, iron, lead, manganese, etc., and chlorinated organic and inorganic salts are the main components of landfill leachate and pose a serious threat to the nearby soil, groundwater, and even surface water. The use of properly constructed landfills and waste recycling, both of which are missing in most developing nations, is the only way to address the health risks associated with open dump sites.

References

1. Abudeif A (2015) Integrated electrical tomography and hydro-chemical analysis for environmental assessment of El-Dair waste disposal site, west of Sohag city, Egypt. Environ Earth Sci 74:5859–5874. https://doi.org/10.1007/s12665-015-4610-5
2. Aisien E, Nwatah C, Aisien F (2010). Biological treatment of landfill leachate from Benin City, Nigeria. Electron J Environ Agric Food Chem 9(11)
3. Alvarez-Vazquez H, Jefferson B, Judd SJ (2004) Membrane bioreactors vs conventional biological treatment of landfill leachate: a brief review. J Chem Technol Biotechnol 79(10):1043–1049. https://doi.org/10.1002/jctb.1072
4. Aziz HA, Daud Z, Adlan MN, Hung Y-T (2009) The use of polyaluminium chloride for removing colour, COD and ammonia from semi-aerobic leachate. Int J Environ Eng 1(1):20–35. https://doi.org/10.1504/IJEE.2009.02644
5. Bashir MJ, Aziz HA, Yusoff MS, Aziz SQ (2012) Color and chemical oxygen demand removal from mature semi-aerobic landfill leachate using anion-exchange resin: An equilibrium and kinetic study. Environ Eng Sci 29(5):297–305. https://doi.org/10.1089/ees.2010.0187
6. Bhalla B, Saini M, Jha M (2012) Characterization of leachate from municipal solid waste (MSW) landfilling sites of Ludhiana, India: a comparative study. Int J Eng Res Appl 2(6):732–745. https://doi.org/10.6088/ijes.002020300034
7. Bhalla G, Kumar A, Bansal A (2011) Assessment of groundwater pollution near municipal solid waste landfill. Asian J Water Environ Pollut 8(1):41–51
8. Chian ES, DeWalle FB (1976) Sanitary landfill leachates and their treatment. J Environ Eng Div 102(2):411–431. https://doi.org/10.1061/JEEGAV.0000476
9. Daud Z, Abubakar MH, Kadir AA, Latiff A, Awang H, Halim AA, Marto A (2016) Optimization of leachate treatment with granular biomedia: Feldspar and Zeolite. Indian J Sci Technol 9(37):91845. https://doi.org/10.17485/ijst/2016/v9i37/91845
10. Daud Z, Abubakar MH, Kadir AA, Latiff AAA, Awang H, Halim AA, Marto A (2017) Adsorption studies of leachate on cockle shells. Geomate J 12(29):46–52. https://doi.org/10.21660/2017.29.54771
11. Daud Z, Nasir N, Aziz Abdul Latiff A, Ridzuan MB, Awang H (2016) Treatment of biodiesel wastewater by coagulation-flocculation process using polyaluminium chloride (PAC) and polyelectrolyte anionic. ARPN J Eng Appl Sci 11(20):11855–11859
12. Foul AA, Aziz HA, Isa MH, Hung Y-T (2009) Primary treatment of anaerobic landfill leachate using activated carbon and limestone: batch and column studies. Int J Environ Waste Manage 4(3–4):282–298. https://doi.org/10.1504/IJEWM.2009.027397
13. Granet C, Courant P, Millot N, Rousseau C, Navarro A (1986) Diagnostic detraitabilite dex lixiviats: definition D'une methodologie. L'Eau et l'Industrie, Janvier

14. Harmsen J (1983) Identification of organic compounds in leachate from a waste tip. Water Res 17(6):699–705. https://doi.org/10.1016/0043-1354(83)90239-7
15. Hui TS (2005) Leachate treatment by floating plants in constructed wetland. Master's Thesis, Universiti Teknologi Malaysia, Malaysia
16. Kjeldsen P, Barlaz MA, Rooker AP, Baun A, Ledin A, Christensen TH (2002) Present and long-term composition of MSW landfill leachate: a review. Crit Rev Environ Sci Technol 32(4):297–336. https://doi.org/10.1080/10643380290813462
17. Ohwoghere-Asuma O, Aweto K (2013) Leachate characterization and assessment of groundwater and surface water qualities near municipal solid waste dump site in Effurun, Delta State, Nigeria. J Environ Earth Sci 3(9):126–134
18. Öman CB, Junestedt C (2008) Chemical characterization of landfill leachates—400 parameters and compounds. Waste Manage 28(10):1876–1891. https://doi.org/10.1016/j.wasman.2007.06.018
19. Renou S, Givaudan J, Poulain S, Dirassouyan F, Moulin P (2008) Landfill leachate treatment: Review and opportunity. J Hazard Mater 150(3):468–493. https://doi.org/10.1016/j.jhazmat.2007.09.077
20. Robinson A (2005) Landfill leachate treatment. Membr Technol 2005(6):6–12. https://doi.org/10.1016/S0958-2118(05)70435-3
21. Robinson H, Gronow J (1995) A review of landfill leachate composition in the UK. Paper presented at the IWM Proceedings
22. Słomczyńska B, Słomczyński T (2004) Physico-chemical and toxicological characteristics of leachates from MSW landfills. Pol J Environ Stud 13(6):627–637
23. Uygur A, Kargı F (2004) Biological nutrient removal from pre-treated landfill leachate in a sequencing batch reactor. J Environ Manage 71(1):9–14. https://doi.org/10.1016/j.jenvman.2004.01.002
24. Welander U (1998) Characterization and treatment of municipal landfill leachates (Doctor). Lund University

Chapter 5
A Review of Landfill Leachate with Environment Impacts: Sustainable Waste Management and Treatment Methods of Vellalore Dump Yard, Coimbatore Corporation

K. Manikanda Bharath, R. Ruthra, Archana Kasinath, and Usha Natesan

Abstract The study examined to identify, characterize, and quantify microplastics in Groundwater samples around the municipal solid waste dumpsites, namely Vellalore dump yard, Coimbatore Corporation in South India. We have collected groundwater samples from bore well (15 nos) in 1-L Silanized amber glass bottles with Teflon-lined caps within 2 km distance from the municipal solid waste disposal sites during the post-monsoon season (November 2019) in South India. To evaluate the identification and quantification of microplastic abundance, characteristics (composite, size, colour, shape and surface morphology), and distribution of groundwater performed using LABOMED zoom stereo microscope, Fourier Transform Infrared Spectroscopy (ATR-FTIR) fitted with Complete Attenuated Reflectance and Scanning electron microscope (SEM) fitted with an energy dispersive X-ray (EDX) analyser. The total amount of microplastic particles was found to contain freshwater sources among the groundwater samples ($n = 15$) ranging from 6 and 80 items/L. Vellalore dump yard sampling sites dominant-coloured particles included white (18%), black (32%), red (19%), blue (16%), green (10%), and yellow (5%) respectively. Also, a Polymer type occurs in the following order nylon (70%), pellets (18%), foam (6%), fragments (3%) fibers/PVC (2%) and polythene (1%). The study

K. M. Bharath (✉) · U. Natesan
National Institute of Technical Teachers' Training and Research Chennai, Anna University, Chennai, Tamilnadu 600113, India
e-mail: krmanibharath93@gmail.com

R. Ruthra
Department of Civil Engineering, St. Joseph's College of Engineering, OMR, Chennai, Tamilnadu 600119, India

A. Kasinath
Department of Water and Wastewater Technology, Faculty of Civil and Environmental Engineering, Gdansk University of Technology, 11/12 Narutowicza St, 80-233 Gdańsk, Poland
e-mail: archana.kasinath@pg.edu.pl

A. Anouzla and S. Souabi (eds.), *A Review of Landfill Leachate*, Springer Water,
https://doi.org/10.1007/978-3-031-55513-8_5

confirmed the microplastic tracers in groundwater contamination near dump yards. In Vellalore dump yard sampling sites, 90% of microplastics were derived from the fragmentation of the buried waste of the plastics predominantly polypropylene (PP), polystyrene (PS) was presented. These methods for identifying and quantifying microplastics in groundwater is likely to provide a scope for further research into the pathways by which microplastic enter the environment.

Keywords Groundwater quality · Microplastic contamination · Solid waste mismanagement · Polymer types · Vellalore dump yard · Hazardous waste

5.1 Introduction

Plastic pollution has emerged as a global crisis with profound environmental consequences [29]. The production and consumption of plastic have surged worldwide, generating a substantial increase in plastic waste [33]. Unlike organic materials, plastic does not naturally decompose, leading to widespread environmental accumulation [34]. Concerns are growing that open landfill sites are releasing microplastics into freshwater ecosystems, though research in this area lags behind investigations into land-based sources [22]. Microplastics, due to their small size and persistence, present significant health risks to humans, organisms, and the environment [20]. This unbridled pursuit of industrial development and resource exploitation has negatively impacted all forms of life on Earth, rendering air, water, and soil increasingly unsuitable for living beings and disrupting ecosystems [21].

To combat plastic pollution, many countries have implemented policies and strict regulations on plastic usage and alternatives. However, this has raised concerns about the survival of all living organisms [6]. Microplastics, extremely small plastic debris resulting from product disposal and breakdown, are now ubiquitous in our daily lives, entering the environment directly and accumulating over time [1]. These minute plastic particles, along with associated toxic elements, are readily ingested and accumulate in various organisms, causing environmental harm and potentially entering the human food chain. The issue of microplastics in the environment should be of great concern [2]. Firstly, they can pose health risks, particularly to children who may inhale them. Secondly, the environment acts as a widespread distributor, spreading plastic particles to undesirable locations such as agricultural fields, national parks, seas, and even remote regions like the Arctic and Antarctica [14]. The presence of plastics in such high proportions is a looming problem [7]. While organic components of solid waste are biodegradable and less problematic, plastic waste poses significant challenges in urban and suburban environments due to its non-biodegradable nature [5].

In India, urban areas, including municipalities, are grappling with high levels of plastic waste, with households being significant contributors. Plastics have replaced traditional materials like leaves, glass, and metals as a cost-effective and efficient means of packaging [3]. In our daily lives, numerous materials, including synthetic and natural fibers in furniture and textiles, contribute to this plastic waste issue. Despite the urgent need for research on environmental plastic pollution, several challenges persist in understanding the degradation of plastic particles and associated health risks [8]. Microplastics, in particular, carry toxic substances, posing significant threats to microorganisms and ecosystems. These persistent pollutants enter the food chain, exacerbating health and environmental concerns. There is a pressing need for enhanced knowledge and a fundamental understanding of the abundance and characteristics of microplastics, as well as the key challenges they present to sustainable strategies [11].

Due to urbanization and population growth, plastic usage continues to rise, resulting in increased plastic waste and environmental concerns. In India, many urban areas, including Chennai, are grappling with high levels of plastic pollution [3]. The Chennai urban heat island is no exception, with plastic litter emerging as a critical issue affecting both society and the environment. The rapid increase in the use of plastic products has led to a substantial surge in plastic waste in metro cities like Chennai. The challenges posed by rapid urbanization, inadequate waste management, and the contamination of air, water, and soil threaten water sources, contribute to environmental pollution, and increase the risk of flooding and climate change [23]. While some studies have provided valuable insights into the occurrence, distribution, and sources of microplastics in Asian wetlands, rivers, beaches, and estuaries, there remains a significant lack of groundwater data [28]. Comparisons with findings from other regions around the world suggest that microplastic contamination in Asia's inland waters could be more severe. Some research indicates that plastic particles are not evenly distributed in groundwater samples and are more abundant near dumping sites. However, investigations into the pathways of microplastics in aquatic taxa, bioaccumulation of contaminants, and toxicological impacts, as well as microplastic ingestion in freshwater ecosystems, are still largely unexplored [24].

In recent decades, the use of plastic products has steadily increased, leading to a significant rise in plastic waste in major Indian cities, particularly in Coimbatore. The shift from traditional packaging methods like leaf wrappers, brown paper, and metal cups to plastic packaging, driven by hygiene concerns, highlights the unique properties of plastics such as versatility, inertness, and flexibility. Consequently, plastics have become the preferred choice for packaging, resulting in substantial profits for companies while burdening the public with the environmental cost of plastic waste cleanup. Currently, the disposal of plastic waste is by incineration and landfill in Coimbatore, which is a threat to the environment. The former is not environmentally friendly and sustainable since this may release carbon dioxide, a major contributor to global warming (greenhouse effect). The latter is not also desirable since plastic is non-degradable. However, very little has been yet done in this area. Hence the earlier the plastic waste menace problem is tackled the better it would be for the environment and sustainable livelihood. The main objective of this study is to establish a cohesive framework that combines plastic waste management strategies with contemporary analytical technology for the purpose of mitigating microplastic contamination and promoting environmental sustainability. Moreover, in order to enhance sustainability by satisfying our collective needs without exacerbating the adverse impact or exhaustion of the remaining ecological resources. Furthermore, In order to illustrate the advantages of prioritizing the rise in recycling of materials and energy over landfilling, this study aims to examine the resulting reduction in environmental impact, diminished health dangers, decreased consumption of energy resources, lower economic costs, and ultimately the promotion of sustainable waste management.

5.2 Study Zone

The study area of Coimbatore is situated in the southern part of India, approximately 494 km southwest of Chennai. Coimbatore is a prominent city in the state of Tamil Nadu, covering an expansive area of approximately 246 sq. km. With a population of over 1.6 million people as of the 2011 census, it is one of the major urban centers in the region. Coimbatore is divided into multiple zones and wards, making it a diverse and densely populated city in its own right. The topography of the Coimbatore region is characterized by a generally flat terrain, with elevations ranging from about 400–600 m above mean sea level. The climate in Coimbatore is tropical, with distinct wet and dry seasons. The lowest recorded temperature in the city is around 15 °C, while the highest can reach up to 40 °C, with relative humidity levels typically fluctuating between 60 and 80% [30]. Coimbatore, much like Chennai, experiences significant influence from seasonal monsoons. The two primary monsoon seasons affecting the area are the southwest monsoon, which typically arrives in June and lasts until September, and the northeast monsoon, which arrives from October to December. The city receives a substantial amount of its annual rainfall during these monsoon periods. On average, Coimbatore receives around 700 mm of rainfall annually, with the southwest monsoon contributing approximately 400 mm, and the northeast monsoon contributing the remaining 300 mm.

Groundwater plays a critical role in meeting the water needs of Coimbatore's growing population. As of the last census in 2011, the city's population stood at approximately 1.6 million people. Coimbatore, like many urban areas in India, generates a significant amount of solid waste. On average, each person in the city produces an estimated 985 g of solid waste daily. It has been reported that Coimbatore generates about 2500 tons of solid waste per day across its various zones, with municipal authorities responsible for managing a substantial portion of this waste.

The Vellalore Dump Yard, also known as the Vellalore Solid Waste Management Plant, is a waste disposal facility located in Vellalore, a suburban of Coimbatore in the Indian state of Tamil Nadu (Fig. 5.1). It is one of the primary locations for waste disposal and management in the Coimbatore Municipal Corporation area. The Vellalore Dump Yard is situated in the Vellalore area on the outskirts of Coimbatore city in Tamil Nadu, India. The facility is primarily used for the disposal of solid waste generated in Coimbatore and its surrounding areas. It serves as a dumping ground for the waste collected from households, commercial establishments, and industries in the region.

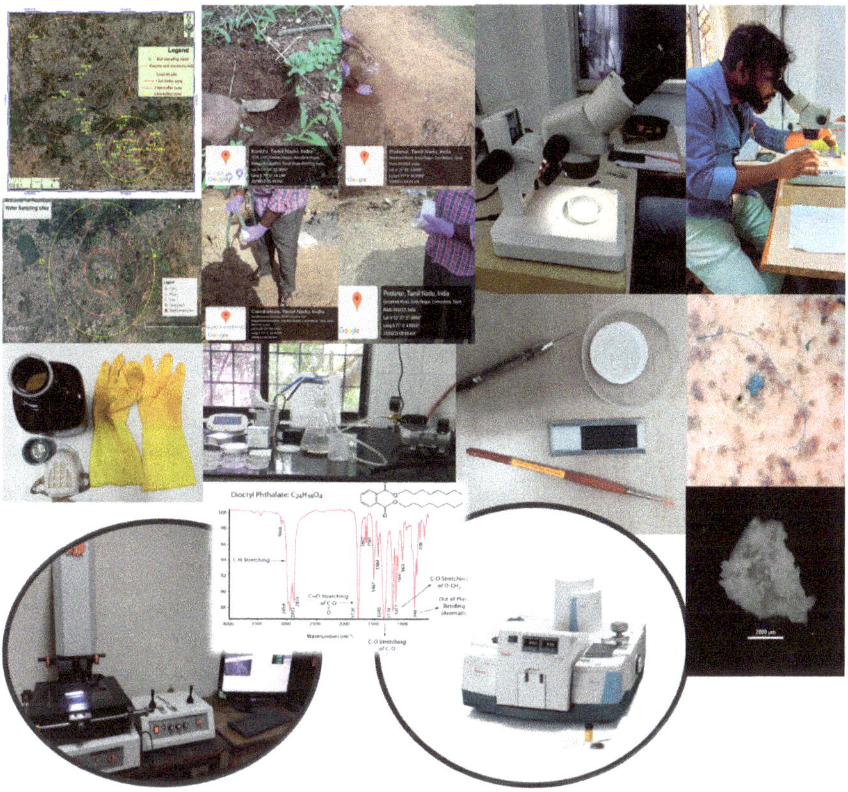

Fig. 5.1 Vellalore Dump Yard investigation and data collection equipment and tools

The Vellalore Dump Yard has faced significant environmental concerns and controversies over the years. Improper waste management practices, including open dumping and burning of waste, have resulted in air and water pollution, odor problems, and adverse health effects for nearby residents. These issues have led to protests and legal actions in an effort to address and mitigate the environmental impact. The unabated disposal of waste has resulted in persistent odors and a general nuisance for the surrounding communities. Residents have raised concerns about the quality of life and the negative impact on their daily lives. The proximity of the dump yard to residential areas has raised significant health concerns. Prolonged exposure to air pollutants and contaminants from the dump yard has been linked to respiratory illnesses and other health issues among local residents.

5.3 Materials and Methods

The methodology of the research includes four main components that are interlocked (geology, hydrology, biotechnology, remote sensing), which will be combined with existing technologies and models to produce more accurate and fine-tuned maps for human health risk in well-established landfills environments [31]. This review will determine the concentration of microplastic in soil, surface, and groundwater in and around the landfill site. Extraction of MP from sieved sediment samples was done as per National Oceanic and Atmospheric Administration (NOAA) protocol. The wet peroxide oxidation (WPO) and density separation [32] using $ZnCl_2$ to separate MPs through floatation. The supernatant was then filtered using Whatman GF/A filter paper (25 mm) and air dried and stored in petridish. The materials and methodology of the microplastic analytical process are shown in the Fig. 5.2.

Preliminary observations will be carried out using stereomicroscope (Labomed CZH4) to estimate the morphology and color. Microplastics were visually identified and measured according to their physical characteristics [18]. The number, shape (fiber, film, pellet, fragment), color (white and colored), and size of the plastics will be noted and discussed.

The collected particles will be further observed and documented using a Scanning Electron Microscope for the characterization of MPs like colour, morphology, and texture. In addition, Raman analyses will be performed for further identification. The surface elemental composition was determined in SEM coupled energy-dispersive X-ray spectra (EDX) [15]. The microplastic classifications and Densities of plastic materials its applications given in the Table 5.1.

5.4 Problems and Mitigation Measures

5.4.1 Improper Waste Management Practices and Their Environmental and Health Implications at Vellalore Dump Yard

The Vellalore Dump Yard, located on the outskirts of Coimbatore, Tamil Nadu, has long been a subject of concern due to its enduring practice of open dumping and waste burning. These waste management practices have not only had severe environmental repercussions but have also emerged as a significant public health hazard for the communities residing in close proximity. This article provides a comprehensive examination of the detrimental effects and consequences arising from these improper waste management practices.

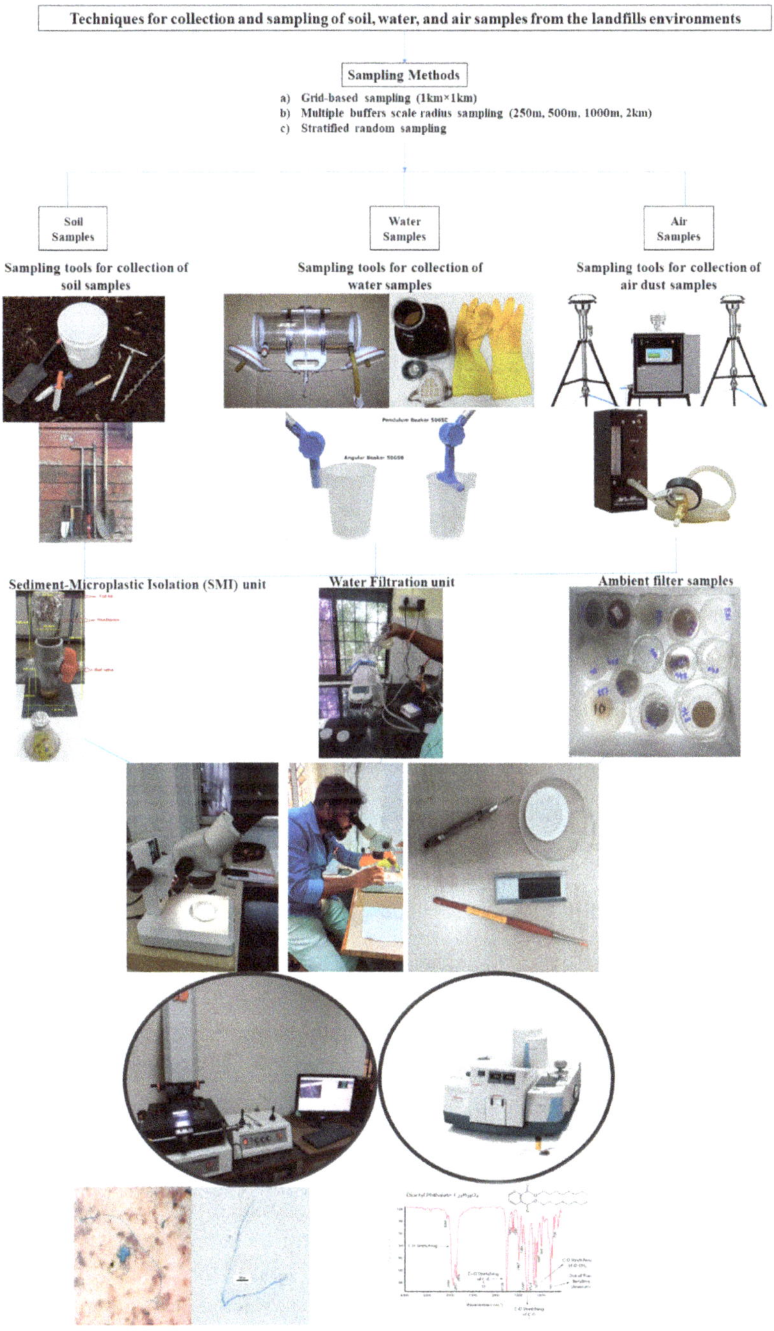

Fig. 5.2 Microplastic extraction and counting methods adopted in Landfills environments

Table 5.1 Several classifications and densities of plastic materials its applications

Plastic class	Density (g/cm^3)	Applications
Expanded polystyrene (styrofoam)	0.01–0.04	Electronics, cups in vending machines, refrigerator liners
Low-density polyethylene	0.89–0.93	Packaging, general-purpose containers, shower curtains, floor tiles
High-density polyethylene	0.94–0.98	Milk containers, detergent bottles, tubing
Polypropylene	0.83–0.92	Packaging, bottle caps, ropes, carpets, laboratory equipment, drinking straws
Polyethylene terephthalate	0.96–1.45	Soft drink bottles, food packaging, thermal insulation, blister packs
Polyamide (nylon)	1.02–1.16	Textiles, toothbrush bristles, fishing lines, automotive
Polystyrene	1.04–1.1	Packaging foam, disposable cups, food containers, CDs, building materials
Polymethyl methacrylate (acrylic)	1.09–1.20	
Polyvinylchloride	1.16–1.58	Pipes, window frames, flooring, shower curtains
Polyester	1.24–2.3	Textiles
Polytetrafluoroethylene	2.1–2.3	Soft drink bottles, food packaging, thermal insulation, blister packs
Polyethylene	0.91–0.96	Supermarket bags, plastic bottles
Acrylonitrile butadiene styrene	1.06–1.08	Musical instruments, printers, computer monitors, drainage pipes, protective equipment

5.4.1.1 Open Dumping

Open dumping at the Vellalore Dump Yard involves the uncontrolled disposal of solid waste in an unrestricted manner. This practice has resulted in the piling up of vast amounts of garbage without any effective containment or containment liners. Over time, this unregulated accumulation of waste has led to several adverse consequences:

- **Land Degradation**: Open dumping has caused significant land degradation in the surrounding areas, as the waste gradually encroaches upon and disrupts natural ecosystems and habitats.
- **Odor Emissions**: The accumulation of waste, coupled with its decomposition, emits foul odors that affect the quality of life for local residents. These odorous emissions contribute to discomfort and distress among the community.
- **Aesthetic Decline**: The unsightly heaps of garbage are not only a source of pollution but also degrade the visual aesthetics of the region, impacting property values and overall well-being.

5.4.1.2 Waste Burning

Waste burning is another alarming practice at the Vellalore Dump Yard. This involves the incineration of waste materials in the open air, releasing a plethora of harmful substances into the atmosphere. The implications of waste burning are multifaceted:

- **Air Pollution**: Burning waste generates a complex mixture of air pollutants, including particulate matter, volatile organic compounds, dioxins, and heavy metals. These pollutants deteriorate air quality and can have dire consequences on respiratory health, particularly for those residing in the vicinity.
- **Health Impacts**: The toxic emissions from waste burning pose acute and chronic health risks to nearby communities. Respiratory problems, cardiovascular issues, and a heightened risk of cancer have been associated with prolonged exposure to these pollutants.
- **Environmental Consequences**: Beyond its effect on human health, waste burning contributes to environmental degradation by releasing greenhouse gases and contributing to climate change.

The persistence of open dumping and waste burning at the Vellalore Dump Yard stands as a stark illustration of the challenges posed by improper waste management. The consequences span a wide spectrum, encompassing environmental degradation, public health hazards, and adverse socio-economic effects. It is imperative that effective and sustainable waste management practices be implemented urgently to mitigate these detrimental impacts and ensure the well-being of the local community and the environment. This research underscores the pressing need for immediate action and policy reforms to address the longstanding issues at the Vellalore Dump Yard.

5.4.2 Air Pollution from Waste Burning at Vellalore Dump Yard

5.4.2.1 Particulate Matter (PM) Emissions

- **Sources**: The open burning of waste materials, which often include plastics, paper, and organic matter, releases fine particulate matter (PM) into the air. PM is categorized based on its size, with PM2.5 (particles with a diameter of 2.5 μm or smaller) and PM10 (particles with a diameter of 10 μm or smaller) being of particular concern.
- **Health Impact**: Inhaling PM can have serious health consequences. PM2.5 particles are small enough to penetrate deep into the lungs and even enter the bloodstream. Prolonged exposure to PM is associated with respiratory diseases such as bronchitis, aggravated asthma, and can lead to cardiovascular problems, including heart attacks and strokes.

- **Visibility**: PM emissions also reduce visibility, affecting not only air quality but also posing safety hazards on roads and for transportation.

5.4.2.2 Volatile Organic Compounds (VOCs)

- **Sources**: Waste burning releases a variety of volatile organic compounds (VOCs) into the atmosphere. These compounds can originate from plastics, paints, solvents, and other materials commonly found in waste.
- **Health Impact**: VOCs can have both short-term and long-term health effects. Short-term exposure can cause eye, nose, and throat irritation, headaches, and nausea. Long-term exposure to certain VOCs is associated with more serious health issues, including damage to the liver, kidneys, and central nervous system. Some VOCs are also classified as carcinogens.
- **Environmental Impact**: VOCs can contribute to the formation of ground-level ozone and smog, which have detrimental effects on the environment and ecosystems.

5.4.2.3 Hazardous Chemicals:

- **Sources**: The waste disposed of at the Vellalore Dump Yard may contain hazardous chemicals such as heavy metals (lead, mercury, cadmium), polycyclic aromatic hydrocarbons (PAHs), and dioxins and furans. These chemicals can be released during combustion.
- **Health and Environmental Impact**: Exposure to these hazardous chemicals can result in a range of health problems, from developmental issues in children to cancer in adults. Additionally, these pollutants can persist in the environment and accumulate in the food chain, impacting wildlife and human health.

5.4.2.4 Respiratory Problems and Health Risks:

- **Exposure**: The poor air quality resulting from waste burning poses significant health risks to nearby residents who are consistently exposed to noxious fumes. Children, the elderly, and individuals with preexisting respiratory conditions are particularly vulnerable.
- **Health Consequences**: Respiratory problems such as coughing, wheezing, and exacerbation of asthma symptoms are common among individuals exposed to the polluted air. Long-term exposure can lead to chronic respiratory diseases and a higher risk of respiratory infections.

In conclusion, the practice of burning waste at the Vellalore Dump Yard has dire consequences for air quality in the vicinity. It releases a complex mixture of harmful pollutants, including particulate matter, volatile organic compounds, and hazardous chemicals, which have far-reaching health implications for local residents and environmental repercussions. Addressing this issue requires urgent measures to curb waste burning and implement proper waste management practices that prioritize public health and environmental protection.

5.4.3 Water Pollution from Contaminants Leaching into Groundwater at Vellalore Dump Yard

5.4.3.1 Contaminant Sources

- **Percolation Through Waste**: The waste disposed of at the Vellalore Dump Yard contains a diverse range of contaminants, including heavy metals, organic chemicals, and hazardous substances. Over time, rainwater can percolate through the accumulated waste, dissolving and carrying these contaminants into the underlying soil and groundwater.

5.4.3.2 Groundwater Contamination

- **Pathways**: The contaminants that leach into the groundwater can follow various pathways, including infiltration through the waste itself and lateral movement through the subsurface. These pathways enable the pollutants to gradually reach and contaminate the local groundwater aquifers.

5.4.3.3 Drinking Water Supply Contamination:

- **Public Wells and Boreholes**: Many communities in the vicinity of the Vellalore Dump Yard rely on groundwater from public wells and boreholes as their primary source of drinking water. The contamination of these water sources can have severe implications for public health.
- **Well Vulnerability**: Contaminants such as heavy metals and organic chemicals can render well water unsafe for consumption. Exposure to contaminated water can lead to acute and chronic health issues, including gastrointestinal problems, organ damage, and an increased risk of certain cancers.

5.4.3.4 Adverse Effects on Aquatic Ecosystems:

- **Surface Water Bodies**: In addition to contaminating groundwater, pollutants can also find their way into nearby surface water bodies, such as rivers, streams, and ponds. This can result in the degradation of aquatic ecosystems.
- **Impact on Flora and Fauna**: Aquatic flora and fauna are particularly vulnerable to changes in water quality. Contaminants can disrupt the balance of these

ecosystems, leading to fish kills, altered species compositions, and reduced biodiversity.

5.4.3.5 Long-Term Environmental Impact

- **Persistence**: Some contaminants, such as heavy metals, can persist in the environment for extended periods, posing a long-term threat to both human health and the environment.
- **Bioaccumulation**: Certain contaminants can bioaccumulate in the food chain, meaning that organisms at higher trophic levels can accumulate higher concentrations of toxins. This can have far-reaching consequences for the health of wildlife and, indirectly, for human populations that rely on these ecosystems for resources.

5.4.3.6 Mitigation and Remediation:

- **Monitoring and Assessment**: Continuous monitoring of groundwater quality in the affected areas is essential to assess the extent of contamination and its impact on drinking water sources.
- **Remediation Measures**: Implementing remediation measures, such as groundwater treatment and wellhead protection zones, can help mitigate the spread of contaminants and protect drinking water supplies.
- **Waste Management Improvements**: Addressing the root cause of the issue requires a shift toward proper waste management practices that prevent the leaching of contaminants in the first place.

In conclusion, water pollution resulting from contaminants leaching into groundwater at the Vellalore Dump Yard presents a multifaceted challenge. It not only jeopardizes local drinking water supplies but also has adverse effects on aquatic ecosystems. Addressing this issue necessitates a comprehensive approach that includes monitoring, remediation, and improved waste management practices to safeguard both human health and the environment.

5.4.4 Odor Nuisance from the Vellalore Dump Yard

5.4.4.1 Odor Sources

- **Decomposing Waste**: The primary source of foul odors at the Vellalore Dump Yard is the decomposition of organic waste materials. As waste degrades, it produces various gases, including hydrogen sulfide (H2S), methane (CH4), and volatile organic compounds (VOCs), which can emit pungent and unpleasant odors.

5.4.4.2 Factors Contributing to Odor Emissions

- **Waste Composition**: The composition of the waste in the dump yard, which includes a mixture of household waste, industrial waste, and other materials, influences the types and intensity of odors produced.
- **Temperature and Weather**: Weather conditions, such as temperature and wind direction, can impact the dispersion and concentration of odors. High temperatures can intensify odor emissions, while wind can carry odors over longer distances.
- **Moisture Levels**: The presence of moisture in the waste can accelerate decomposition and increase odor production.
- **Waste Disturbance**: Activities such as waste dumping and compacting can disturb waste layers, releasing trapped gases and exacerbating odor issues.

5.4.4.3 Impact on Quality of Life

- **Health Concerns**: Prolonged exposure to foul odors can lead to various health issues, including headaches, nausea, and respiratory discomfort. It can also worsen existing respiratory conditions like asthma [12].
- **Discomfort**: The constant presence of unpleasant odors affects the quality of life for residents in the surrounding areas. It can lead to discomfort, annoyance, and distress among the local population.
- **Social and Economic Impact**: Odor nuisance can have social and economic implications, including reduced property values, difficulty attracting businesses or new residents, and impacts on tourism and local businesses.

5.4.4.4 Community Complaints

- **Frequent Complaints**: The persistent odor issues at the Vellalore Dump Yard have resulted in frequent complaints from residents in the nearby communities.
- **Community Activism**: Local residents and community groups have actively advocated for solutions to address the odor problem and improve their living conditions.

5.4.4.5 Mitigation Measures

- **Odor Control Technologies**: Various odor control technologies, such as landfill gas collection and treatment systems, can help capture and neutralize odorous gases before they are released into the atmosphere.
- **Waste Management Practices**: Implementing better waste management practices, including waste segregation, can reduce the production of odorous gases.
- **Buffer Zones**: Establishing buffer zones and vegetative barriers around the dump yard can help mitigate the dispersion of odors to nearby residential areas.

5.4.4.6 Regulatory and Compliance Measures

- **Environmental Regulations**: Government agencies may have regulations in place to address odor emissions from landfill and waste management facilities. Compliance with these regulations is essential to minimize odor nuisance.

In conclusion, the odor nuisance originating from the Vellalore Dump Yard is a significant concern for local residents, impacting their quality of life and well-being. Addressing this issue requires a combination of technological solutions, improved waste management practices, and regulatory compliance to mitigate the impact of foul odors and provide relief to the affected communities.

5.4.5 Health Concerns Due to Proximity to Vellalore Dump Yard

5.4.5.1 Respiratory Illnesses

- **Airborne Pollutants**: Residents living near the dump yard are at an increased risk of respiratory illnesses due to prolonged exposure to airborne pollutants. The burning of waste and the release of particulate matter (PM) and volatile organic compounds (VOCs) can irritate the respiratory system [10].
- **Conditions**: Common respiratory problems include coughing, wheezing, exacerbation of asthma symptoms, bronchitis, and upper respiratory tract infections. Prolonged exposure may contribute to the development of chronic respiratory diseases [27].

5.4.5.2 Skin Problems

- **Contact with Contaminated Water**: In areas where groundwater is contaminated due to leaching from the dump yard, residents may be exposed to contaminated water during daily activities such as bathing and washing clothes.
- **Skin Irritation**: Contaminants in water can lead to skin problems, including irritation, rashes, and dermatitis.

5.4.5.3 Gastrointestinal Issues

- **Contaminated Water and Food**: If groundwater or local crops are contaminated with hazardous substances from the dump yard, residents may ingest these contaminants through water consumption or the consumption of contaminated produce [19].

- **Gastrointestinal Distress**: Ingesting contaminated substances can result in gastrointestinal problems, including diarrhea, nausea, vomiting, and abdominal pain.

5.4.5.4 Increased Cancer Risk

- **Carcinogenic Substances**: Some of the pollutants found in waste, such as certain heavy metals and organic chemicals, are known or suspected carcinogens. Prolonged exposure to these substances can increase the risk of cancer among nearby residents [26].

5.4.5.5 Neurological and Developmental Effects

- **Exposure to Hazardous Chemicals**: Children and pregnant women are particularly vulnerable to the effects of exposure to hazardous chemicals, which can have neurological and developmental impacts [25].

5.4.5.6 Mental Health Implications

- **Psychological Stress**: Living in proximity to a dump yard, with associated health concerns and unpleasant odors, can lead to psychological stress and reduced quality of life. Residents may experience anxiety, depression, and a sense of helplessness [9].

5.4.5.7 Vulnerable Populations

- **Children and the Elderly**: Children and the elderly are often more susceptible to the health effects of pollution due to weaker immune systems and preexisting health conditions.

5.4.5.8 Monitoring and Medical Care

- **Health Surveillance**: Local health authorities should conduct health surveillance to monitor residents' health in proximity to the dump yard and provide timely medical care and interventions as needed.
- **Public Awareness**: Raising public awareness about the potential health risks associated with living near a dump yard is crucial so that residents can take preventive measures and seek medical attention if necessary.

In conclusion, the proximity to the Vellalore Dump Yard raises serious health concerns among nearby residents. Prolonged exposure to air pollutants and the potential contamination of groundwater can result in a range of health issues, including respiratory illnesses, skin problems, gastrointestinal distress, and an increased risk of cancer. Addressing these health concerns requires comprehensive monitoring, health-care access, and public awareness initiatives to protect the well-being of affected communities.

5.4.6 *Lack of Waste Segregation at Vellalore Dump Yard*

5.4.6.1 Ineffective Waste Separation

- **Mixing of Materials**: The dump yard often receives mixed waste streams, including household waste, industrial waste, construction debris, and organic matter, without proper separation. This results in a chaotic jumble of materials that are difficult to manage or recycle effectively.
- **Loss of Valuable Resources**: Without waste segregation, valuable resources such as metals, plastics, paper, glass, and organic waste that could be reused, recycled, or composted are often lost in the general waste stream.

5.4.6.2 Environmental Impact

- **Reduced Recycling Rates**: The lack of waste segregation hinders recycling efforts. Recyclable materials that could be diverted from landfills end up being buried, contributing to the depletion of natural resources and increased landfill space requirements.
- **Wasted Organic Matter**: Organic waste, such as food scraps and yard waste, when mixed with general waste, decomposes in landfills, producing methane, a potent greenhouse gas that contributes to climate change.

5.4.6.3 Economic Implications

- **Lost Economic Opportunities**: Proper waste segregation can create economic opportunities by facilitating recycling and composting industries. When recyclable and organic materials are disposed of in landfills, these economic potentials remain untapped.

5.4.6.4 Operational Challenges

- **Waste Sorting Difficulties**: The presence of mixed waste at the dump yard makes waste sorting and disposal more challenging and costly. It requires additional labor and resources to separate materials manually, which may not be fully effective.

5.4.6.5 Health and Safety Concerns

- **Hazards for Waste Workers**: Waste workers who manually sort waste at the dump yard are exposed to health and safety risks when dealing with unsegregated waste. This includes the potential for injury and exposure to hazardous materials.

5.4.6.6 Aesthetic and Environmental Pollution

- **Visual Impact**: The presence of mixed waste contributes to the unsightly appearance of the dump yard, impacting the visual aesthetics of the area and diminishing the overall quality of life for nearby residents.
- **Environmental Pollution**: The decomposition of unsegregated organic waste generates odors and releases pollutants into the air and groundwater, contributing to environmental pollution.

5.4.6.7 Need for Education and Awareness

- **Community Education**: Encouraging waste segregation at the source, such as at households and businesses, is essential. Community education programs can help raise awareness about the importance of proper waste separation and its positive environmental and economic effects.

5.4.6.8 Policy and Infrastructure Improvements

- **Regulatory Frameworks**: Implementation and enforcement of waste segregation regulations can help address this issue. Penalties for improper disposal of segregated waste can incentivize compliance.
- **Infrastructure Upgrades**: Investment in infrastructure, including waste sorting and recycling facilities, can enhance the efficiency of waste segregation and recycling processes.

In conclusion, the lack of waste segregation at the Vellalore Dump Yard results in a multitude of challenges, including reduced recycling rates, environmental pollution, economic losses, and operational difficulties. Addressing this issue requires a multifaceted approach that includes community education, regulatory enforcement, and improvements in waste management infrastructure to encourage proper waste separation and promote more sustainable waste management practices.

5.4.7 Ecological Disruption Due to Vellalore Dump Yard

5.4.7.1 Habitat Destruction

- **Encroachment**: The establishment and expansion of the dump yard have led to the physical encroachment on natural habitats, including wetlands, grasslands, and woodlands. These areas were once crucial for local wildlife species.

5.4.7.2 Soil Contamination

- **Leaching of Contaminants**: The dumping of waste at the site can result in the leaching of contaminants into the surrounding soil. These contaminants can affect soil quality and fertility, making it less suitable for plant growth.

5.4.7.3 Water Contamination

- **Groundwater and Surface Water**: As mentioned earlier, contaminants from the dump yard can leach into groundwater, potentially affecting local water sources. This can disrupt aquatic ecosystems and impact aquatic species.
- **Contaminated Runoff**: Rainwater runoff from the dump yard may carry pollutants into nearby streams, rivers, or ponds, affecting aquatic life downstream.

5.4.7.4 Biodiversity Decline

- **Loss of Native Species**: The disruption of natural habitats and the presence of pollutants can lead to a decline in native flora and fauna. Some species may be directly harmed by pollutants, while others may lose their habitat or food sources.
- **Invasive Species**: Dump yards can create conditions that favor the establishment of invasive species, which can outcompete native species and further reduce biodiversity.

5.4.7.5 Impact on Migratory Species

- **Migration Routes**: Dump yards located along the migration routes of birds and other wildlife can disrupt their annual journeys, leading to exhaustion, reduced breeding success, and potentially affecting the populations of these species.

5.4.7.6 Ecosystem Functionality

- **Disrupted Ecosystem Services**: Natural ecosystems provide vital services, such as pollination, water purification, and nutrient cycling. The disruption caused by the dump yard can impair these services, affecting the overall health and resilience of the local environment.

5.4.7.7 Long-Term Consequences

- **Cumulative Effects**: The ecological disruption caused by the dump yard can have long-term, cumulative effects on local ecosystems. These effects may extend far beyond the dump yard's operational lifespan.

5.4.7.8 Conservation Efforts

- **Habitat Restoration**: Efforts to mitigate the ecological disruption may involve habitat restoration projects that aim to reclaim degraded areas and create suitable habitats for native species.
- **Wildlife Corridors**: The establishment of wildlife corridors or protected areas in the vicinity can help maintain connectivity between fragmented habitats, allowing species to move and maintain genetic diversity.
- **Monitoring and Research**: Ongoing monitoring and research are essential to assess the impact of the dump yard on local ecosystems and inform conservation efforts.

In conclusion, the presence of the Vellalore Dump Yard has resulted in significant ecological disruption, including habitat destruction, soil and water contamination, a decline in biodiversity, and potential harm to migratory species. Addressing these ecological issues requires a concerted effort involving habitat restoration, wildlife corridor creation, and ongoing research to mitigate the long-term consequences on the local environment.

5.4.8 Public Outcry and Protests Against Vellalore Dump Yard

5.4.8.1 Environmental and Health Concerns

- **Growing Awareness**: Over time, as the environmental and health issues linked to the dump yard became more evident, local residents, environmental activists, and concerned citizens became increasingly aware of the negative consequences of its operations.
- **Health Impact**: The visible impact of pollution, the prevalence of foul odors, and the rising health concerns have galvanized the community to take action.

5.4.8.2 Community Mobilization

- **Local Residents**: Residents in the vicinity of the dump yard, who have directly experienced the adverse effects of its operations, have been at the forefront of the protest movement.
- **Environmental Activists**: Local and regional environmental groups and activists have played a pivotal role in organizing and mobilizing communities against the dump yard.

5.4.8.3 Key Issues and Demands

- **Improved Waste Management**: Protesters and activists have called for immediate improvements in waste management practices, including waste segregation, recycling, and the reduction of waste burning.
- **Relocation of the Dump Yard**: A prominent demand has been the relocation of the dump yard to a more suitable location, away from residential areas, to protect public health and the environment.
- **Environmental Compliance**: Demonstrators have urged authorities to ensure that the dump yard complies with environmental regulations and standards.

5.4.8.4 Protests and Demonstrations

- **Rallies**: Public protests, rallies, and demonstrations have been organized to draw attention to the issues surrounding the dump yard.
- **Petitions**: Citizens have initiated petitions and signature campaigns to formally present their concerns and demands to local authorities and government bodies.

5.4.8.5 Legal Actions

- **Court Petitions**: Some activists and concerned citizens have taken legal action by filing petitions and lawsuits against the dump yard. These legal actions aim to hold responsible parties accountable and seek remedies for environmental and health damages.

5.4.8.6 Media Coverage

- **Media Attention**: The issues related to the Vellalore Dump Yard have received substantial media coverage, including print, broadcast, and online outlets. This coverage has helped to raise awareness and garner public support.

5.4.8.7 Collaboration with Environmental Organizations

- **NGOs and Environmental Groups**: Local environmental organizations and non-governmental organizations (NGOs) have joined forces with the affected communities to provide support, resources, and expertise in addressing the issues associated with the dump yard.

5.4.8.8 Government Response

- **Policy Revisions**: In response to the protests and public outcry, local and regional authorities may consider revising waste management policies and regulations to address the concerns raised by the community.
- **Dialogue**: Government officials may engage in dialogues and negotiations with community representatives to find mutually acceptable solutions.

In conclusion, the environmental and health concerns associated with the Vellalore Dump Yard have sparked significant public outcry and protests. Local residents, activists, and environmental groups have united to demand improved waste management practices, environmental compliance, and, in many cases, the relocation of the dump yard to safeguard public health and protect the environment. The collective efforts of these concerned individuals and organizations underscore the importance of community engagement in addressing pressing environmental issues.

5.4.9 Legal Actions and Petitions Regarding Vellalore Dump Yard

1. **Petitions to Authorities**
   - **Formal Appeals**: Concerned citizens, community groups, and environmental activists have submitted formal petitions to various government authorities at the local, state, and national levels. These petitions outline the issues associated with the dump yard and present specific demands for improved waste management and environmental compliance.
2. **Environmental Impact Assessment (EIA)**
   - **Demand for EIA**: Some legal actions and petitions have called for a comprehensive Environmental Impact Assessment (EIA) of the Vellalore Dump Yard's operations. An EIA evaluates the environmental and social consequences of a project or facility and is often a requirement under environmental regulations.
3. **Court Petitions and Lawsuits**
   - **Seeking Remedies**: Concerned individuals and organizations have taken legal action by filing petitions and lawsuits against the dump yard's operators, local authorities, and government agencies. These legal actions seek remedies

for environmental violations and health hazards caused by the dump yard's operations.
- **Compliance with Regulations**: Court petitions often emphasize the importance of ensuring compliance with existing environmental regulations and standards.

4. **Injunctions and Compliance Orders**
- **Court Orders**: In response to legal actions, courts may issue injunctions or compliance orders that require specific actions to be taken by the dump yard operators or local authorities. These orders can mandate changes in waste management practices, pollution control measures, or even the closure and relocation of the facility.

5. **Public Interest Litigation (PIL)**
- **Community Advocacy**: Public interest litigation (PIL) is a legal mechanism that allows concerned citizens and organizations to bring legal actions in the interest of the public and the environment. PIL has been used in some cases to advocate for improved waste management practices and the protection of local ecosystems.

6. **Advocacy and Awareness Building**
- **Legal Support**: Environmental and legal advocacy organizations have supported affected communities in their legal actions and petitions by providing legal expertise, resources, and guidance.
- **Media and Public Awareness**: Legal actions and petitions have often been accompanied by media campaigns and public awareness initiatives to garner support and inform the public about the issues surrounding the dump yard.

7. **Collaborative Approaches**
- **Negotiations**: Legal actions may also lead to negotiations and dialogues between community representatives, environmental activists, and government authorities. These discussions aim to find solutions that balance environmental protection and public health with practical waste management needs.

8. **Accountability and Transparency**
- **Demand for Accountability:** Legal actions and petitions underscore the importance of accountability and transparency in waste management practices. They seek to hold responsible parties accountable for environmental violations and health hazards.

In conclusion, legal actions and petitions have been instrumental in addressing the issues at the Vellalore Dump Yard. They serve as a formal and structured means for concerned individuals and communities to advocate for better waste management, compliance with environmental regulations, and the protection of public health and the environment. These legal efforts highlight the importance of the rule of law in ensuring that environmental concerns are addressed and that the rights and well-being of communities are protected.

5.4.10 *Mitigation Measures and Environmental Policies and Guidelines that Can Be Implemented or Suggested for the Vellalore Dump Yard*

1. **Waste Segregation and Recycling**
   - **Implementation**: Encourage and enforce waste segregation at source, where households and businesses separate recyclables, organic waste, and non-recyclable materials.
   - **Recycling Infrastructure**: Invest in recycling facilities and promote recycling programs to divert recyclable materials from landfills.
   - **Composting**: Establish composting facilities to manage organic waste, reducing methane emissions from decomposition.
2. **Proper Waste Management**
   - **Landfill Upgrades**: Upgrade the dump yard to modern landfill standards with lined cells and landfill gas collection systems to minimize groundwater contamination and methane emissions.
   - **Daily Cover**: Implement a daily cover system to reduce odor emissions and prevent scavenging by animals.
3. **Waste-to-energy (WtE) Facilities**
   - **WtE Plants**: Consider the construction of Waste-to-Energy (WtE) facilities that can convert non-recyclable waste into energy, reducing the volume of waste sent to landfills and providing an additional source of energy.
4. **Environmental Impact Assessment (EIA)**
   - **Mandatory EIAs:** Enforce the requirement for an Environmental Impact Assessment (EIA) before establishing or expanding dump yards. EIAs should assess potential environmental and health impacts and suggest mitigation measures.
5. **Air Quality Monitoring**
   - **Continuous Monitoring**: Install air quality monitoring stations in and around the dump yard to track pollutants, such as particulate matter (PM) and volatile organic compounds (VOCs). This data can inform timely actions to address air pollution issues.
6. **Groundwater Protection**
   - **Groundwater Monitoring**: Implement a comprehensive groundwater monitoring program to detect and address any contamination promptly.
   - **Wellhead Protection Zones**: Establish wellhead protection zones around nearby drinking water wells to safeguard them from contamination.
7. **Buffer Zones and Land Use Planning**
   - **Buffer Zones**: Create buffer zones around the dump yard to limit its impact on neighboring communities. These zones can include green belts and vegetative barriers to absorb odors and pollutants.
   - **Land Use Policies**: Enforce land use policies that prevent the development of residential or sensitive land uses near dump yards.

8. **Public Awareness and Education**
   - **Community Engagement**: Engage with the local community through educational programs, workshops, and public meetings to raise awareness about waste management practices and environmental protection.
   - **Reporting Mechanisms**: Establish mechanisms for residents to report environmental concerns, odors, and violations for prompt action.
9. **Regulatory Enforcement**
   - **Stringent Regulations:** Strengthen environmental regulations and compliance standards for dump yards, including waste management, air quality, and groundwater protection.
   - **Penalties:** Impose penalties for violations to incentivize adherence to regulations.
10. **Closure and Relocation Plans**
    - **Closure Procedures**: Develop a comprehensive closure plan for the dump yard, outlining the steps and procedures for its eventual closure in a way that minimizes long-term environmental impact.
    - **Relocation**: Explore options for relocating the dump yard to a more suitable location away from residential areas.
11. **Sustainable Waste Management Practices**
    - **Zero Waste Initiatives**: Promote the adoption of "Zero Waste" principles, aiming to reduce waste generation, increase recycling and composting rates, and minimize landfilling.
    - **Extended Producer Responsibility (EPR)**: Implement EPR programs, where producers are responsible for managing the waste generated by their products, encouraging more sustainable product design.
12. **Stakeholder Collaboration**
    - **Multi-stakeholder Committees:** Establish multi-stakeholder committees involving government agencies, environmental organizations, waste management experts, and affected communities to oversee dump yard operations and propose improvements.
13. **Long-Term Monitoring and Reporting**
    - **Environmental Impact Assessments**: Require periodic Environmental Impact Assessments to evaluate the ongoing environmental and health impacts of the dump yard and recommend further mitigation measures as needed.
    - **Transparency**: Ensure transparency in reporting environmental data and compliance status to the public.

These comprehensive measures, policies, and guidelines can help address the environmental and health challenges associated with the Vellalore Dump Yard while promoting sustainable waste management practices and safeguarding the well-being of local communities and ecosystems.

5.5 Conclusions

In developing countries, only limited work has been done on microplastic occurrence in water and soil ecosystems. Plastic particles are one of the parts of our day-to-day life and affect our lifestyle. In India and developing countries, the problem of waste management is increasing the consumption or use of plastics. The main source of microplastics in groundwater are surface runoff, wastewater effluent, open dumping, and open burning, which is the main specific source for sampling sites.

Proper waste management scheme is needed to implement the pollution of microplastics and mitigate these problems in developing countries. We can follow basic remediation methods such as the use of bio-based or decomposable polymers, engineering tools and biotechnological tools, etc. We believe that further research and review of differences are of great significance to prioritize the adoption of solutions and the creation of innovative strategies that would be useful in the future. Considerable progress has been made in recent years in the knowledge of Micro and Nano Plastic contamination of freshwater ecosystems, but this new field of water research still preserves several key challenges. These initial investigations call for further research efforts to better understand the source and fate of microplastics in freshwater ecosystems. There is an immediate need for extensive and in-depth research to overcome the research gap to allow for a more detailed risk assessment of microplastics in groundwater and to promote national policies to address these problems.

Recommendation

Establishing four ultramodern recycling plants, in the major cities of the country in particular urban and sub-urban zones. Education of the populace about the consequences of indiscriminate littering will be an ongoing concern of this research and skill training. Creating a direct value in plastic waste so as to attract scavengers to collect these plastic wastes from the ground for recycling. This will also attract itinerant waste buyers to start moving from house to house to buy plastic waste.

Acknowledgements The authors express their gratitude to the Director of the National Institute of Technical Teachers Training and Research (NITTTR), under the Ministry of Education, Government of India, located in Taramani, Chennai-600 113, India, for their assistance in facilitating the work and providing the necessary software for data analysis.

References

1. Bakir A, Rowland SJ, Thompson RC (2014) Transport of persistent organic pollutants by microplastics in estuarine conditions. Estuar Coast Shelf Sci 140:14–21. https://doi.org/10.1016/j.ecss.2014.01.004
2. Bellas J, Martínez-Armental J, Martínez-Cámara A, Besada V, Martínez-Gómez C (2016) Ingestion of microplastics by demersal fish from the Spanish Atlantic and Mediterranean coasts.

Mar Pollut Bull 109(1):55–60. https://doi.org/10.1016/j.marpolbul.2016.06.026. Epub 2016 Jun 8. PMID: 27289284
3. Bharath KM, Natesan U, Ayyamperumal R, Kalam SN (2020) Microplastics as an emerging threat to the freshwater ecosystems of Veeranam lake in south India: A multidimensional approach. Chemosphere 264(Pt 2):128502–128502
4. Bharath KM, Muthulakshmi AL, Usha N (2022) Microplastic contamination around the landfills: distribution, characterization and threats—a review. Curr Opin Environ Sci Health 100422
5. Brennan RB, Clifford E, Devroedt C, Morrison L, Healy MG (2017) Treatment of landfill leachate in municipal wastewater treatment plants and impacts on effluent ammonium concentrations. J Environ Manage 188:64–72
6. Brennan RB, Healy MG, Morrison L, Hynes S, Norton D, Clifford E (2016) Management of landfill leachate: The legacy of European Union Directives. Waste Manage 55:355–363
7. Browne MA, Crump P, Niven SJ, Teuten E, Tonkin A, Galloway T, Thompson R (2011) Accumulation of microplastic on shorelines worldwide: sources and sinks. Environ Sci Technol 45:9175–9179. https://doi.org/10.1021/es201811s
8. Cole M, Lindeque P, Halsband C, Galloway TS (2011) Microplastics as contaminants in the marine environment: a review. Mar Pollut Bull 62:2588–2597. https://doi.org/10.1016/j.marpolbul.2011.09.025
9. Derraik JGB (2002) The pollution of the marine environment by plastic debris: a review. Mar Pollut Bull 44:842–852. https://doi.org/10.1016/S0025-326X(02)00220-5
10. Dris R, Gasperi J, Mirande C, Mandin C, Guerrouache M, Langlois V, Tassin B (2017) A first overview of textile fibers, including microplastics, in indoor and outdoor environments. Environ Pollut 221:453–458. https://doi.org/10.1016/j.envpol.2016.12.013
11. Duggan J (2005) The potential for landfill leachate treatment using willows in the UK—a critical review. Resour Conserv Recycl 45(2):97–113
12. Duis K, Coors A (2016) Microplastics in the aquatic and terrestrial environment: sources (with a specific focus on personal care products), fate and effects. Environ Sci Eur 28:2. https://doi.org/10.1186/s12302-015-0069-y
13. Frias JPGL, Gago J, Otero V, Sobral P (2016) Microplastics in coastal sediments from Southern Portuguese shelf waters. Mar Environ Res 114:24–30. https://doi.org/10.1016/j.marenvres.2015.12.006
14. Ghosh A, Kumar S, Das J (2023) Impact of leachate and landfill gas on the ecosystem and health: research trends and the way forward towards sustainability. J Environ Manage 336:117708
15. Gopikumar S, Raja S, Robinson YH, Shanmuganathan V, Chang H, Rho S (2021) A method of landfill leachate management using internet of things for sustainable smart city development. Sustain Cities Soc 66:102521
16. Horton AA, Svendsen C, Williams RJ, Spurgeon DJ, Lahive E (2017) Large microplastic particles in sediments of tributaries of the River Thames, UK e abundance, sources and methods for effective quantification. Mar Pollut Bull 114:218–226. https://doi.org/10.1016/j.marpolbul.2016.09.004
17. Huffer T, Hofmann T (2016) Sorption of non-polar organic compounds by microsized plastic particles in aqueous solution. Environ Pollut 214:194–201. https://doi.org/10.1016/j.envpol.2016.04.018
18. Kamaruddin MA, Yusoff MS, Rui LM, Isa AM, Zawawi MH, Alrozi R (2017) An overview of municipal solid waste management and landfill leachate treatment: Malaysia and Asian perspectives. Environ Sci Pollut Res 24:26988–27020
19. Kole PJ, Lohr AJ, Van Belleghem FGAJ, Ragas AMJ (2017) Wear and tear of € tyres: a stealthy source of microplastics in the environment. Int J Environ Res Public Health 14:1265. https://doi.org/10.3390/ijerph14101265
20. Kumar V, Sharma N, Umesh M, Chakraborty P, Kaur K, Duhan L et al (2023) Micropollutants characteristics, fate, and sustainable removal technologies for landfill leachate: a technical perspective. J Water Process Eng 53:103649

21. Luo H, Zeng Y, Cheng Y, He D, Pan X (2020) Recent advances in municipal landfill leachate: a review focusing on its characteristics, treatment, and toxicity assessment. Sci Total Environ 703:135468
22. Manjula G, Rajagopalan V, Nayahi NT (2019) Environmental assessment of solid waste dump site. J Environ Biol 40(4):758–762
23. Manoj N, Gandhimathi A (2020) Environmental impact assessment of ground water quality around Vellalore dumping yard. Int Res J Eng Technol 7(2)
24. Napper IE, Thompson RC (2016) Release of synthetic microplastic plastic fibres from domestic washing machines: effects of fabric type and washing conditions. Mar Pollut Bull 112:39–45. https://doi.org/10.1016/j.marpolbul.2016.09.025
25. Nizzetto L, Futter M, Langaas S (2016) Are agricultural soils dumps for microplastics of urban origin? Environ Sci Technol 50:10777–10779. https://doi.org/10.1021/acs.est.6b04140
26. Postacchini L, Ciarapica FE, Bevilacqua M (2018) Environmental assessment of a landfill leachate treatment plant: impacts and research for more sustainable chemical alternatives. J Clean Prod 183:1021–1033
27. Pratheba S, Johnpaul V, Balasundaram N (2020) Study on environmental contamination and assessment of the impact of municipal solid waste on soil at the Vellalore dumping yard, Coimbatore City. In: IOP conference series: materials science and engineering, vol. 1006, no. 1. IOP Publishing, p. 012026
28. Ramakrishnan A, Blaney L, Kao J, Tyagi RD, Zhang TC, Surampalli RY (2015) Emerging contaminants in landfill leachate and their sustainable management. Environ Earth Sci 73:1357–1368
29. Remmas N, Manfe N, Zerva I, Melidis P, Raga R, Ntougias S (2023) A critical review on the microbial ecology of landfill leachate treatment systems. Sustainability 15(2):949
30. Shanthi P, Sundari PM, Meenambal T (2013) Evaluating the physico-chemical characteristics of municipal solid waste in Coimbatore City, Tamilnadu. Nat Environ Pollut Technol 12(1):69
31. Siddiqi SA, Al-Mamun A, Baawain MS, Sana A (2022) A critical review of the recently developed laboratory-scale municipal solid waste landfill leachate treatment technologies. Sustain Energy Technol Assess 52:102011
32. Teng C, Zhou K, Peng C, Chen W (2021) Characterization and treatment of landfill leachate: a review. Water Res 203:117525
33. Vaverková MD (2019) Landfill impacts on the environment. Geosciences 9(10):431
34. Wijekoon P, Koliyabandara PA, Cooray AT, Lam SS, Athapattu BC, Vithanage M (2022) Progress and prospects in mitigation of landfill leachate pollution: Risk, pollution potential, treatment and challenges. J Hazard Mater 421:126627

Chapter 6
Impact of Landfill Leachate on Ground Water Quality: A Review

Anshu Gupta, Akanksha Verma, and Paulraj Rajamani

Abstract The most widely utilized method of solid waste disposal around the world is the use of Municipal Solid waste (MSW) landfills. Upto 95% of the waste collected worldwide is disposed off in landfills. The main issues related with the landfills are the production of methane gas and liquid leachate. Among these issues, generation of leachate is of growing concern. In developing countries landfilling is the commonly used method and most of the landfills do not have any liners at the base or proper top covers, which as a result of leachate generation from solid waste results in the potential problems of ground water/surface water contamination. Besides landfilling is considered the most economical method of waste disposal, in recent decades several problems of social and environmental attention are associated with landfills. Amongst the environmental problems related to MSW landfilling, ground water contamination is the most prominent. From physical, chemical, and biological point of view the leachate generated from MSW disposal sites is considered as one of the highly contaminated resources. Sanitary landfill that is present in urban areas is often related to potential environmental pollution of ground water and surface water. Landfills pose potential risk to human health as well due to the toxic impacts caused by the leachates when released into the environment and the potential of landfills to generate leachates even after 100 years of closure. It has been reported earlier that even small amounts of leachate can pollute large volume of groundwater that can make them unusable for domestic and many other purposes. In this chapter authors have reviewed the leachate composition from MSW landfills, contamination of ground and surface water sources by different pollutants present in the leachates. The effect of depth of water source and its distance from landfill along with other factors affecting the contamination of water sources is also reviewed. Authors have

A. Gupta
Department of Environmental Sciences, Jammu and Kashmir Higher Education Department (JKHED), UT of Jammu and Kashmir, Jammu, India

A. Verma · P. Rajamani (✉)
School of Environmental Sciences, Jawaharlal Nehru University, New Delhi, India
e-mail: paulrajr@yahoo.com

A. Verma
Shyam Lal College (Evening), University of Delhi, Delhi, India

A. Anouzla and S. Souabi (eds.), *A Review of Landfill Leachate*, Springer Water,
https://doi.org/10.1007/978-3-031-55513-8_6

suggested remedial measures and that proper management of solid waste can help to decrease the contamination of through leachate production.

Keywords Groundwater contamination · Solid waste landfill · Leachate · Emerging contaminants · Microplastics

6.1 Introduction

Landfills are considered the most widely practiced method for disposal of municipal solid waste (MSW) and 95% of the total MSW collected worldwide is disposed of in landfills [1]. In most of the developing countries, it is the widely accepted method for the disposal of solid waste due to its economic advantages. At present most of the solid waste in the country is disposed unscientifically. It has adverse effects on not only on the ecosystem but also on the human health and environment. However poorly designed landfill can create contamination of ground water, soil, and air [2]. The decomposition of solid waste in the landfill, along with rainwater penetration is responsible for the formation of a dark liquid with unpleasant odour known as leachate that contains dissolved organic and inorganic compounds, heavy metals, suspended particles and hazardous substances [3, 4]. In the under-developed and developing countries, hazardous industrial waste is also co-disposed with municipal waste and no provision of separate secured hazardous landfills exists. Even if there are no hazardous wastes placed in municipal landfills, the leachate is still reported as a significant threat to groundwater [5]. A few studies have been carried out in Delhi in order to assess the effect of leachate contamination on the nearby ground water quality and it has been found that landfills are the sources of contamination to the environment surrounding the landfills [2, 6, 7]. Gupta and Rajamani [8] found very high values of LPI were obtained for all the three landfill sites in Delhi, Bhalaswa having the highest followed by Ghazipur and Okhala indicating the comparative leachate pollution potential. Due to the absence of collection and treatment facilities, leachate is posing a potential threat to ground and surface water. Across the world, landfills have been highlighted as one of the most significant hazards to groundwater resources, not only in India but also in other parts of the world. It is estimated that more than 90 percent of the Municipal Solid Waste (MSW) generated in India is dumped directly on land in an unsatisfactory way. Solid waste disposed of in landfills, or open dumps is exposed to groundwater underflow, precipitation-induced infiltration, and any other form of water infiltration that may occur [9].

6.1.1 Composition of Leachate

Leachate is a complex chemical mixture that contains dissolved organic matter, inorganic salts, organic trace impurities, and heavy metals (Fig. 6.1). Each of these substances is found in varying concentrations as a result of the physical, chemical, and microbiological processes occurring in the deposited waste [10]. Composition of leachate is affected by the type of trash deposited, usage of the landfill, availability of oxygen, prevailing hydrogeological conditions, and the age of the landfill [11]. Landfill leachate consists of four groups of pollutants: Dissolved organic matter, Inorganic macro-components, Heavy metals and Xenobiotic-organic compounds [3, 4].

In general, the composition of MSW Landfill Leachate is complex but there are some chemicals which are common to all landfill effluents. These include high levels of Biological Oxygen Demand (BOD), Chemical oxygen Demand (COD), Ammonia (NH_4^+), Na, and Cl. Other pollutants of concern include biological organisms that include bacteria, fungi and viruses, and a variety of organic chemicals [12]. The leachates can also be contaminated with bacteria including aerobic, psychrophilic, mesophilic bacteria, coliform and fecal coliform, and spore forming bacteria, including clostridium perfringens and with filamentous fungi [13]. The

Fig. 6.1 Composition of landfill leachate

leachates from landfills have been found to contain hundreds of different chemicals, including pharmaceuticals, fluorinated surfactants, phosphate-based flame retardants, phthalate acid esters, bisphenol A, personal care products, polybrominated diphenyl ethers, and chlorinated dioxins and furans [14].

During stabilization of landfilled wastes, non-conservative constituents of leachate (primarily organic in nature) tend to decompose and stabilize with time, whereas conservative constituents will remain long after waste stabilization occurs. Conservative constituents include various heavy metals, chloride, and sulfide. Metals often are precipitated within the landfill and are infrequently found at high concentrations in leachate, except for iron [2]. The concentrations of metals in the leachate, which can vary widely from the microgram to the milligram per litre concentration [3].

Contamination from municipal landfill leachate has become a serious problem for the aquatic environment because the leachates contaminate both surface as well as ground water, which is either due to faulty design or construction of the landfill site [15].

Municipal solid waste landfill dump sites play a significant role in polluting groundwater reserves to a significant extent [16]. However, a lack of sanitation and improper waste management have the potential to degrade the quality of groundwater, resulting in higher levels of pollution in the groundwater.

Following are the major contaminants present in leachates:

Contaminants of emerging concern (CEC): CEC include predominantly unregulated chemicals such as micropollutants, pharmaceuticals and personal care products (PPCPs) and endocrine-disrupting compounds (EDCs) which are occurring in the environment [17]. Similarly, fifty-eight CEC compounds were detected in landfill leachate samples from Greece with highest average concentrations of Bisphenol A, valsartan and 2-OH-benzothiazole. Pharmaceuticals, industrial chemicals and agrochemicals were among the predominantly detected CEC [18]. According to [17] Phthalate esters and PPCPs are the two most frequently investigated CEC classes in Chinese landfill leachate. Nine CEC have reported wide range of concentration starting from 0.03 to 4500 μg/L.

Inorganic macro-compounds: Inorganic macro compounds generally show higher concentrations in landfill leachate, whereas the concentration can depend on landfill age for some macro components. $NH4^+$, NO^{3-}, NO^{2-}, $SO_4{}^{2-}$, Cl^-, $PO_4{}^{3-}$, sodium (Na^+), potassium (K^+), calcium (Ca^{2+}), magnesium (Mg^{2+}), iron (Fe^{2+}) and hydrogen carbonate ($HCO_3{}^-$) are among the most predominant inorganic compounds [19]. The high pH in methanogenic leachate induces the sorption of some cations (such as calcium, magnesium and iron) into organic materials while forming precipitates with anions. Therefore, lower concentrations of the cations remain in the leachate. Some macro components including Cl^-, Na^+, and K^+ barely undergo sorption, precipitation, and complexation into other substances due to the washing out by leaching [20].

Xenobiotic organic compounds (XOCs): Xenobiotic organic compounds (XOCs) are presented in low concentrations, usually less than 1 mg/L for individual compounds. Examples for XOCs are benzene, toluene, phenols, chlorinated aliphatics, phthalates and halogenated hydrocarbons (tetrachloroethylene and trichloroethylene) [21]. The primary sources of these pollutants are household and industrial chemicals, together with pesticides and fertilizers. Waste composition, landfill technology and age are the factors that affect the level of these pollutants in leachate.

Heavy metals: Heavy metals form an important constituent of Landfill leachate. Leachates containing heavy metals are the result of the wastes that are disposed of in landfills, including lamps, batteries, paints, and packaging materials for chemicals and pharmaceuticals. They have the ability to form complexes with organic molecules, salts, and alkalinity, which are frequently present in leachates [22–24].

Landfill leachate can be divided into three categories according to landfill age (Fig. 6.2). According to the classification of landfills based on age, those less than five years old are considered young, those between five and ten years old are considered middle-aged, and those more over ten years old are considered old [27, 28]. Thus, the age difference between the two landfills may be one of the causes of the variance in the concentration of heavy metals in the leachate at various sites. Even though the buried wastes have a comparable composition, the change in pH in older landfills might cause the concentration of heavy metals to differ from that in younger landfills. In addition, compared to new landfills, the old landfills' decreased biodegradability is an indirect factor that alters the quantity of heavy metals in leachate [29]. Due to metals' poor solubility at alkaline pH, there are not many heavy metals in the leachates from landfills in Brazil. Likewise, because of the low pH value at the beginning of operation, heavy metal concentrations in landfill leachates tend to be high. Over time, as the wastes stabilize and the landfill enters the methanogenic phase, the high pH value limits the solubility of the metals [22]. Leachate is one of the sources of heavy metal contamination because of the sluggish leaching of heavy metals caused by the leachate's acidic nature [30].

Heavy metals present in leachates from both municipal solid waste landfills and hazardous waste disposal sites pose a severe threat to public health. For instance, long-term exposure to arsenic can result in cardiovascular disease, hypertension, internal malignancies, as well as infertility and reproductive failure, bone demineralization, gastrointestinal cancer and neurological issues. On the *Lemna gibba*, heavy metals can also prevent the growth and synthesis of photosynthetic pigments [31, 32].

The influence of the quantity of cigarette butts in the waste mass on the leachate's quality was investigated by [33], they observed a correlation between the amount of cigarette butts in the buried garbage and the increase in the content of heavy metals in the leachate.

The elevated concentration of Fe in the leachate suggests that more iron and steel trash is being deposited at the landfill. Zn in the leachate indicates that trash from batteries and fluorescent bulbs is dumped at the landfill [34].

The sources of lead and copper emissions include spent batteries, pipes, and paint [29]. According to [35], high Pb levels can also disturb the neurological system and induce brain damage. When such water is utilised for drinking, a greater amount of

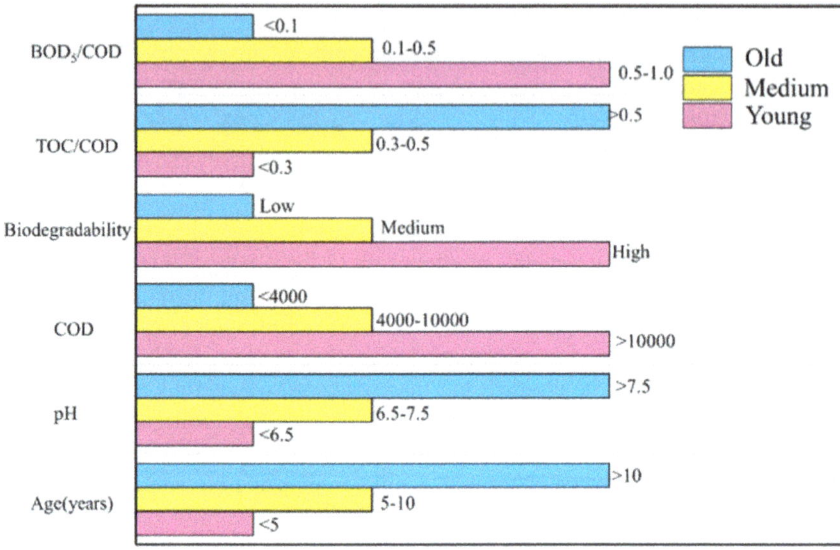

Fig. 6.2 Changes in landfill leachate classification with age. *Source* [25, 26]

Cd might result in cramps in the kidneys and the artery walls, as well as nausea, vomiting, and diarrhea [31]. Leachate percolation through the unlined landfill is thought to be the primary source of heavy metals in the dump site and the nearby groundwater.

Organic pollutants: Landfill leachates have a high concentration of organic material, expressed as BOD_5. This means that most of the organic material present in these landfills is made up of recalcitrant substances (like humic and fulvic acids), which are difficult to break down. This can be explained by the fact that the biodegradability of waste material declines over time and that these landfills are older [22].

According to [36], the ratio between BOD_5 and COD levels for old landfills is extremely low, a ratio of 0.022 has been documented for the Muribeca landfill of Brazil. Another research by [37], has shown that landfills at the older stage, are in methanogenic phase, had BOD to COD ratios of less than 0.1. Additionally, it was discovered that the landfills having high pH values (an average of 8.11), may be due to the existence of high alkalinity levels given that the landfills had progressed to the methanogenic phase [22]. The older landfills produce leachates that are alkaline in character and range in pH from 8.0 to 8.5. While more neutral pH values are anticipated in leachates that have already undergone some stabilization, mature landfills produce leachates with pH values greater than 7 [1]. The toxicity of leachate is the result of high concentrations of ammoniacal nitrogen or the overall high conductivity of the leachate. The amount of dissolved inorganic elements and the total ion concentration are both indicated by conductivity [4].

It is generally recognised that leachate can endanger the public's health and the quality of the water. Most of the pollutants found in the leachate have been reported to be hazardous and able to cause mutations in living things. The high concentration of hazardous organic compounds, such as pesticides, insecticides, or solvents, as well as unused medications, bisphenol A, phenol, biphenyls, organic phosphates, and polychlorinated biphenyls, that are present in the leachate obscured the effects of nutrient components and inhibited plant growth [1, 38, 39].

Industrial trash dumped in landfills from surrounding industrial areas is a substantial source of polyaromatic compounds that leak from top soil and percolate down to join leachate. Polycyclic aromatic hydrocarbons (PAHs), which have two or more rings, frequently have high environmental stability. They have teratogenic, mutagenic, and carcinogenic effects on animals [40, 41].

Persistent organic pollutants (POPs), endocrine-disrupting chemicals (EDCs), pharmaceuticals, phenols, polychlorinated biphenyls, and polyaromatic hydrocarbons (PAHs) and personal care products (PPCPs) all have been found to contaminate surface water bodies [23, 24, 42]. According to [43], these organic pollutants found in water bodies can contaminate water sources and endanger human health in addition to posing a hazard to natural aquatic ecosystems. Another possible source of environmental pollutants that has received little attention is landfill leachates processed at municipal wastewater treatment facilities [44].

Other notable organic compounds found in landfill leachates include carbamazepine, DEET (Diethyltoluamide), PFOA (Perfluorooctanoic acid), primidone, perfluorochemicals (PFCs) and gemfibrozil. Wastewater and surface/source drinking water are frequently found to contain carbamazepine, which is used to treat epileptic seizures. In USA landfill leachates, it was reported. leachates sample have the presence of primidone, an anticonvulsant, DEET used to repel biting pests such as mosquitoes and ticks. An antibiotic called sulfamethoxazole is frequently found in wastewater samples, and surface water [14].

Microplastics: According to Plastics Europe (2016), 332 million tonnes of plastic were produced worldwide in 2015, according to estimates. 2015 saw the production of almost 6,300 million tonnes of plastic garbage, of which 9% was recycled, 12% was burned, and the remaining 79% was dumped in landfills or released into the environment. By 2050, landfills and the environment are expected to contain almost 12,000 million tonnes of plastic trash if no action is taken [45].

The amount of plastic garbage produced globally that may be stored in landfills, which are utilised for waste disposal, ranges from 21 to 42% [46]. Although the harm caused by big plastics is well known, a growing understanding of microplastics (those less than 5 mm in diameter) and their potential for danger has led to greater worries [47]. Micro-plastics can absorb organic contaminants in addition to entering the soil because of their propensity to do so [45]. Due to the inclusion of plastic microbeads in some widely used personal care products, leachates may also contain microplastics. The degradation of macroplastics in landfills can also produce microplastics. Landfilled macroplastics will continue to break down into microplastics due to the fluctuating temperatures (reaching as high as 60–90 °C) and pH (4.5–9), deep-seated fires, physical stress, and compaction, and (although limited) oxygen. Once landfilled, plastics undergo aerobic biodegradation (at an initial stage), but quickly make the transition to anaerobic conditions, with the consequent influence of acid formation and methane fermentation from organic solid wastes [48, 49].

Microplastics are known to act as carriers of hazardous substances because of their hydrophobic nature [50], therefore if such uncollected/untreated plastic particles which may transport such contaminants find their way to natural ecosystems, the negative impacts would be exacerbated. In fact, packaging plastics comprised of high- and low- density polyethylene (HDPE and LDPE), PET, and PVC make up a sizable portion of the plastic trash that is landfilled, but other materials like textiles and tyres are also included [51]. Emphasizing that landfilling is not the final destination for plastic waste but rather a possible source of microplastics into the ecosystem, 99.4% of microplastics were also associated with plastic trash buried in landfills that might remain in the environment for decades (PE, PP, PET) [52].

Their physicochemical characteristics determine their size-dependent transit, sorption, bioavailability, and fate. Due to their large surface areas, nano-plastics have the potential to cause the release of substances (such as monomers, additives, and other environmental pollutants) into the environment [53]. It also makes it a possible source of plastic pollution in landfills as they are lower scale particles than microplastics. Which can have negative impacts on the environment and on individuals.

6.2 Leachate and Ground Water Pollution

Landfill leachate is generated due to the penetration of precipitation through the waste mass and due to the biodegradation of waste. Leachate can cause serious pollution to ground water aquifer as well as adjacent surface waters as it migrates away from landfill [55]. Since a large number of landfills, whether active or abandoned, have been operated with minimal consideration for groundwater contamination. One of the main environmental effects of leachate is groundwater pollution. The escape of surface water sources may be relatively easy to control than groundwater sources, which is very difficult to control and clean up [56]. The contamination of ground water caused by landfill leachate is regarded as the biggest threat to the environment and public health associated with garbage disposal [11].

Several landfill sites are known to cause contamination of the nearby aquifers and it is because of the potential pollution source of leachate originating from the base of landfill sites. The natural resource user and the natural environment get potential risk because of such contamination of ground water [55, 57]. It has been reported that small amount of leachate may cause severe pollution to the ground water aquifer and to the adjacent surface waters [55, 58].

Across the world, landfills have been highlighted as one of the most significant hazards to groundwater resources, not only in India but also in other parts of the world. Solid waste disposed off in non-engineered landfills, or open dumps is exposed to groundwater underflow, precipitation-induced infiltration, and any other form of water infiltration that may occur [9]. During rain, deposited solid waste absorbs

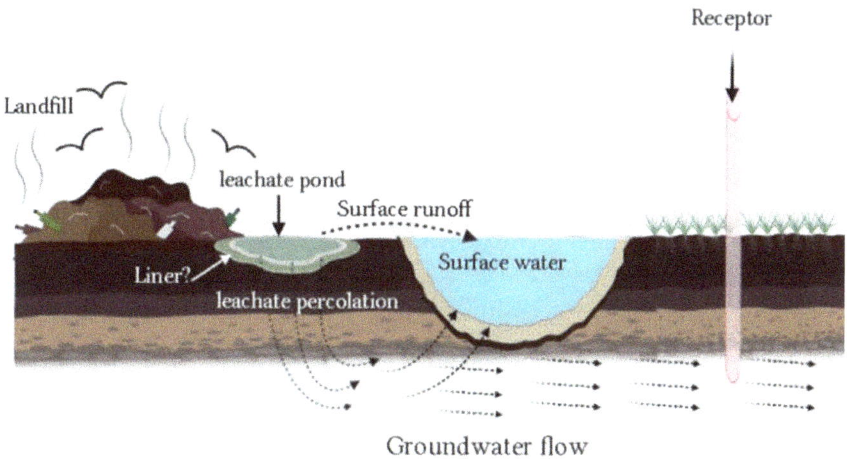

Fig. 6.3 Illustration of landfill leachate (with improper lining/no lining) mobilization in groundwater and surface water. *Source* [54]

water, and byproducts of waste breakdown and trickle into the water via waste deposition. The liquid containing innumerable organic and inorganic compounds is called 'leachate' that accumulates at the bottom of the landfill and percolates through the soil further reaching the groundwater [57] (Fig. 6.3).

Leachate percolating through the unlined trash dump site can be considered the primary potential source of heavy metals in the landfill and nearby groundwater [31]. The interaction of landfill waste with groundwater and the movement of contaminants into the aquatic ecosystem may have been facilitated by higher content of very mobile leachate effluents like Cd. Under the impact of solid waste leaching, the pollutants that entered the groundwater were susceptible to transformation processes. Groundwater flow aids in the diffusion and dispersion of leached contaminants in an aquifer system [6].

The most common pathway for leachate to mobilize to the aquatic environments is from the bottom of the landfill through the unsaturated soil layers to the ground water [60, 61] (Fig. 6.4). Since most of the landfills in developing countries, especially in Bangladesh, were constructed without engineered liners and proper leachate collection systems [37, 62, 63], leachate could be moved from groundwater to surface water through hydraulic connections.

There is direct and indirect association between health impacts and handling, treatment, and disposal methods of waste [64]. In 1990, Wrensch et al. investigated and verified that contaminated well near San Jose, California, disposal site had an adverse effect on spontaneous abortions, birth defects, and children heath concerns such as leukemia [65, 66]. In 2002, Jarup et al. found cancer risk, leukemia in children as well as in adults who were living around 2 km from landfill sites in Great Britain [67]. there was an increased concentration of contaminants in groundwater near the Bhalaswa landfill in Delhi [2]. The local residents suffered from a number of illnesses,

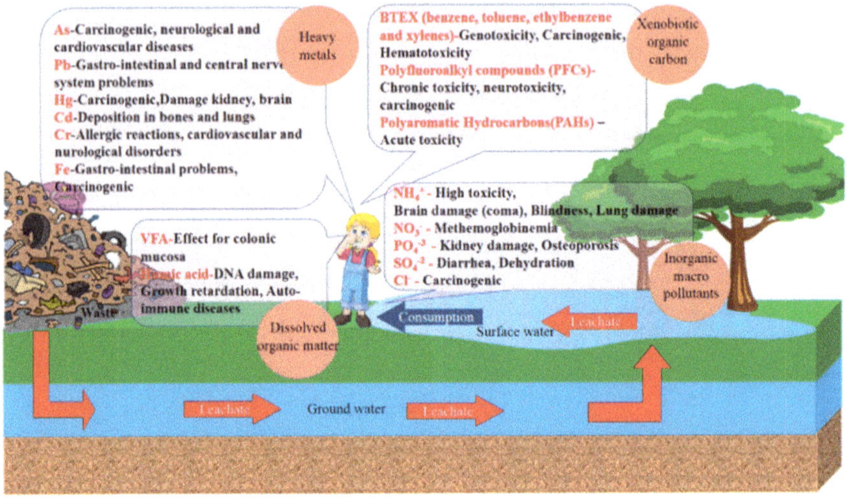

Fig. 6.4 Short term and long-term hazardous impacts of landfill leachates. *Source* [59]

especially gastro-intestinal diseases, musculoskeletal pain, skin and eye irritation, and respiratory problems. Of the sample population in Bhalaswa resettlement colony, 21.1% of women and 31.9% of men suffered from diarrhea and vomiting. This could indicate occurrence of faecal contamination of the drinking water. On the other hand, 62.6% of people suffered from gas and ache problems. The percentage of people was also found to be significant,20% of men and 18.2% of women in Bhalaswa resettlement colony had skin problems such as itching and skin rash. This may be due to regular contact with the polluted groundwater for the domestic use such as bathing, washing utensils, and clothes.

6.3 Influence of Depth and Distance from Landfill on Ground Water Quality

According to Negi et al. [68] the concentration of ammoniacal nitrogen (9.8 mg/L), chemical oxygen demand (128 mg/L), chloride (115 mg/L), sodium (98 mg/L) and potassium (42.2 mg/L) was found relatively higher in water samples that have lower depth (30 ft) and distance (1 km) from the landfill. The concentration of measured parameters decreases with increase in depth and distance confirming that the leachate is the potential source of groundwater contamination. [68]. According to [34], the concentrations of Cl^-, NO_3^-, SO_4^{2-}, NH_4^+ were found to be in considerable levels in the groundwater samples particularly near to the landfill sites, likely indicating that groundwater quality is being significantly affected by leachate percolation. Further they were proved to be the tracers for groundwater contamination near Semur and

Vendipalayam dump yards. The presence of contaminants in groundwater particularly near the landfill sites warns its quality and thus renders the associated aquifer unreliable for domestic water supply and other uses [34]. They further found that the quality of the groundwater was found to improve with the increase in depth and distance of the well from the landfill site. Although, the concentrations of few contaminants do not exceed drinking water standard even then the groundwater quality represent a significant threat to public health.

The very high EC and TDS observed in the groundwater suggest a downward transfer of leachate into groundwater as reported earlier by [57], Al-Khaldi [69] and [70]. High concentrations of TDS decrease the palatability of water and may also cause gastrointestinal irritation in humans and laxative effects particularly upon transits [71]. Andremi et al. [72] have reported that the occurrence of Na^+ above the WHO tolerance levels in wells closest to the landfill is an indication of possible leachate flow into groundwater. They further stated that the significant indirect correlations between Na^+, EC, TDS with distance from the dumpsite shows that the concentrations of contaminants in groundwater normally decreases with increase in distance from the pollution source. These parameters could therefore be used as indicators of inorganic groundwater pollution. Based on chloride and sulphate analysis of groundwater it is possible to establish that solid waste dumping site is a source of pollution of groundwater in the area. Chloride and sulphate in ground water increase due to municipal and industrial waste dumped on the ground surface. Elevated concentration of chloride (Cl^-) and sulphate ($SO_4{}^-$) in surface and ground water are common in all countries, and can serve as indicators of ground water pollution [73].

6.4 Conclusion

The disposal of solid waste that is followed in most of the developing countries is posing serious hazards to environmental as well as human health. Most landfills are being operated as mere dump sites that have become cause of serious groundwater pollution in the vicinity as the composition of leachate is very high for most of the pollutants having concentrations beyond the permissible levels. There is a need for new regulations and techniques for controlling the pollution due to leachates, since the high concentrations of organic and inorganic pollutants (heavy metals) lead to significant groundwater contamination from leachate percolating from the landfill site.

Therefore, stress should be put on the necessity to improve on waste management practices and also construct properly engineered sanitary landfill sites to curb the pollution of groundwater in the vicinity of landfills. In addition, there is need to provide ground water monitoring shallow wells around the dumpsite so that more information about the quality changes of leachate due to the waste age can be gathered.

References

1. Gupta A, Paulraj R (2017) Leachate composition and toxicity assessment: an integrated approach correlating physicochemical parameters and toxicity of leachates from MSW landfill in Delhi. Environ Technol 38(13–14):1599–1605
2. Jhamnani B, Singh SK (2009) Groundwater contamination due to Bhalaswa landfill site in New Delhi. Int J Environ Sci Eng 1(3):121–125
3. Christensen TH, Kjeldeson P, Bjerg (2001) Biogeochemistry of landfill leachate plumes. Appl Geochem 16:659–718
4. Gupta A, Gaharwar US, Verma A, Rajamani P (2022) Municipal solid waste landfill leachate induced cytotoxicity in root tips of Vicia faba: Environmental risk posed by non-engineered landfill
5. Sabahi EA, Samsudin AR, Yaacob WZW, Nozaey FA, Alshaebi F (2009) 'Leachate composition and groundwater pollution at municipal solid waste landfill of Ibb City, Yemen. Sains Malaysiana 38(3):295–304
6. Srivastava SK, Ramanathan AL (2008) Geochemical assessment of groundwater quality in vicinity of Bhalswa landfill, Delhi, India, using graphical and multivariate statistical methods. Environ Geol 53:1509–1528
7. Zafar M, Alappat BJ (2004) Environmental mapping of water quality of the River Yamuna in Delhi with landfill locations'. Manage Environ Qual Int J 15(6):608–621
8. Gupta A, Rajamani P (2016) Evaluation of the leachate composition and contamination potential of municipal solid waste landfill sites in Delhi. Int J Environ Waste Manage 18(4):285–302
9. Chatterjee R (2010) Municipal solid waste management in Kohima city-India. J Environ Health Sci Eng 7(2):173–180
10. Aziz SQ, Aziz HA, Yuso MS, Bashir MJ, Umar M (2010) Leachate characterization in semi-aerobic and anaerobic sanitary landfills: a comparative study. J Environ Manage 91(12):2608–2614
11. Przydatek G, Kanownik W (2019) Impact of small municipal solid waste landfill on groundwater quality. Environ Monit Assess 191:1–14
12. Jones DL, Williamson KL, Owen AG (2006) Phytoremediation of landfill leachate. Waste Manage 26:825–837
13. Matejczyk M, Płaza GA, Nałecz-Jawecki G, Ulfig K, Markowska-Szczupak A (2011) Estimation of the environmental risk posed by landfills using chemical, microbiological and ecotoxicological testing of leachates. Chemosphere 82:1017–1023. https://doi.org/10.1016/j.chemosphere.2010.10.066
14. Clarke BO, Anumol T, Barlaz M, Snyder SA (2015) Investigating landfill leachate as a source of trace organic pollutants. Chemosphere 127:269–275
15. Gupta A, Paulraj R (2016) Relationship between physicochemical parameters and toxicity of leachate from municipal solid waste landfill site in Delhi. In: Geostatistical and geospatial approaches for the characterization of natural resources in the environment: challenges, processes and strategies. Springer International Publishing, pp 361–367
16. Groundwater I. (2012) A valuable but diminishing resource. The World Bank
17. Qi C, Huang J, Wang B, Deng S, Wang Y, Yu G (2018) Contaminants of emerging concern in landfill leachate in China: a review. Emerging Contaminants 4(1):1–10
18. Nika MC, Ntaiou K, Elytis K, Thomaidi VS, Gatidou G, Kalantzi OI, Thomaidis NS, Stasinakis AS (2020) Wide-scope target analysis of emerging contaminants in landfill leachates and risk assessment using risk quotient methodology. J Hazard Mater 394:122493
19. Robinson H (2007) The composition of leachates from very large landfills: an international review. Commun Waste Resour Manag 8:19–32
20. Asadi M (2008) Investigation of heavy metals concentration in landfill leachate and reduction by different coagulants. In: Proceedings of the 7th international conference on environmental engineering. Faculty of Environment Engineering. Vilnius Gedim. Tech. Univ., pp 484–488

21. Suter MJF, Riediker S, Zipper C, Kohler HPE, Giger W (1997) Polar organic compounds in landfill leachates. Analusis 7:M23–M25
22. Costa AM, Alfaia RGDSM, Campos JC (2019) Landfill leachate treatment in Brazil—an overview. J Environ Manage 232:110–116
23. Verma A, Gupta A, Rajamani P (2023a) Application of wastewater in agriculture: benefits and detriments. In: River conservation and water resource management. Springer Nature Singapore, Singapore, pp 53–75
24. Verma A, Gupta A, Gaharwar US, Rajamani P (2023) Effect of wastewater on physiological, morphological, and biochemical levels and its cytotoxic potential on *Pisum sativum*. Int J Environ Sci Technol: 1–18
25. Renou S, Givaudan JG, Poulain S, Dirassouyan F, Moulin PJJOHM (2008) Landfill leachate treatment: Review and opportunity. J Hazard Mater 150(3):468–493
26. Zhao R, Wang X, Chen X, Liu Y (2019) Impacts of different aged landfill leachate on PVC corrosion. Environ Sci Pollut Res 26:18256–18266
27. Pasalari H, Farzadkia M, Gholami M, Emamjomeh MM (2019) Management of landfill leachate in Iran: valorization, characteristics, and environmental approaches. Environ Chem Lett 17:335–348
28. Ghosh P, Thakur IS, Kaushik A (2017) Bioassays for toxicological risk assessment of landfill leachate: a review. Ecotoxicol Environ Saf 141:259–270
29. Beinabaj SMH, Heydariyan H, Aleii HM, Hosseinzadeh A (2023) Concentration of heavy metals in leachate, soil, and plants in Tehran's landfill: Investigation of the effect of landfill age on the intensity of pollution. Heliyon 9(1)
30. Kanmani S, Gandhimathi R (2013) Assessment of heavy metal contamination in soil due to leachate migration from an open dumping site. Appl Water Sci 3:193–205
31. Boateng TK, Opoku F, Akoto O (2019) Heavy metal contamination assessment of groundwater quality: a case study of Oti landfill site, Kumasi. Appl Water Sci 9(2):33
32. Verma A, Gaharwar US, Priyadarshini E, Rajamani P (2022) Metal accumulation and health risk assessment in wastewater used for irrigation around the Agra Canal in Faridabad, India. Environ Sci Pollut Res: 1–15
33. Torkashvand J, Godini K, Norouzi S, Gholami M, Yeganeh M, Farzadkia M (2021) Effect of cigarette butt on concentration of heavy metals in landfill leachate: health and ecological risk assessment. J Environ Health Sci Eng 19:483–490
34. Nagarajan R, Thirumalaisamy S, Lakshumanan E (2012) Impact of leachate on groundwater pollution due to non-engineered municipal solid waste landfill sites of erode city, Tamil Nadu, India. Iran J Environ Health Sci Eng 9:1–12
35. Bulut Y, Baysal Z (2006) Removal of Pb (II) from wastewater using wheat bran. J Environ Manag 78:107–113
36. Assou M, El Fels L, El Asli A, Fadiki H, Souabi S, Hafidi M (2016) Landfill leachate treatment by a coagulation-flocculation process: effect of the introduction order of the reagents. Desalin Water Treat 57:21817–21826
37. Kjeldsen P, Barlaz MA, Rooker AP, Baun A, Ledin A, Christensen TH (2002) Present and long-term composition of MSW landfill leachate: a review. Crit Rev Environ Sci Technol 32(4):297–336
38. Buszka PM, Yeskis DJ, Kolpin DW, Furlong ET, Zaugg SD, Meyer MT (2009) Waste-indicator and pharmaceutical compounds in landfill-leachate-affected ground water near Elkhart, Indiana, 2000–2002. Bull Environ Contam Toxicol 82:653–659
39. Gupta A, Rajamani P (2015) Toxicity assessment of municipal solid waste landfill leachate collected in different seasons from Okhala landfill site of Delhi. J Biomed Sci Eng 8(06):357
40. Lau EV, Gan S, Ng HK (2010) Extraction techniques for polycyclic aromatic hydrocarbons in soils. Int J Anal Chem
41. Verma A, Meena R, Maurya A, Gaharwar US, Rajamani P (2019) Identification, quantification, and in-vitro genotoxicity of major polyaromatic hydrocarbons produced by sugarcane fly ash emitted from sugarmill. J Environ Prot 10(10):1244–1261

42. Kolpin DW, Furlong ET, Meyer MT, Thurman EM, Zaugg SD, Barber LB, Buxton HT (2002) Pharmaceuticals, hormones, and other organic wastewater contaminants in U.S. Streams, 1999–2000: a national reconnaissance. Environ Sci Technol 36:1202–1211
43. Holzer J, Goen T, Rauchfuss K, Kraft M, Angerer J, Kleeschulte P, Wilhelm M (2009) One-year follow-up of perfluorinated compounds in plasma of German residents from Arnsberg formerly exposed to PFOA-contaminated drinking water. Int J Hyg Environ Health 212:499–504
44. Eggen T, Moeder M, Arukwe A (2010) Municipal landfill leachates: a significant source for new and emerging pollutants. Sci Total Environ 408:5147–5157
45. Afrin S, Uddin MK, Rahman MM (2020) Microplastics contamination in the soil from urban landfill site, Dhaka, Bangladesh. Heliyon 6(11)
46. Nizzetto L, Futter M, Langaas S (2016) Are agricultural soils dumps for microplastics of urban origin Environ. Sci Technol 50(20):10777–10779
47. Avio CG, Gorbi S, Regoli F (2017) Plastics and microplastics in the oceans: from emerging pollutants to emerged threat. Mar Environ Res 128:2–11
48. Mahon AM, O'Connell B, Healy MG, O'Connor I, Officer R, Nash R, Morrison L (2017) Microplastics in sewage sludge: effects of treatment. Environ Sci Technol 51(2):810–818
49. Sundt P, Schulze PE, Syversen F (2015) Sources of microplastic pollution to the marine environment. Norwegian Environ Agency: 495
50. Hartmann NB, Rist S, Bodin J, Jensen LH, Schmidt SN, Mayer P, Meibom A, Baun A (2017) Microplastics as vectors for environmental contaminants: exploring sorption, desorption, and transfer to biota. Integr Environ Assessment Manage 13(3):488–493
51. Plastics Europe, Plastics—The Facts (2019) An analysis of European plastics production, demand and waste data. https://www.plasticseurope.org/application/files/9715/7129/9584/FINAL_web_version_Plastics_the_facts2019_14102019.pdf. Accessed 15 Oct 2020
52. Silva AL, Prata JC, Duarte AC, Soares AM, Barceló D, Rocha-Santos T (2021) Microplastics in landfill leachates: the need for reconnaissance studies and remediation technologies. Case Stud Chem Environ Eng 3:100072
53. Silva TA, Girardclos S, Stutenbecker L, Bakker M, Costa A, Schlunegger F (2018) The sediment budget and dynamics of a delta-canyon-lobe system over the Anthropocene timescale: the Rhone river delta, Lake Geneva (Switzerland/France). Sedimentology 66:838–858
54. Parvin F, Tareq SM (2021) Impact of landfill leachate contamination on surface and groundwater of Bangladesh: a systematic review and possible public health risks assessment. Appl Water Sci 11(6):100
55. Li G, Sang N, Guo D (2006) Oxidative damage induced in hearts, kidneys and spleens of mice by landfill leachate. Chemosphere 65:1058–1063
56. Abiriga D, Vestgarden LS, Klempe H (2020) Groundwater contamination from a municipal landfill: effect of age, landfill closure, and season on groundwater chemistry. Sci Total Environ 737:140307
57. Mor S, De Visscher A, Ravindra K, Dahiya RP, Chandra A, Van Cleemput O (2006) Induction of enhanced methane oxidation in compost: temperature and moisture response. Waste Manage 26(4):381–388
58. Pivato A, Gaspari L (2006) Acute toxicity test of Leachates from traditional and sustainable landfills using luminescent bacteria. Waste Manag 26:1148–1155
59. Wijekoon P, Koliyabandara PA, Cooray AT, Lam SS, Athapattu BC, Vithanage M (2022) Progress and prospects in mitigation of landfill leachate pollution: risk, pollution potential, treatment, and challenges. J Hazard Mater 421:126627
60. Fadhullah W, Kamaruddin MA, Ismail N et al (2019) Characterization of landfill leachates and its impact to groundwater and river water quality: a case study in Beris Lalang waste dumpsite, Kelantan. Pertanika J Sci Technol 27:633–646
61. Mishra S, Tiwary D, Ohri A, Agnihotri AK (2019) Impact of municipal solid waste landfill leachate on groundwater quality in Varanasi, India. Groundw Sustain Dev 9:100230. https://doi.org/10.1016/j.gsd.2019.100230
62. Alam R, Ahmed Z, Howladar MF (2020) Evaluation of heavy metal contamination in water, soil and plant around the open landfill site Mogla Bazar in Sylhet. Groundw Sustain Dev Bangladesh. https://doi.org/10.1016/j.gsd.2019.100311

63. DNCC (2016) Dhaka North city corporation waste report 2016–2017.111. https://dncc.portal.gov.bd/sites/default/files/files/dncc.portal.gov.bd/annual_reports/efedb2ad_8950_461e_960f_d1bfac6d10e3/Waste%20Report%202016-2017.pdf
64. De S, Debnath B (2016) Prevalence of health hazards associated with solid waste disposal—a case study of Kolkata, India. Procedia Environ Sci 35:201–208
65. Wrensch M, Swan S, Lipscomb J, Epstein D, Fenster L, Claxton K, Murphy PJ, Shusterman D, Neutra R (1990) Pregnancy outcomes in women potentially exposed to solvent-contaminated drinking water in San Jose, California. Am J Epidemiol 131(2):283–300
66. Wrensch M, Swan SH, Lipscomb J, Epstein DM, Neutra RR, Fenster L (1992) Spontaneous abortions and birth defects related to tap and bottled water use, San Jose, California, 1980–1985. Epidemiology: 98–103
67. Jarup L, Briggs D, De Hoogh C, Morris S, Hurt C, Lewin A, Maitland I, Richardson S, Wakefield J, Elliott P (2002) Cancer risks in populations living near landfill sites in Great Britain. Br J Cancer 86(11):1732–1736
68. Negi P, Mor S, Ravindra K (2020) Impact of landfill leachate on the groundwater quality in three cities of North India and health risk assessment. Environ Dev Sustain 22:1455–1474
69. Al-Khadi S (2006) Assessment of groundwater contamination vulnerability in the vicinity of Abqaiq landfill-A GIS Approach, Dissertation, King Fahd University of Petroleum and Minerals, Saudi Arabia
70. Longe EO, Enekwechi LO (2007) Investigation on potential groundwater impacts and influence of local hydrogeology on natural attenuation of leachate at a municipal landfill. Int J Environ Sci Technol 4:133–140
71. World Health Organization (WHO) (1997) Guideline for drinking water quality, 2nd edn., vol 2. In: Health criteria and other supporting information. World Health Organization, Geneva, 9 p
72. Aderemi AO, Oriaku AV, Adewumi GA, Otitoloju AA (2011) Assessment of groundwater contamination by leachate near a municipal solid waste landfill. Afr J Environ Sci Technol 5(11):933–940
73. Venkatesan G, Swaminathan G (2009) Review of chloride and sulphate attenuation in ground water nearby solid-waste landfill sites. J Environ Eng Landsc Manag 17(1):1–7

Chapter 7
Assessing the Impact of Landfill Leachate on Surface and Ground Water in Bangladesh: A Comparison with Other South Asian Regions

Mohammad Toha, Sadia Sikder, and Md. Mostafizur Rahman

Abstract This book chapter thoroughly evaluates landfill leachate's significant influence on surface and groundwater systems in Bangladesh, alongside a comparative examination of similar patterns in other South Asian regions. The increase in urbanization and population expansion has led to a significant rise in garbage production, making landfills a necessary but challenging method of trash disposal. The emergence of leachate, a liquid byproduct resulting from the percolation of water via decomposing garbage, has raised considerable environmental concerns due to its potential to contaminate water resources. The chapter explores the complex relationship between landfill leachate and aquatic ecosystems, emphasizing the concerning decline in surface water quality due to the infiltration of leachate contaminants into rivers, lakes, and streams. The groundwater in Bangladesh, which serves as a crucial drinking water source, is confronted with a comparable peril. Leachate elements have the potential to infiltrate aquifers, endangering the health of residents. The results emphasize the necessity of taking proactive actions, such as implementing solid regulatory frameworks, utilizing sophisticated leachate treatment technology, and engaging with the community to reduce the adverse effects caused by leachate contamination. This chapter highlights the importance of understanding the complex relationship between landfill leachate and water resources. It emphasizes the urgent need to address this issue in Bangladesh and provides valuable insights for other South Asian regions that are working towards waste management and the protection of their water resources.

Keywords Landfill leachate · Surface water · Groundwater · Water contamination · Waste management

M. Toha · S. Sikder · Md. Mostafizur Rahman (✉)
Department of Environmental Science, Bangladesh University of Professionals, Mirpur, Cantonment, Dhaka 1216, Bangladesh
e-mail: rahmanmm@juniv.edu

Md. Mostafizur Rahman
Laboratory of Environmental Health and Ecotoxicology, Department of Environmental Sciences, Jahangirnagar University, Savar, Dhaka 1342, Bangladesh

A. Anouzla and S. Souabi (eds.), *A Review of Landfill Leachate*, Springer Water,
https://doi.org/10.1007/978-3-031-55513-8_7

7.1 Introduction

Leachate from a landfill is a liquid of absorbed materials, some of which may be soluble solids and others unacceptable. It develops when precipitation soaks into a landfill, seeps through various trash types (including municipal and industrial), and evaporates. The leachate created is typically highly toxic due to the enormous variety of chemicals and organic and inorganic compounds that might be present in waste material and their subsequent breakdown. Landfill leachate can contain toxic heavy metals, including lead, arsenic, cadmium, mercury, etc., and xenobiotic chemicals like halogenated organic compounds, which have become a significant concern for quite some time. Many municipal solid waste landfill leachates also contain microplastics, making them a potential health threat. Polyethylene and polypropylene are two of the most common varieties; both have grown steadily in popularity over the years thanks to their durability and ease of use since their development several decades ago. Pollution of groundwater, soil, and air can be prevented through consistent and timely monitoring of the landfill and its leachate. Due to poor management and a lack of monitoring, several landfill regions have seen significant unintentional bursts caused by gases released due to decomposition and microbiological activity.

The main issue with municipal landfills is the possibility of pollution from leachate being mobilized through the subsoil and into surface and groundwater. Water containing leachate from the landfill also drains into lowlands and surface water bodies during the wet season, further polluting the local habitats. As a result, this hazardous aqueous effluent from a landfill site can pose dangers to both surface and groundwater, endangering aquatic biota, plants, and people. Uncontrolled and unengineered open landfill sites are the norm rather than the exception in most developing countries. Toxic leachate from improper waste disposal pollutes drinking water and crops, disproportionately affecting these countries' urban and semi-urban poor.

The geographical, meteorological, and socioeconomic similarities between South Asian nations exacerbate the complex dynamics of leachate pollution. Leachate has penetrated and polluted both surface water bodies and groundwater resources due to the rising volumes of solid waste generated and the inadequacy of waste management procedures. As a result of its large population, Bangladesh produces and discards over 8000 tons of solid waste per day [9]. Approximately 334 million metric tons of trash are generated annually in South Asia, of which 174 million metric tons (or 57%) are organic. According to Kaza et al. [16], the amount of garbage generated in South Asia is projected to be somewhere between 466 and 661 million metric tons in 2030 and 2050, respectively, and the proportion of waste generated in metropolitan areas is expected to be relatively high. In highly populated regions, where landfill space is in high demand, yet land is at a premium, this problem becomes especially acute [3, 8, 22]. Concerns regarding water quality decline, human health risks, and ecological effects have resulted from leachate and its related contaminants being present in the aquatic environment [1, 7, 17, 23, 26]. This book chapter aims to examine the effects of landfill leachate in Bangladesh and compare them with the experiences of other South Asian countries. By doing so, potential solutions can be discovered, modified, and applied to alleviate the negative consequences of leachate on surface and groundwater resources. In pursuing sustainable development and responsible management of the environment, it is imperative to acknowledge and tackle the obstacles presented by landfill leachate. This undertaking is essential in safeguarding the welfare of current and future generations.

7.2 Landfills and Leachate Generation

The use of landfills as designated locations for the disposal of solid waste produced by communities and enterprises is an essential function of modern waste management systems, which include these facilities as essential components [1, 22]. These locations are significant in controlling waste materials, preventing their uncontrolled spread, and minimizing the potential harmful effects of waste materials on the environment and public health. Leachate is a complicated type of liquid waste produced as a byproduct of the decomposition process inside landfills [14, 15]. The production of leachate is a continuous process that will go on for as long as the waste is broken down into smaller and smaller pieces inside the landfill [21]. The built-up leachate is a significant difficulty since it must be managed appropriately to prevent contamination of surface water bodies and groundwater aquifers. The inability to manage leachate efficiently can result in pollution of the surrounding environment, adverse effects on public health, and the destruction of aquatic ecosystems.

7.2.1 Mechanism of Leachate Generation

Waste soil contains three different kinds of water: free, bound, and intracellular. Intracellular water is water found within organic matter cells; as waste soil undergoes compression and degradation during the landfill process, this water is liberated as free water and bound water in the soils various components. Because it is so difficult to move the limited amount of bound water that does exist in the waste soil's pores as hydrocolloids, this water is thought to exist in the waste soil in the form of field water holding capacity. There are two different types of free water: capillary water, which is present in the unsaturated zone and is subject to matrix suction and gravity, and gravity water, which descends and eventually releases itself as leachate. Research into waste soil's water-holding and hydraulic qualities often centers on the flow of capillary water, which, when gravity plays a dominant role, goes downward and is released as leachate-like gravity water. Suction plays a deciding role; capillary water stops moving and remains in the waste soil as field water holding capacity. This occurs because the substrate suction increases in the unsaturated zone, preventing the ongoing passage of capillary water [29]. The mechanism of landfill leachate depicts in Fig. 7.1.

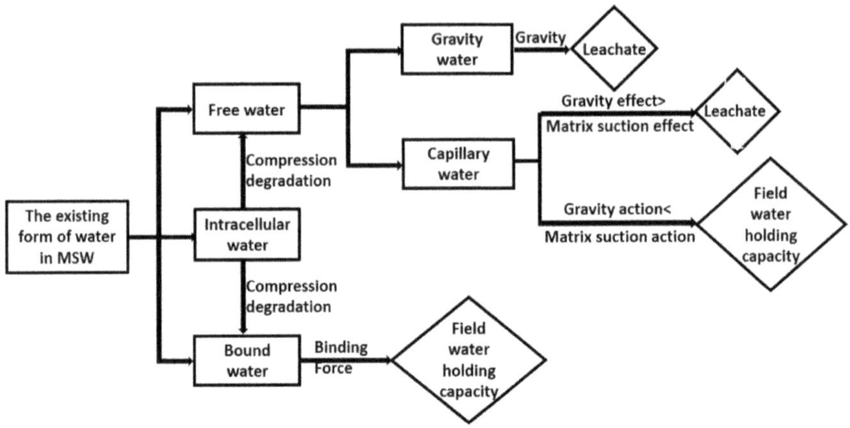

Fig. 7.1 Mechanism of landfill leachate generation

7.2.2 *Factors Influencing Leachate Composition*

The formation of leachate within landfills can result in diverse pollutants, which can vary in quantity and composition according to several influencing variables. Potential variables in this context include the type of waste, its geographical location, and the strategies employed for its management. Several key factors influence the composition of leachate, which are of significant importance.

7.2.2.1 Waste Composition

The composition of leachate in landfills is significantly impacted by the specific categories of garbage placed within these facilities. The increase in the biological oxygen demand (BOD) and chemical oxygen demand (COD) of leachate is attributed to the decomposition of organic waste, such as food and yard waste, which liberates soluble organic compounds. Hazardous waste is responsible for releasing toxic compounds into the environment, leading to potential contamination of groundwater due to the leaching of heavy metals from specific waste categories. The degradation of polymers and additives contributes to a further decline in leachate quality. The rates at which decomposition occurs, as indicated by the composition of leachate, are determined by the microbial activity that is attracted to the waste materials. Consequently, the intricate amalgamation of toxins is directly impacted by the composition of trash, thereby emphasizing the imperative of efficient waste management in mitigating adverse consequences for the environment and human well-being.

7.2.2.2 Age of Landfill

The duration of landfill operations significantly influences the composition of leachate. The leachate from recently established landfills exhibits a substantial concentration of organic matter due to the ongoing waste decomposition process, resulting in elevated levels of biochemical oxygen demand (BOD) and chemical oxygen demand (COD). The leachate derived from older landfills is characterized by a reduced presence of organic matter, which can be attributed to the comparatively sluggish decomposition rate. However, it is worth noting that such leachate may contain elevated levels of degradation byproducts. The dynamic nature of leachate throughout time is evident in its varying composition, necessitating the implementation of phase-specific strategies for landfill management.

7.2.2.3 Climate and Precipitation

The prevailing climatic conditions and precipitation patterns in the surrounding area significantly impact the leachate composition generated from landfills. A rise in precipitation leads to a higher moisture level, resulting in elevated leachate production through an augmented rate of liquid percolation across waste layers. The intensified interaction generates leachate with higher concentrations of pollutants, facilitating the removal of a broader range of contaminants, including organic compounds and heavy metals. The rates of waste breakdown are influenced by temperature, which impacts the quantity of organic matter discharged into leachate. Therefore, it is imperative to consider regional environmental factors when implementing landfill management strategies, as variations in climate and precipitation directly influence the generation of leachate, both in terms of its quality and quantity.

7.2.2.4 Landfill Design and Management Practices

Landfills' design and management practices notably impact the composition of landfill leachate. Various elements, including waste compaction, drainage systems, and cover materials, influence the liquid infiltration process and the subsequent creation of leachate. Implementing an effective landfill liner system is crucial in mitigating the potential risks associated with leachate migration, preventing it from contributing significantly to environmental contamination. The modification of leachate contents can occur due to alterations in waste exposure to moisture and microbial activity, which are brought about by implementing strategic waste layering and cover installation techniques. Implementing efficient leachate collection and treatment systems can potentially mitigate the release of pollutants. The diversity and concentration of pollutants in leachate are significantly impacted by the interaction between design choices and management approaches, underscoring their pivotal importance in determining the environmental outcomes of landfill activities.

7.2.2.5 Microbial Activity

Microbial activity is pivotal in determining the chemical composition of leachate within landfills. The waste breakdown process is facilitated by microbes, leading to the subsequent release of organic matter and chemicals into leachate. The decomposition rates of various substances are influenced by the composition of microbial communities, which are influenced by the types of waste and environmental conditions. The enzymatic activity of leachate influences its chemical composition, leading to alterations in both organic and inorganic constituents. Leachate composition is subject to continuous change due to differences in ambient factors, such as temperature, humidity, and waste composition, directly influencing microbial activity.

7.3 Characteristics of Landfill Leachate

The leachate generated by landfills is a turbid aqueous solution resulting from the interaction between water and decomposing waste materials. The wide range of leachate properties can be attributed to various factors such as the type of trash, the layout of the landfill, prevailing meteorological conditions, and the age of the landfill. Various conventional parameters, including chemical oxygen demand (COD), biological oxygen demand (BOD5), pH, total dissolved solids (TSS), ammonium nitrogen (NH4-N), total Kjeldahl nitrogen (TKN), and heavy metals, can be employed for the characterization of landfill leachate. The age of the landfill was seen to have a substantial impact on the amounts of organic matter and ammonia. The composition of leachate generated from municipal landfills is typically characterized by the compounds enumerated in Table 7.1.

Depending on the waste decomposition processes and interactions with the surrounding materials, the pH of landfill leachate can range from severely acidic to slightly alkaline [20]. The pH of leachate from newly constructed landfill sites is between 5.0 and 6.5, while that from established landfill sites is between 7.8 and 8.64 [28]. It is worth noting that landfilled leachate can raise the pH of drinking water and contribute to the development of trihalomethane (THM), a carcinogenic compound. When the BOD5/COD ratio of landfill leachate is less than 0.1, it is considered toxic because it indicates the presence of large amounts of poorly biodegradable COD and because this COD concentration can significantly alter the physiochemical properties of groundwater and introduce organic contamination into the water. Since ammonia-N may stay in the water for a long time, it is a significant contaminant in leachate that endangers humans and fish. The [10] recommends that the content of ammonia-N in inland surface water not exceed 50 mg/L [12]. Even if only minimal amounts of heavy metals are present in leachate, they can significantly affect the quality of groundwater and surface water. Toxic metal concentrations at Bangladeshi landfill sites vary from 0.45 to 0.30 mg per liter (mg/L) for lead, 0.2 to 2 mg (mg) for cadmium, 0.07 to 1.5 mg (mg/L) for copper, and 0.33 to 4.5 mg

Table. 7.1 Typical landfill leachate composition

Parameters	Range (mg/l)	Upper limit (mg/L)
Total alkalinity ($CaCO_3$)	730–15,050	20,850
Calcium	240–2330	4080
Chloride	47–2400	11,375
Magnesium	4–780	1400
Sodium	85–3800	7700
Sulfate	20–730	1826
Specific Conductance	2000–8000 μmhos cm^{-1}	9000 μmhos cm^{-1}
TDS	1000–20,000	55,000
COD	100–51,000	99,000
BOD	1000–30,300	195,000
Iron	0.1–1700	5500
Total Nitrogen	2.6–945	1416
Potassium	28–1700	3770
Chromium	0.5–1	5.6
Manganese	BDL-400	1400
Copper	0.1–9	9.9
Lead	BDL-1	14.2
Nickel	0.1–1	7.5

(mg/L) for nickel, depending on the year. Different areas and countries around the world may have varying landfill leachate properties. Table 7.2 summarizes the main features of Southeast Asian countries' landfill leachate.

7.4 Impact of Landfill Leachate on Water Bodies

The combination of water and waste materials within a landfill generates a heavily contaminated liquid known as leachate. The water's dissolved and suspended forms contain various substances, including organic and inorganic chemicals, heavy metals, pathogens, and other pollutants. The interaction between landfill leachate and surface and groundwater can substantially harm water quality and the ecosystem. The phenomenon of landfill leachate pollution in surface and groundwater can be elucidated through the collaborative operation of many processes [24]. Moisture, such as precipitation or condensation, infiltrates the layers of waste within a landfill and accumulates toxic substances during its percolation. The water, referred to as leachate, infiltrates the landfill's waste materials and undergoes dissolution and suspension of various pollutants. Leachate formation occurs within a landfill's lower region as it permeates through the permeable layers of waste above it. The potential for leachate

Table 7.2 Landfill leachate characteristics in Southeast Asian countries (mg/l except pH)

Country	Lanfill's Name	Zn	Mn	Pb	Cd	Cu	Cr	Fe	Ni	DO	BOD	COD	pH	TDS	References
Bangladesh	Aminbazar	–	–	–	0.30	1.14	0.18	14.7	2.32	–	–	–	–	–	[12]
Bangladesh	Matuail	0.04	–	BDL	0.005	0.15	0.7	–	1.05	0.9	–	1630	6.9	734	[5]
Bangladesh	Moglabazar	0.12	0.88	0.30	0.05	0.16	–	0.65	–	–	–	–	7.9	–	[2]
Bangladesh	Rowfabad	2.5	2.12	0.03	0.09	0.65	2	7.25	–	0.8	216	430	6.5	2700	[13]
India	Brahmapuram	0.27	–	0.02	–	0.1	0.12	60	0.53	–	39,500	57,300	6.6	14,490	[4]
India	Ramna MSW	BDL	–	BDL	BDL	0.33	1.77	4.5	BDL	–	1335	8332	8.82	2322	[19]
Malaysia	AmpangJajar	0.01	–	0.3	–	0	0	3	0	–	48	599	7.5	2543	[25]
Malaysia	Kuala Septang	0.2	–	0.4	–	0.75	0.05	5	0.3	–	85	990	8.1	12,568	[25]
Malaysia	BerisLalang	1.5	–	1	BDL	5	–	0.06	0.04	0.70	1640	1965	8.31	–	[11]

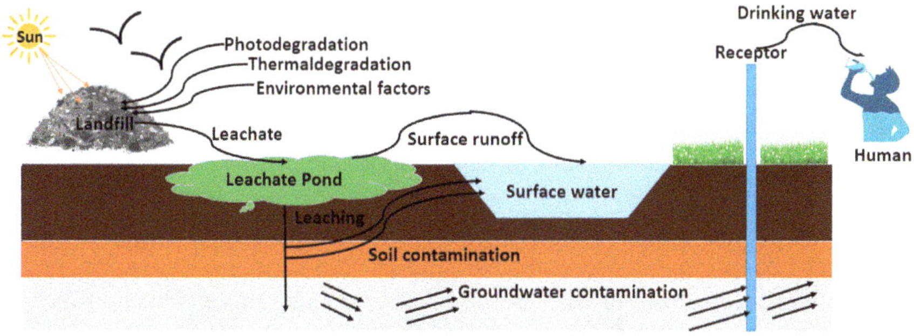

Fig. 7.2 Landfill leachate mobilization into surface and groundwater

to permeate a landfill's liner and pollute the surrounding soil exists when suitable management measures are not implemented. Leachate facilitates the transportation of various contaminants, such as metals, organic compounds, pathogens, and nutrients, as it permeates the waste material. The leachate can potentially transport these pollutants as it traverses the soil. The potential for contamination of surface and groundwater by leachate originating from landfills is evident in Fig. 7.2.

7.4.1 Impact on Surface Water

Landfill leachate poses a significant danger to surface water quality and ecosystems. Water percolation occurs within the confines of a landfill, whereby water infiltrates through the various waste materials, thereby absorbing dissolved and suspended pollutants. This results in the formation of a liquid substance known as leachate. Leachate presents several environmental hazards when it enters surface water systems such as rivers, streams, and lakes.

7.4.1.1 Water Contamination

The contamination of surface water bodies resulting from leachate discharge from landfills can initiate a cascade of adverse biological and environmental consequences. Various pollutants, including heavy metals, organic compounds, pathogens, nutrients, and hazardous chemicals, are introduced into surface and groundwater through precipitation or the infiltration of trash deposits in landfills. Landfill leachate refers to a hazardous liquid generated from the decomposition of waste materials, subsequently infiltrating adjacent surface water bodies such as rivers, streams, and lakes. The introduction of landfill leachate into water systems results in the rapid dissemination of its harmful constituents, leading to severe degradation of water quality and adverse impacts on the surrounding ecosystems. The presence of contaminants in leachate can result in notable alterations in the chemical composition of water, such as fluctuations in pH levels and the introduction of novel chemicals that are typically absent in aquatic environments.

Consequently, the aquatic organisms that have undergone evolutionary adaptations to flourish within a specific pH range and environmental circumstances may have difficulties or face potential extinction due to the modifications. Furthermore, the surplus nutrients, namely nitrogen and phosphorus, introduced by the pollutants present in the leachate have the potential to induce harmful algal blooms [6, 18]. Algal blooms have the potential to develop into dense layers on the surface of water bodies, resulting in decreased oxygen levels due to the decomposition of algae and the subsequent obstruction of sunlight. Hypoxia, resulting from oxygen deprivation, leads to the formation of areas in marine ecosystems known as dead zones, characterized by the absence of viable marine organisms. The endangerment of fish and other aquatic species poses a significant threat to the food chain and the ecological stability of the water body. The potential impact of landfill leachate on surface water bodies extends beyond ecological issues, encompassing significant concerns for human health. The contamination of water supplies by leachate presents significant risks to human well-being and economic sustenance, as it renders water unsuitable for human use, recreational activities, and agricultural purposes.

7.4.1.2 Eutrophication

The primary cause of eutrophication in surface water bodies is landfill leachate, which initiates a series of biological transformations that have negative implications for water quality and aquatic ecosystems' overall health. The infiltration of groundwater or precipitation through landfill waste contaminates these water sources with a diverse range of hazardous compounds, such as nitrates and phosphates. Landfill leachate is the aqueous solution that arises from the dissolution of specific nutrients in waste materials. When the leachate comes into contact with water bodies such as rivers, lakes, or streams, it introduces an excessive quantity of nutrients, resulting in pollution. Algae, specifically, exhibit a prompt increase in biomass as a response to adding additional nutrients. The proliferation of algae and subsequent formation of algal blooms, leading to the alteration of water colour to green or brown, occur due to elevated nitrogen and phosphorus concentrations. The algae's fast proliferation confers a competitive edge and disrupts the fragile ecological equilibrium. Decomposing deceased algal blooms results in the consumption of substantial quantities of dissolved oxygen within the water, hence inducing hypoxia as oxygen concentrations progressively decline over the water column. Mass fish mortalities and the degradation of the aquatic environment can be attributed to diminished oxygen concentrations in the water, posing a significant threat to the survival of several aquatic organisms. When an ecosystem undergoes degradation, it results in an imbalance wherein a limited number of species emerge as dominant due to their superior ability to adapt to altered environmental conditions.

Furthermore, biodiversity loss is exacerbated due to the creation of "dead zones" where no aquatic species can survive due to the lack of oxygen. Hence, eutrophication has significant impacts on aquatic ecosystems and human communities reliant on these bodies of water due to the excessive nutrient load stemming from landfill leachate. The introduction of additional nutrients into aquatic ecosystems and the consequential disturbance of the eutrophication cycle can be alleviated by implementing suitable waste management strategies, effective containment of leachate, and using advanced treatment methodologies.

7.4.1.3 Altered Water Quality

Landfill leachate significantly impacts surface water bodies, initiating a series of events that disrupt the intricate equilibrium of aquatic ecosystems. Rainwater or groundwater can become contaminated with heavy metals, organic compounds, and various other contaminants as they percolate through landfill waste. The conglomerate, referred to as landfill leachate, is subsequently discharged into nearby water bodies, resulting in numerous new constituents that fundamentally modify the characteristics of the receiving water. The introduction of contaminants can lead to alterations in pH, temperature, turbidity, and chemical composition, significantly impacting the aquatic ecosystem. Aquatic creatures have undergone evolutionary adaptations to inhabit specific ranges of environmental circumstances, and abrupt alterations in these settings can impose physiological stress and ultimately lead to their mortality. Particularly susceptible species may encounter difficulties in adapting to the alterations, potentially resulting in a decline in their population or even leading to their extinction. Changes in water quality can impede the growth of beneficial aquatic vegetation and facilitate the proliferation of invasive or opportunistic species, thus disrupting ecological balance to a greater extent. The altering water quality poses a significant threat to aquatic organisms' reproductive success, behaviour, and overall health. The adverse impact of turbidity-induced light attenuation on photosynthesis and, subsequently, the entire food chain is well-documented. The potential for oxygen depletion and harm to aerobic organisms is increased due to the presence of organic substances introduced by leachate, which can stimulate the proliferation of oxygen-consuming bacteria during decomposition. The alteration in water quality resulting from the introduction of landfill leachate has extensive ecological implications, placing aquatic organisms at risk and compromising the overall equilibrium of the aquatic ecosystem. In light of the extensive ramifications associated with alterations in water quality, it is imperative to adopt stringent waste management protocols and employ effective leachate treatment methodologies to mitigate the influx of pollutants into our aquatic ecosystems. It is imperative to undertake these precautionary measures to uphold the ecological well-being of these systems, ensuring their ability to sustain biodiversity and furnish indispensable resources for humanity and the environment.

7.4.1.4 Reduced Biodiversity

The presence of landfill leachate in surface water bodies has been observed to significantly impact the decline of aquatic life since it initiates a chain reaction of ecological disturbances. Introducing leachate into nearby water bodies can disrupt the equilibrium of the aquatic ecosystem by transporting a diverse range of contaminants. The pollutants encompass heavy metals, organic molecules, and toxic chemicals. The presence of contaminants inside leachate has the potential to cause mortality in organisms and diminish the abundance of vulnerable species by means of direct toxic effects. The introduction of leachate can also lead to modifications in habitat structure and the availability of resources, posing detrimental consequences for species that depend on stable environmental conditions to ensure their survival and reproductive success. Eutrophication, oxygen depletion, and the creation of areas with limited biodiversity, sometimes called "dead zones," are consequences of elevated nutrient levels in leachate, particularly nitrogen and phosphorus. These excessive nutrients serve as catalysts for the proliferation of algae and aquatic plants.

Consequently, the hostile nature of this habitat leads to a significant decline in the population of diverse aquatic creatures. Potential consequences of a shift towards species that exhibit more tolerance towards pollution and modified surroundings include alterations in trophic relationships, disturbances in nutrient cycling, and overall dysfunction within the ecosystem. The impacts extend beyond aquatic creatures and have far-reaching consequences across the food web, posing a threat to the overall equilibrium of the ecosystem and the interdependencies between predators and prey. In light of the intricate nature of aquatic ecosystems, it is imperative to establish and execute effective waste management strategies, stringent regulatory measures, and state-of-the-art leachate treatment techniques in order to mitigate the discharge of detrimental pollutants into the water reservoir. The preservation of the intricate network of aquatic biodiversity is vital to ensure the long-term well-being of aquatic ecosystems, hence safeguarding their sustainability for future generations.

7.4.2 Impact on Groundwater

Leachate, the liquid that emerges from landfills due to water percolating through the deposited garbage, is a significant concern in waste management. The leachate has the potential to adversely impact the quality of groundwater and the overall environmental conditions.

7.4.2.1 Deterioration of Groundwater Quality

The potential degradation of groundwater quality due to landfill leachate initiates a cascade of interconnected environmental and health-related issues. Rain and moisture can accumulate many hazardous substances, such as heavy metals, volatile organic compounds, viruses, and nutrients when they infiltrate layers of disposed waste materials. The leachate comprises a hazardous combination of pollutants that infiltrate the soil and then permeate the groundwater. The contaminants mentioned above are derived from the process of organic matter decomposition, industrial waste, and the disposal of chemical substances. Upon contact, the polluted leachate has the potential to contaminate groundwater, leading to alterations in the chemical composition, pH levels, and nutritional equilibrium of the water. Organic compounds such as benzene and chlorinated solvents have been linked to the development of cancer and other chronic illnesses.

Additionally, heavy metals like lead, mercury, and cadmium have the potential to contaminate groundwater, hence presenting significant health risks to individuals who consume it. The presence of leachate in water systems presents a significant concern for water quality due to its potential to disseminate pathogens, posing a public health risk. The elevated nutritional content of leachate can also contribute to eutrophication, thereby facilitating the proliferation of algal blooms that have detrimental effects on aquatic ecosystems, such as suffocating fish and depleting oxygen levels in water bodies. The potential long-term consequence involves disseminating contaminated groundwater plumes from landfills, posing a threat to the potable water supply of populations residing in proximate regions. The imperative for improved waste management and environmentally sustainable disposal options is underscored by the detrimental impacts on ecosystems, risks to public health, and the substantial costs associated with landfill leachate remediation.

7.4.2.2 Human Health Risks

The contamination of groundwater caused by landfill leachate presents significant risks to human health and can have extensive implications. Heavy metals such as lead and cadmium in leachate have been associated with significant neurological abnormalities, developmental deficiencies in children, and organ damage in adults when these metals are consumed through the ingestion of drinking water. Individuals residing close to groundwater sources that have been contaminated face an elevated susceptibility to cancer as a result of their exposure to leachate. This leachate consists of hazardous compounds such as benzene and chlorinated solvents, which possess carcinogenic properties. The presence of bacterial, viral, and parasite pathogens inside leachate has the potential to contaminate drinking water sources and contribute to the occurrence of waterborne disease outbreaks. These outbreaks tend to significantly impact marginalized communities, exacerbating existing health disparities and manifesting symptoms such as diarrhoea, pneumonia, and mortality. The presence of endocrine-disrupting chemicals in leachate has the potential to infiltrate the groundwater reservoir, giving rise to a range of health complications such as reproductive disorders, impaired growth, and imbalances in the immune system. Prolonged exposure to groundwater that has been contaminated presents significant health hazards to the affected communities, resulting in an elevated susceptibility to chronic ailments, reduced life expectancy, and compromised overall well-being.

In order to safeguard human well-being and mitigate the exacerbation of a potential public health emergency, it is imperative to tackle the issue of groundwater pollution caused by landfill leachate. Adopting robust waste management protocols, stringent regulatory enforcement, and the implementation of efficacious water treatment strategies can achieve this.

7.4.2.3 Ecological Damage

The impact of landfill leachate on groundwater systems is substantial, leading to a cascade of adverse consequences that pose a significant danger to the integrity of ecosystems. Various detrimental substances, such as heavy metals, persistent organic pollutants, and microbial diseases, have the potential to infiltrate groundwater as leachate permeates the soil and reaches the water table. The presence of these contaminants in aquatic environments threatens the well-being of aquatic organisms due to their detrimental effects on water quality and disruption of the intricate equilibrium within aquatic ecosystems. The accumulation of heavy metals in aquatic flora and fauna tissues has been observed to result in several detrimental effects, including impaired growth and development, aberrant behavioral patterns, and reduced reproductive capacity. The metabolic activities of microorganisms result in the consumption of organic compounds, reducing oxygen levels and posing a detrimental impact on aquatic ecosystems.

Moreover, eutrophication is fueled mainly by minerals, such as nitrogen and phosphorus, transported by leachate. These nutrients stimulate the proliferation of abundant algal blooms, impeding sunlight penetration and depleting oxygen concentrations. Consequently, this detrimental phenomenon leads to the mortality of aquatic species and disrupts the stability of food chains. Leachate poses a significant threat to biodiversity, disrupts crucial biological relationships, and further destabilizes already vulnerable ecosystems by introducing a toxic mixture. The ecological harm from landfills extends beyond their immediate vicinity when toxins are transported through groundwater, impacting downstream water bodies and the habitats they support. The emergence of contaminated groundwater through springs and seeps has the potential to alter the surrounding soil's chemical composition, posing a threat to terrestrial flora and fauna. In general, the impact of landfill leachate on groundwater has extensive ecological consequences, resulting in the degradation of ecosystem resilience, the compromise of biodiversity, and the necessity for substantial measures to mitigate the harm inflicted on the natural environment.

7.4.2.4 Migration of Contaminants

The dispersion of pollutants extends beyond the confines of the landfill site when groundwater becomes contaminated by landfill leachate, resulting in significant environmental ramifications. The soil is infiltrated by leachate water, which introduces a mixture of pollutants such as heavy metals, organic compounds, and pathogens. The poisons are transported through the lower flow of groundwater and disseminated by tracking aquifers and geological formations. Various factors, including groundwater flow rates, soil and rock strata permeability, and topographical gradients, facilitate the movement of groundwater. The emissions originating from a landfill possess the capacity to disperse over considerable distances, polluting adjacent groundwater and surface water sources, thereby posing a hazard to the ecosystems reliant upon them. The ascent of contaminated groundwater to the surface in the form of rivers or lakes results in the dissemination of pollutants across aquatic ecosystems, perpetuating the detrimental pollution cycle. The widespread movement of toxins has a significant impact on the deterioration of water quality in various places and the human and ecological communities that depend on these water sources. To address this issue, it is imperative to employ efficient strategies for managing landfills, establish comprehensive networks for monitoring groundwater, and implement containment and remediation methods. These actions are crucial in mitigating the dissemination of pollutants, safeguarding ecosystems, and preserving water sources.

7.4.3 Case Studies of Specific Landfill Sites

From the landfill's base through the unsaturated soil layers to the groundwater is the most prevalent route for leachate to move to aquatic ecosystems. As a result of hydraulic linkages, leachate can be transferred from groundwater to surface water in many poor nations, including Bangladesh, where most landfills were built without adequate liners. The weak design of landfill sites and poor management in subtropical countries like Bangladesh allow monsoon foods to be introduced straight to surface water.

Most water quality parameters of surface water within a 1 km radius of the Matuail landfill were below the safe limit of inland surface water [10], except Cr (1.03 mg/L) and DO (2.3 mg/L) (Azim et al. 2020). This study was conducted by Azim et al. [5]. Leachate discharge to water bodies is thought to pose a significant risk of oxygen depletion in the surface water, which can have an influence on the stream bottom fauna and fora and generate ammonia toxicity. On the other hand, fishing is an everyday activity in the lowlands adjacent to the Matuail dump. Therefore, such heavy metals can bioaccumulate in fish and then be consumed by people. Although Zn and Pb concentrations surpassed the standard threshold for drinking water set by the World Health Organization [27], it was observed that the overall concentration of metal ions in groundwater at the Matuail landfill site was generally relatively low due to the decreasing solubility of metal. Unfortunately, to our knowledge, no study has been found on surface and groundwater contamination near the Amin Bazar landfill site, even though it has poor waste management and leachate treatment, is located in a residential area, and is within the designated food flow zone area. Except for Pb (0.33 mg/L in surface water and 0.17 mg/L in groundwater), the concentration of toxic metals in the surface and groundwater near the Mogla Bazar landfill site is below the safe limit for inland surface water and drinking water set by the DoE,

Bangladesh, and WHO (2020). Residents in the area rely on groundwater as their primary source of drinking water. Thus, it is crucial that no hazardous heavy metals seep into it. High Fe, Cr, and Cd concentrations were found in surface water at the Rowfabad landfill site, as reported by Hossain et al. [13]. Maximum permitted levels (MPL) for Fe, Cd, and Cr in drinking water [10, 27] are exceeded in the groundwater near the Rowfabad Landfill. E. coli, a fecal indicator bacterium, was also detected in groundwater samples (15/100 mL in the winter and 71/100 mL in the rainy season) by Hossain et al. [13]. The physicochemical properties of the surface and groundwater near Bangladeshi landfills are shown in Table 7.3.

7.5 Remedial Measures

Leachate created from landfill is harming surface water quality through surface runoff and groundwater quality through percolation into the subsurface. Therefore, it is unavoidable to take required actions to prevent additional contamination from landfill leachate. Following remedial procedures can be taken to limit or avoid the contamination of surface and groundwater by landfill leachate:

- Implementing a thorough leachate collection system within the landfill to capture and channel leachate away from nearby soil and water bodies.
- Setting up an on-site leachate treatment plant to treat collected leachate before release, employing procedures such biological treatment, coagulation, flocculation, and advanced oxidation.
- Emphasizing good landfill construction that includes impermeable liners (clay or synthetic materials) to avoid leachate migration into the soil and groundwater.
- Installing gas collecting devices to limit the development of pressure within the landfill, which can force leachate out into adjacent areas.
- Regularly monitoring leachate levels and quality in both the landfill and local water bodies to detect any contamination early.
- Constructing containment barriers such as berms and dikes around the landfill area to prevent leachate runoff during heavy rainfall.
- Implementing appropriate landfill cover systems to prevent precipitation infiltration, which might cause leachate. This can include employing impermeable coverings or vegetation.
- After landfill closure, placing a final capping layer to decrease water infiltration and minimize leachate generation.
- Establishing buffer zones surrounding landfills with plants that can absorb excess moisture and toxins, limiting the risk for leachate migration.
- Using native plants and trees with deep root systems to promote soil stability and avoid erosion.
- Educating the public on correct waste disposal and the potential risks of landfill leachate contamination, supporting responsible waste management practices.

Table 7.3 Physico-chemical characteristics of surface and groundwater surrounding the landfill sites in Bangladesh along with the recommended values (mg/l except pH)

Landfill sites	Type of water body	Zn	Mn	Pb	Cd	Cu	Cr	Fe	Ni	DO	BOD	COD	pH	TDS	References
Matuail	Surface	0.116	–	0.03	0.005	0.2	1.03	–	0.4	2.3	–	50	6.5	427	[5]
Mogla Bazar	Surface	0.01	0.03	0.33	0.00	0.01	–	0.03	–	–	–	–	6.3	–	[2]
Rowfabad	Surface	0.14	0.015	0.005	0.067	0.25	1.07	4.8	–	2.35	21	69	7.2	–	Hossain et al. [13]
Inland surface water standard		5	5	0.10	0.05	0.5	0.1	2	1	4.5–8	50	200	6–9	2100	[10]
Matuail	Ground	5	–	0.05	0.001	0.01	0.05	–	0.02	–	–	3.88	6.8	502	[5]
Mograbazar	Ground	0.02	–	0.17	–	0.02	–	0.04	–	–	–	–	7.1	–	[2]
Rowfabad	Ground	0.5	0.12	0.007	0.04	0.02	0.09	3.26	–	2.45	0.4	44	6.7	–	Hossain et al. [13]
WHO guideline for drinking water		3	0.2	0.01	0.003	2	0.05	0.3	0.02	NM	NM	NM	6.5–9.2	500	[27]
ECR standard for drinking water		5	0.1	0.05	0.005	1	0.05	0.3–1	0.1	6	6	4	6.5–8.5	1000	[10]

- Strengthening rules and enforcement to ensure that landfills follow best practices in containment, collection, and treatment of leachate.
- Promoting trash reduction, recycling, and composting to lower the volume of garbage that ends up in landfills, hence reducing the possibility for leachate formation.
- Encouraging the adoption of contemporary waste-to-energy technology that can transform garbage into energy while reducing environmental concerns.
- Conducting detailed hydrogeological research to determine the local groundwater flow patterns and potential pathways for leachate migration.
- Using this knowledge to implement appropriate mitigation measures based on the individual characteristics of the location.
- Developing a detailed emergency response plan that covers procedures for containing and addressing leachate spills, leaks, or other events.
- Training landfill employees and local authorities to rapidly respond to any leachate-related emergencies.
- Establishing a long-term monitoring and maintenance program to ensure the effectiveness of adopted measures and to resolve any potential concerns that may occur over time.
- Regularly analyzing and modifying remedial methods based on new technology, research, and evolving environmental requirements.

7.6 Conclusion

In Bangladesh, landfill leachate has seriously contaminated the surface and groundwater systems. The necessity of tackling this issue to protect public health and maintain the ecosystem is emphasized because leachate contains various toxins. There should be a strong focus on the importance of strict laws, efficient waste management techniques, and cutting-edge leachate treatment technologies. Comprehensive strategies prioritizing sustainable garbage disposal, appropriate site selection, and constant monitoring must incorporate mitigation measures. To ensure the long-term health of Bangladesh's water resources in the face of rising urbanization and water shortages, landfill leachate management has emerged as a pressing issue needing cooperation between government agencies, businesses, and communities.

References

1. Abunama T, Moodley T, Abualqumboz M, Kumari S, Bux F (2021) Variability of leachate quality and polluting potentials in light of leachate pollution index (LPI)–a global perspective. Chemosphere 282:131119
2. Alam R, Ahmed Z, Howladar MF (2020) Evaluation of heavy metal contamination in water, soil and plant around the open landfill site Mogla Bazar in Sylhet Bangladesh. Groundwater Sustain Dev 10:100311

3. Ali SA, Ahmad ATEEQUE (2019) Analysis of chemical and heavy metal concentrations of leachates and impact on groundwater quality near Dhapa dumping ground Kolkata. Asian Profile 47(1):19–37
4. Arunbabu V, Indu KS, Ramasamy EV (2017) Leachate pollution index as an effective tool in determining the phytotoxicity of municipal solid waste leachate. Waste Manage 68:329–336
5. Azim MD, Rahman MM, Khan RH, Kamal ATMM (2011) Characteristics of leachate generated at landfill sites and probable risks of surface and groundwater pollution in the surrounding areas: a case study of Matuail landfill site, DHAKA. J Bangladesh Acade Sci 35(2):153–160
6. Banch TJ, Hanafiah MM, Alkarkhi AF, Salem AM (2019) Statistical evaluation of landfill leachate system and its impact on groundwater and surface water in Malaysia. Sains Malays 48(11):2391–2403
7. Bhagwat A, Ojha CSP, Pant A, Kumar R (2023) Interaction among heavy metals in landfill leachate and their effect on the phytoremediation process of Indian Marigold. J Hazard Toxic Radioactive Waste 27(1):04022039
8. Chaudhary R, Nain P, Kumar A (2021) Temporal variation of leachate pollution index of Indian landfill sites and associated human health risk. Environ Sci Pollut Res 28:28391–28406
9. DNCC (2016) Dhaka North City Corporation Report 2016–2017 (Internet), pp 1–11. Available from: https://dncc.portal.gov.bd/sites/default/fles/fles/dncc.%20portal.gov.bd/annual_reports/efedb2ad_8950_461e_960f_d1bfa%20c6d10e3/Waste%20Report%202016-2017.pdf
10. ECR (1997) The environmental conservation rules, 1997. Bangladesh Dep Environ Minist for Gov Peoples' Repub Bangladesh, pp 179–227
11. Fadhullah W, Kamaruddin MA, Ismail N, Sansuddin N, Abdullah H (2019) Characterization of landfill leachates and its impact to groundwater and river water quality: a case study in beris lalang waste dumpsite Kelantan. Pertanika J Sci Technol 27(2):633–646
12. Hoque MA, Haque MA, Mondal MSA (2014) Seasonal effects on heavy metal concentration in decomposed solid waste of DNCC and DSCC landfill sites. Civ Eng Archit 2(1):52–56
13. Hossain ML, Das SR, Hossain MK (2014) Impact of landfill leachate on surface and ground water quality. J Environ Sci Technol 7(6):337–346. https://doi.org/10.3923/jest.2014.337.346
14. Huang FY, Zhao Y, Zhou XY, Li H, Zhang X, Su JQ (2022) Dissemination of antibiotic resistance genes from landfill leachate to groundwater. J Hazard Mater 440:129763
15. Irfeey AMM, Najim MM, Alotaibi BA, Traore A (2023) Groundwater Pollution Impact on Food Security. Sustainability 15(5):4202
16. Kaza S, Yao L, Bhada-Tata P, Van Woerden F (2018) What a waste 2.0: a global snapshot of solid waste management to 2050. World Bank Publications
17. Lindamulla L, Nanayakkara N, Othman M, Jinadasa S, Herath G, Jegatheesan V (2022) Municipal solid waste landfill leachate characteristics and their treatment options in tropical countries. Current Pollution Reports 8(3):273–287
18. Ma S, Zhou C, Pan J, Yang G, Sun C, Liu Y, Zhao Z (2022) Leachate from municipal solid waste landfills in a global perspective: characteristics, influential factors and environmental risks. J Cleaner Prod 333:130234
19. Mishra S, Tiwary D, Ohri A, Agnihotri AK (2019) Impact of municipal solid waste Landfill leachate on groundwater quality in Varanasi India. Groundwater Sustain Dev 9:100230
20. Mukherjee S, Mukhopadhyay S, Hashim MA, Sen GB (2014) Contemporary environmental issues of landfill leachate: assessment and remedies. Crit Rev Environ Sci Technol 45(5):472–590
21. Natesan U, Vaikunth R, Kumar P, Ruthra R, Srinivasalu S (2021) Spatial distribution of microplastic concentration around landfill sites and its potential risk on groundwater. Chemosphere 277:130263
22. Negi P, Mor S, Ravindra K (2020) Impact of landfill leachate on the groundwater quality in three cities of North India and health risk assessment. Environ Dev Sustain 22:1455–1474
23. Rajoo KS, Karam DS, Ismail A, Arifin A (2020) Evaluating the leachate contamination impact of landfills and open dumpsites from developing countries using the proposed leachate pollution index for developing countries (LPIDC). Environ Nanotechnol Monit Manage 14:100372

24. Tang ZW, Hamid FS, Yusoff I, Chan V (2023) A review of PFAS research in Asia and occurrence of PFOA and PFOS in groundwater, surface water and coastal water in Asia. Groundwater Sustain Dev 100947
25. Umar M, Aziz HA, Yusoff MS (2010) Variability of parameters involved in leachate pollution index and determination of LPI from four landfills in Malaysia. International J Chem Eng
26. Vaccari M, Tudor T, Vinti G (2019) Characteristics of leachate from landfills and dumpsites in Asia, Africa and Latin America: an overview. Waste Manage 95:416–431
27. WHO (2012) Guideline for drinking water quality, 4th ed. Incorporating the 1st addendum. [Internet]. Available from: https://www.who.int/publications/i/item/%209789241549950
28. Zakaria SNF, Aziz HA (2018) Characteristic of leachate at Alor Pongsu landfll site, Perak, Malaysia: a comparative study. IOP Conf Ser Earth Environ Sci 140:012013
29. Zhang P, Chai J, Cao J, Qin Y, Dang M, Geng K et al (2022) Landfill leachate generation mechanism study: a review. Int J Environ Sci Technol 20(8):9271–9290

Chapter 8
Landfill Leachate and Ecotoxicity

Alyne Moraes Costa, Sarah Dario Alves Daflon, and Juacyara Carbonelli Campos

Abstract Leachate is wastewater generated in sanitary landfills, resulting from the degradation of solid waste, with a high polluting potential for water bodies. The ecotoxicological assessment of this effluent is essential to determine the impact of its pollutants on the aquatic ecosystem. In this context, the objective of this study was to evaluate the acute toxic effects caused by some pollutants measured by the parameters total ammonia nitrogen (TAN), total alkalinity, humic substances (HS), and chloride present in the leachates from different landfills in Rio de Janeiro State (Brazil). The test organisms evaluated represented different trophic levels of the food chain: *Vibrio fischeri*, *Danio rerio*, *Daphnia similis*, activated sludge microorganisms and *Artemia* sp. Ecotoxicity results for the selective removal of pollutants from the leachate, such as air-stripping and membrane filtration experiments, are presented. The obtained data were evaluated by principal components analysis methodology. This study highlights the importance of toxicity assessment for landfill leachates and choosing suitable test organisms.

Keywords Landfill leachate · Ecotoxicity · Physicochemical parameters · Treatment · Environmental impact

8.1 Introduction

Urban solid waste management is a significant environmental, economic and social problem. Every year, more than one billion tons of solid waste are produced worldwide, and it is estimated that this volume will only increase over the years, reaching around 3.4 billion tons by the year 2050 [1]. Karak et al. [2] state that solid waste production is overgrowing and outstripping the global population growth.

The final disposal of solid waste in landfills has become the most widely used method in most developing countries [1], such as Brazil, due to the simplicity and low cost of operation [3].

A. M. Costa · S. D. A. Daflon · J. C. Campos (✉)
School of Chemistry, Universidade Federal do Rio de Janeiro, Rio de Janeiro, RJ, Brazil
e-mail: juacyara@eq.ufrj.br

A. Anouzla and S. Souabi (eds.), *A Review of Landfill Leachate*, Springer Water,
https://doi.org/10.1007/978-3-031-55513-8_8

The confinement of waste favors the formation of leachate, one of the most significant environmental threats from landfills [4, 5]. Landfill leachate has a variable and heterogeneous composition containing pollutants: organic matter (biodegradable and recalcitrant), ammonia nitrogen, inorganic salts, and heavy metals, among others [6–10]. Due to the complexity of its composition, leachate is considered a potentially toxic wastewater associated with groundwater and surface water contamination.

According to Marttinen et al. [11] and Prestes et al. [12], leachate ecotoxicity results from its numerous contaminants, their additive, synergistic and antagonistic effects and their physicochemical properties. Therefore, the need to determine each pollutant's contribution to this wastewater's ecotoxicity has become one of the primary measures to understand the risks of contamination to the aquatic ecosystem.

Landfill leachate ecotoxicity is generally related to ammonia nitrogen, alkalinity, organic matter and chloride [13–15]. Ammonia nitrogen has been identified as the most toxic pollutant in leachate, found in high concentrations even in landfills that have ceased operations for more than 50 years [16]. When found in abundance in the environment, ammoniacal nitrogen can stimulate algae growth, deplete dissolved oxygen, and harm aquatic organisms [17, 18].

Traditionally, the assessment of landfill leachate quality is based on identifying pollutants through physicochemical analysis. However, these analyses alone cannot determine the complex interaction between contaminants and their biological effects. Ecotoxicity tests have therefore become essential to complement physicochemical analyses and directly assess the impact of leachate on living organisms [5, 19].

Ecotoxicity tests using test organisms covering different trophic levels are the most appropriate for wastewaters with various contaminants, such as leachate [13, 20]. Analyzing the action of leachate pollutants on different organisms and their respective sensitivities is a way of estimating the potential environmental risk related to the aquatic ecosystem as a whole, as well as a way of obtaining information on the efficiency of the treatment system and seeking additional strategies for its remediation [19].

8.2 Principles of Aquatic Ecotoxicology

The term ecotoxicology emerged in the late 1960s and was first defined by René Truhaut as the study of anthropogenic toxic effects on ecological systems [21]. Over the years, more comprehensive definitions of this term have emerged, and the definition proposed by [22], still used today, describes ecotoxicology as the science that evaluates the toxic effects of chemical and physical agents on living organisms, especially on populations and communities within defined ecosystems, including the fates and interactions of these agents in the environment. Boudou and Ribeyre [23] further add that ecotoxicology studies the disturbances induced in the short, medium and long term by contamination factors (physical, chemical and biological) in ecosystems from the direct or indirect effects of anthropogenic activity.

The aquatic environment comprises rivers, lakes, estuaries, seas and oceans that harbor ecosystems of unique characteristics, products of the interaction between biotic and abiotic components [24, 25]. While aquatic ecosystems are the primary receptacles of contaminants, aquatic ecotoxicity tests are frequently used to assess environmental impacts. Wastewater quality determination based on ecotoxicity assays may be more relevant for assessing the ecological impact of contamination than the mandatory wastewater quality classification based on physicochemical parameters [26].

According to Moiseenko [27], contaminants are susceptible to different processes affecting their fate and behavior, such as migration, transformation, sedimentation and interaction, generating impacts at the species, population, community or even ecosystem level. Ecotoxicological tools can complement and improve the assessment of natural water bodies' quality, including in situations of climate change [28].

Ecotoxicological tests can detect harmful effects at the molecular and cellular levels using biomarkers [29, 30] and through measures that are most used in this field, such as lethality, alteration in the development, growth, reproduction and behavior of organisms [31]. In addition, they can be classified into acute and chronic, according to the concentration of the toxic agent and the time of exposure of the organisms to the pollutants [32].

Acute tests evaluate the rapid and severe effects of chemical agents in a short exposure period with regard to the life cycle of organisms, which can vary from 24 to 96 h [21, 32]. According to Birge et al. [33], acute tests are simple to perform, low cost and rapid response, and therefore, they were the first to be developed, constituting the basis of ecotoxicological data.

The main variables evaluated in the acute test are the lethality and immobility of the organisms. In this evaluation, the use of statistical tools is fundamental to obtain the following responses: EC50 (median effective concentration), i.e., the concentration of the toxic agent that causes a toxic effect to 50% of the test organisms; and LC50 (median lethal concentration), which represents the concentration of the toxic agent that causes lethality to 50% of the test organisms [34].

Chronic tests measure the effects of pollutants over a prolonged and significant period in relation to the organisms' life cycle. Exposure of the organism to the contaminant occurs at sublethal levels, i.e., it generally does not lead to mortality. However, it generates long-term physiological and behavioral effects.

8.2.1 The Test Organisms

The test organisms used in ecotoxicity assays can range from plants to animals, from prokaryotes to eukaryotes, as long as they can emit a response upon contact with the contaminant [35, 36]. However, some essential criteria should be taken into account for the choice of test organisms, such as the organism should play an important role in its trophic level, occupy a significant position in the food chain, have a wide distribution, exhibit uniform and constant genetics, present sensitivity to the

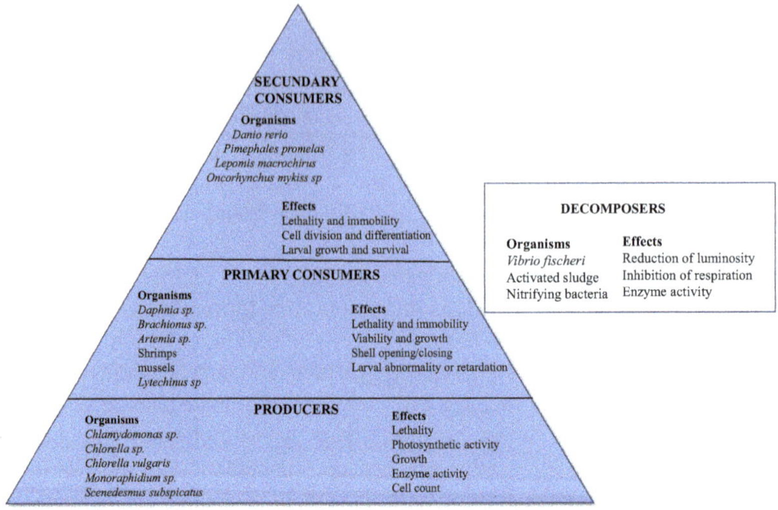

Fig. 8.1 The test organisms commonly used for toxicity assessment of water and wastewater and usual responses to the toxicant. Adapted from [35, 39]

contaminant, and finally, have deeply known data on its physiology, taxonomy and its role in the environment [37, 38]. Figure 8.1 shows some test organisms commonly used in water and wastewater toxicity assessment.

The choice of test organisms influences the success and interpretation of the results, i.e., if the species used in the assay has little or no sensitivity to the contaminant, false positive results may be generated, and the actual effect of the toxicant in the environment will not be measured [40]. In addition, using a single species is usually linked to economic factors and ease of cultivation in the laboratory [38]. According to Newman [31], determining a single species' sensitivity cannot predict population-level effects. Therefore, it is essential to use a battery of test organisms from different trophic levels to obtain an adequate response to the action of the toxic agent in the environment.

The test organisms used in this study are detailed below, as well as their sensitivity to some contaminants and the advantages and disadvantages of their use in trials.

8.2.1.1 *Vibrio fischeri*

The test organism *Vibrio fischeri*, also known as *Aliivibrio fischeri* or *Photobacterium phosphoreum,* is a non-pathogenic bioluminescent marine bacterium belonging to the Vibrionaceae family, commonly used in acute ecotoxicity tests [41, 42]. The luminescence emitted by the bacteria is a result of the presence of the enzyme luciferase, which is intrinsically related to cellular respiration, so that if the cellular metabolism of the bacteria undergoes any alteration, a reduction in the luminous emission will be evidenced [43–45].

The advantages of the ecotoxicity assay with *Vibrio fischeri* are fast response (5–30 min), high sensitivity, need for a smaller sample volume, as well as being highly reproducible in laboratories [46–48]. However, this method presents some disadvantages, and one of the main ones involves the issue that *Vibrio fischeri* is a marine organism, and to perform tests with freshwater samples, it becomes necessary to add saline solution, altering the physicochemical composition of the sample, as well as reducing the solubility of some organic compounds, making the solution turbid [39].

The bioluminescence inhibition assay with the bacterium *Vibrio fischeri* is generally one of the most sensitive acute ecotoxicity assays compared to other bacterial assays, such as nitrification inhibition, respirometry, ATP luminescence and enzyme inhibition [49–51].

A limitation in the sensitivity of bioluminescent bacteria is that they show higher sensitivity to organic compounds and lower sensitivity to heavy metals [47, 52]. In addition, low concentrations of inorganic nutrients (K^+, Ca^{2+}, Na^+) and heavy metals (Cu^{+2}, Zn^{+2}, Cd^{+2}, Cr^{+3}, Mg^{+2}, Fe^{+3}, Ni^{+2}, Co^{+2}) can induce a stimulating effect on luminescent bacteria, called hormesis, increasing the variation of emitted light, which may indicate false toxicity results [53, 54].

8.2.1.2 Microorganisms of Activated Sludge

Activated sludge is a complex ecosystem composed of large biodiversity of organisms, such as heterotrophic and nitrifying bacteria, protozoa, metazoans and fungi [55, 56].

The microbial community of activated sludge, in general, can be divided into decomposers (bacteria and fungi), which feed on organic matter and consumers (protozoa and metazoans), which feed on dispersed bacteria and other microorganisms [57, 58]. According to Nielsen et al. [59], about 80–95% of the microbial population of activated sludge is formed by bacteria. Bacteria are the main group of microorganisms fundamental for the functioning of the activated sludge system, as they are responsible for removing organic matter and forming agglomerates called biological flocs [60]. Figure 8.2 shows the typical structure of a biological floc.

Biological floc is formed by a complex structure of aggregates of microorganisms, mainly by floc-forming and filamentous bacteria [62]. Floc-forming bacteria are responsible for forming the gelatinous matrix, and filamentous bacteria are responsible for structuring the biological floc. The gelatinous matrix is formed through the excretion of polysaccharides by bacteria and has the function of adhesion, which allows the formation of large microbial aggregates [58].

The diversity of microbial species in activated sludge will depend on some variables, such as the presence of macronutrients (nitrogen, phosphorus, potassium, calcium, magnesium, iron), micronutrients (chromium, cobalt, manganese, nickel, selenium, zinc, among others), temperature, pH and the level of dissolved oxygen.

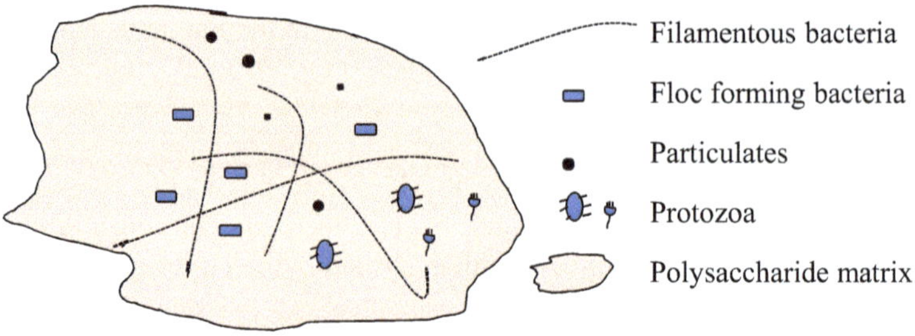

Fig. 8.2 Composition of biological floc in activated sludge systems. Adapted from [61]

The ecotoxicity test that estimates the toxic effect of contaminants on activated sludge microorganisms is called the respiration inhibition test or respirometry [55, 63]. Under defined experimental conditions, the respirometry technique measures dissolved oxygen (DO) consumption by microorganisms for the degradation of substrates [64].

The oxygen uptake rate (OUR) is determined by the rate at which oxygen is consumed (per unit time) by the microorganisms as a means of understanding the metabolic activity of the biomass.

The ratio of OUR to biomass mass is called the specific oxygen consumption rate (SOUR). The SOUR parameter is more consistent in evaluating toxicity in microorganisms, considering the concentration of volatile suspended solids (VSS) in contact with the toxic components [65].

In respirometric tests, oxygen consumption occurs in two phases, called endogenous and exogenous. Endogenous respiration consists of constant oxygen consumption as a way to meet the needs of cell maintenance. In exogenous respiration, oxygen consumption occurs at a high rate to perform the degradation of substrates present in the liquid. The graphical representation of the rate of oxygen consumption by microorganisms can be seen in Fig. 8.3.

According to Ricco [66] and Kungolos [67], the respirometry technique is simple, fast and reproducible, highly applicable in activated sludge systems, being the first parameter to be used for the evaluation of biological activity and the verification of effluent toxicity.

This method is described in detail in OECD 209 [68] [OECD Guidelines for the Testing of Chemicals Activated Sludge, Respiration Inhibition Test (Carbon and Ammonium Oxidation)].

8.2.1.3 *Artemia* sp.

The organism *Artemia* sp. is a microcrustacean frequently used in research and aquatic ecotoxicology tests in order to investigate the toxicity of environmental samples, mixtures, chemicals and nanoparticles [69–72].

Fig. 8.3 Graphical representation of oxygen uptake (OUR) versus time. Adapted from [65]

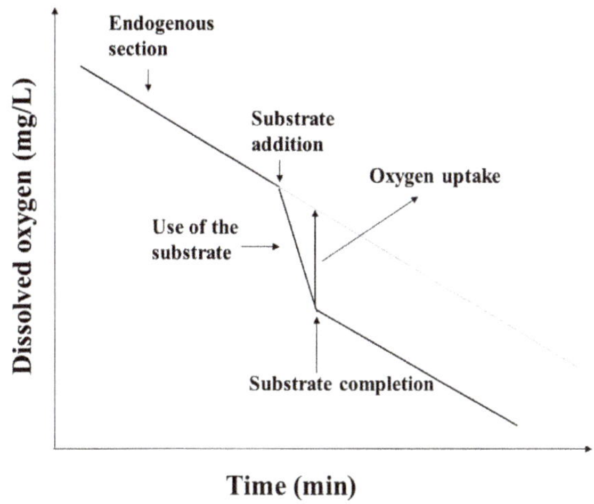

Artemia sp. is one of the main species from brackish to hypersaline ecosystems, tolerating salinity between 5–250 and measuring about 8–12 mm in length [73, 74]. Their geographical distribution occurs in North America, South America and Asia, and they play an essential role in the food chain as a primary consumer [73].

This species of microcrustacean perform sexual and parthenogenic reproduction; however, its reproductive mode can be altered according to the conditions of the environment, mainly related to salinity and temperature [75]. The reproductive cycle of *Artemia* sp. can be visualized in Fig. 8.4.

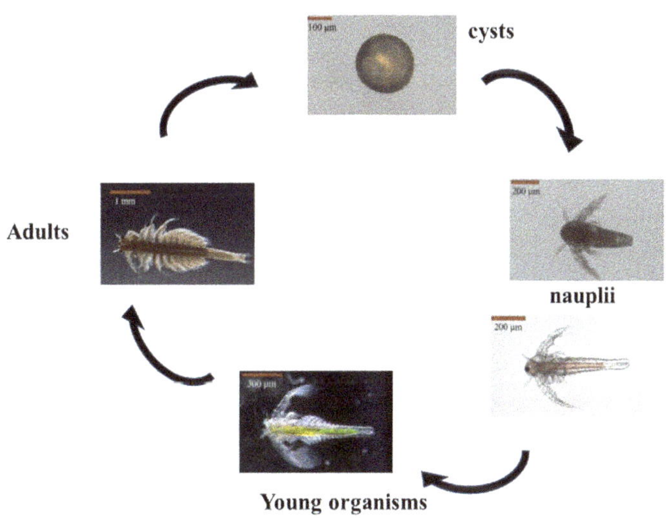

Fig. 8.4 Development stages of *Artemia* sp. Adapted from [76] with permission of Elsevier (license # 5586250566687)

The main advantages of using the microcrustacean *Artemia* sp. as a test organism in ecotoxicology assays are rapidity of the test, low cost, availability of durable commercial cysts, i.e., not requiring cultivation of the organism, and easy handling and maintenance in the laboratory [70, 73, 77].

8.2.1.4 *Daphnia similis*

Cladocera has more than 700 species described, where the family Daphniidae and the genera Daphnia and Ceriodaphnia stand out [78]. The genus *Daphnia* is composed of freshwater planktonic microcrustaceans, widely used in bioassays due to its high sensitivity, ease of cultivation, short life cycle, importance in the food chain (primary consumer) and its method of reproduction (parthenogenesis) [79, 80].

According to Rodgher et al. [81], *Daphnia magna* is universally used in bioassays. However, other species of the genus have been showing great importance as alternative organisms in ecotoxicological assessment, such as the species *Daphnia similis.*

In Brazil, *Daphnia similis* is recommended as a test organism by the ABNT NBR 12,713 standard [82], and although it is not native to the Brazilian territory, it presents the natural characteristics of tropical regions.

Daphnias sp. measure around 1–5 mm, live for approximately 60 days, and their feeding is based on the filtration of suspended material in the water, mainly algae of the genera *Scenedesmus* and *Chlamydomonas* [83]. They are important organisms in the food chain and serve as the basis for feeding invertebrate and vertebrate predators [84].

A healthy population of *Daphnias* sp., in general, is formed mainly by females that perform asexual reproduction, called parthenogenesis. This reproduction form is fundamental in ecotoxicology due to the formation of identical individuals and constant sensitivities [83]. However, under stressful conditions, such as low water quality, extreme temperatures and food scarcity, these organisms will produce male embryos, and the cycle of sexual reproduction begins [85, 86]. Figure 8.5 shows the life cycle of *Daphnia* sp.

According to Ebert et al. [83], an adult female *Daphnia* sp. can produce eggs every three to four days during its entire life cycle. The eggs are incubated in a chamber in its abdomen and released after three days. The neonate presents the appearance of an adult organism and goes through four to six larval stages until it becomes an adult and reproduces [88].

Daphnia sp. is an organism that shows sensitivity to various chemicals, mainly metals, pesticides, nitrogen compounds and solvents [89]. Recent research revealed toxicological negative effects of microplastics on *Daphnia* species [90].

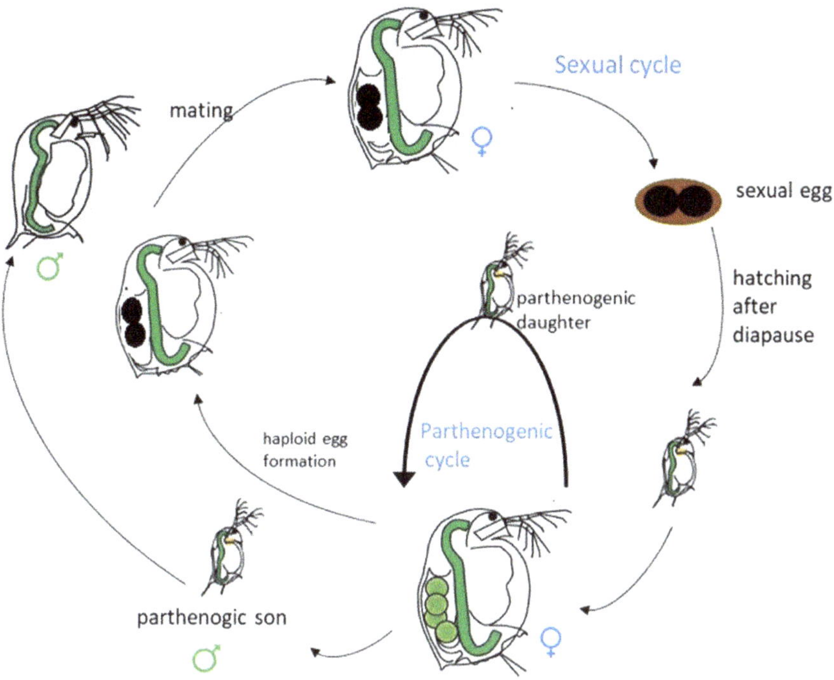

Fig. 8.5 The life cycle of *Daphnia* sp. Adapted from [87] with permission of Elsevier (license # 5586261383020)

8.2.1.5 *Danio rerio*

Danio rerio, also known as zebrafish, is a tropical freshwater fish, a member of the Cyprinidae family [91]. A native of South Asia, it is commonly found in flooded rice field regions in India, Bangladesh, Nepal, Myanmar and Pakistan [91, 92]. The pattern of black and white stripes along the body is the most striking feature of the species, as can be visualized in Fig. 8.6.

This species is gregarious, forming small schools of 5–20 organisms living in still or slow-moving waters [94]. Their diet is mainly based on microcrustaceans, rotifers, and insect larvae, but they can also feed on algae and organic materials [95, 96].

In the wild, the reproduction of *Danio rerio* occurs during the rainy months, when the highest availability of food occurs [92]. Mating occurs in still, shallow waters

Fig. 8.6 *Danio rerio* details. Adapted from [93] with permission of Elsevier (license # 5586641430041)

with abundant vegetation during the early morning. Reproduction is sexual, and fertilization is external, where females scatter between 100–200 eggs on a substrate and male gametes are released simultaneously into the water [97, 98]. The eggs are demersal, i.e., dense, tend to sink, lack adhesive capacity, and hatch within 2–3 days after fertilization [98, 99].

The development of *D. rerio* comprises larval, juvenile and adult stages. Adult organisms are around 90 days old, have an average size of 2–3 cm and continue to grow throughout their life (approximately 3 years), reaching a size of 4–5 cm [100].

Danio rerio is a widely studied organism, either as an embryonic model for genetic studies (replacing higher vertebrate organisms such as rodents) [101] or in environmental ecotoxicology for monitoring organic pollutants, heavy metals and endocrine disruptors [102].

Different characteristics contribute to the use of *D. rerio* in ecotoxicological assays, such as easy maintenance in the laboratory, rapid development, large number of offspring, the transparency of embryos (facilitating the visualization of development and organogenesis) [103, 104].

Danio rerio embryo-larval assays have been used for evaluating the effects of several substances, such as pesticides and pharmaceuticals because zebrafish presents a rapid reproduction rate and genetic similarity (70%) to humans [105, 106].

8.3 Landfill Leachate: Characterization Parameters Related to Ecotoxicity

When municipal solid waste is deposited in landfills, simultaneous and interrelated physical, chemical and biological changes occur. According to McBean et al. [107], physical decomposition consists of sorption and diffusion processes, as well as structural changes that occur in waste in contact with water; chemical decomposition comprises changes resulting from pH change, oxidation, reduction, dissolution, precipitation, complexation, among other chemical reactions; and biological decomposition is considered the most important, as it involves the processes of degradation of organic matter and the formation of leachates and gases in landfills.

The model described by [108] divides the stages of waste degradation into four distinct phases: aerobic phase, non-methanogenic anaerobic phase, unstable methanogenic anaerobic phase and stable or permanent methanogenic phase.

The aerobic phase comprises the initial stage, in which the consumption of oxygen present in the empty spaces between the waste by aerobic microorganisms occurs, resulting in the production of CO_2 and an increase in temperature [4]. According to [109], leachate formation is somewhat limited in the aerobic phase as field capacity has generally not yet been reached. As oxygen availability ends, anaerobic processes begin after the waste has been covered and compacted in the landfill, generally occurring for only a few days [110].

The second phase of the landfill waste degradation process is anaerobic, where the microorganisms involved use nitrate and sulfate as the final electron acceptor [6]. This phase consists of different stages: hydrolysis, acidogenesis and acetogenesis. In the acidogenesis stage, the fermentative or acid phase, significant production of organic acids occurs. The monomers obtained in the hydrolysis stage are degraded by the fermentative bacteria, resulting in intermediate compounds (propionate and butyrate), among other complex organic acids [111]. According to Ehrig [112], in this phase, the leachate is characterized by a high Biochemical Oxygen Demand (BOD_5) and Chemical Oxygen Demand (COD).

The third phase, unstable methanogenic, is reached after months or even years of landfill operation when measurable amounts of methane can be detected [4]. In this phase, the leachate presents neutral or slightly alkaline pH values, a low concentration of biodegradable organic material and a high concentration of sulfide due to the reduction of the different forms of sulfur present in the leachate [7].

Finally, the fourth phase, called stable methanogenic, the maturation phase, occurs the limitation of nutrients and substrates, and consequently, there is low biological activity and a significant decrease in gas production [113]. In the leachate, there are fractions of refractory organic material, that is, difficult to degrade, such as humic and fulvic acids, low concentrations of BOD_5, and pH values in the alkaline range (7–8) [4, 114] and high concentrations of ammoniacal nitrogen [115].

According to El-Fadel et al. [116], leachate formation is directly associated with the field capacity of the waste layers of a landfill. Field capacity is the maximum moisture saturation in a porous medium without percolation. Therefore, leachate formation occurs when moisture exceeds the landfill's field capacity. According to Kulikowska and Klimiuk [117], landfills function as large-scale reactors, where biochemical and physicochemical reactions, such as dissolution, precipitation, adsorption, dilution, and volatilization, influence the composition of the leachate.

The composition of the leachate is variable and heterogeneous. In general, four groups of pollutants are present, such as dissolved organic matter (COD, TOC, volatile acids and refractory organic compounds, such as humic and fulvic acids), inorganic macro components (Ca^{2+}, Mg^{2+}, Na^+, K^+, NH_4^+, Fe^{2+}, Mn^{2+}, Cl^-, SO_4^{2-}, HCO_3^-), heavy metals (Cd, Cr, Pb, Ni and Zn) and xenobiotic organic compounds (aromatic hydrocarbons, phenols, pesticides, among others) [7, 8, 10].

The age of the landfill is a determining factor in leachate composition, as several parameters change as the landfill goes through its stabilization phases. According to Christensen et al. [7], leachate has an acidic pH and a high concentration of easily degradable organic matter and volatile acids in young landfills. In mature landfills, where the methanogenic phase predominates, significant methane production, pH values in the alkaline range, and organic material in the humic and fulvic fractions are observed [117].

In recent decades, the polluting potential of landfill leachate has been the subject of several studies related to groundwater and surface water pollution [118–122]. The environmental impacts related to leachate are one of the reasons for the increasingly restrictive requirements by legislation and efforts in the area of technological innovation related to the planning and operation of landfills [123].

In the literature, the acute ecotoxicity of landfill leachates has been related to the presence of some parameters that represent pollutants present in leachates [13–15]. In the studies by [8], statistically significant correlations between the physicochemical and ecotoxicological parameters of leachates generated from experimental landfill cells showed that organic matter, in the form of COD, alkalinity and ammonia nitrogen and salinity (chloride) were the pollutants responsible for the acute toxicity of the organisms *V. fischeri* and *D. rerio*. Thus, these parameters and their relationships with toxicity are described below.

8.3.1 Ammonia Nitrogen

In landfills, total ammonia nitrogen (TAN) originates from the proteinaceous material contained in the landfill waste during anaerobic decomposition [47]. According to the UK Environment Agency [124], only 10% of the ammoniacal nitrogen contained in the waste is released rapidly through the leachate and the remaining portion is adsorbed onto the waste particles. For this reason, high concentrations of TAN are found over a long time in landfill leachate.

The contribution of ammonia nitrogen to the toxicity of landfill leachate has been evidenced by different authors in the literature, such as those described below.

Silva et al. [125] studied different treatments for the removal of toxicity from the leachate of the Gramacho landfill located in the State of Rio de Janeiro (Brazil). The air stripping treatment was the most efficient for removing TAN from the leachate. After this treatment, the authors verified the reduction of toxicity by the organisms *Vibrio fischeri* and *Daphnia similis* and the absence of toxicity for *Artemia* sp. Concluding that ammoniacal nitrogen has a significant share in the toxicity of the leachate studied. The same observation was made by [126], when evaluating the leachate from the same landfill, found reductions of 98% of ammonia using air stripping treatment, the acute toxicity of the leachate with *Danio rerio* reduced by 30% compared to the raw leachate.

Different authors verified fish ecotoxicity related to ammonia in the literature [127–130]. Studies by [18] found a greater sensitivity of the species *Danio rerio* to ammonia in its un-ionized state compared to microcrustaceans.

The equilibrium between the forms of ammonia (ionized and un-ionized) is defined by pH and temperature, as represented in Eq. 8.1 [131].

$$\mathrm{NH_4^+}_{(aq)} \Leftrightarrow \mathrm{NH_{3(g)}} + \mathrm{H^+}_{(aq)} \quad \mathrm{K_{a(25°C)}} = 5.62 \times 10^{-10}\ \mathrm{mol/L} \tag{8.1}$$

At 25 °C, for pH values lower than 8.0, the reaction equilibrium is shifted towards ammonia in ionized form, but for pH values above 11.0, the equilibrium shifts towards free ammonia [61]. Free ammonia, considered the most toxic form of aquatic life, is related to the mortality of organisms [132].

According to Randall and Tsui [130], the toxicity related to ammonia forms in the aquatic environment varies according to pH, but the influence of temperature variables and salinity are also important. For the same pH value, the proportion of free ammonia will be higher than the liquid medium's temperature, shifting the reaction's equilibrium [130]. In addition, the higher the salinity, the lower the proportion of free ammonia.

According to Emerson et al. [133], the fraction of free ammonia concerning the total ammonia nitrogen can be calculated, as presented in Eq. 8.2. From this, Whitfield [134] developed for saline waters a model that considers the pKa value as a function of temperature and ionic strength, as presented in Eqs. 8.3, 8.4 and 8.5.

$$[NH_3] = \frac{[NH_3] + [NH_4{}^+]}{1 + 10^{pKa\text{-}pH}} \tag{8.2}$$

$$pKa(T) = pKa^s + 0.0324 \times (298 - (273 + T)) \tag{8.3}$$

$$pKa^s = 9.2516 + 0.153 \times F \tag{8.4}$$

$$F = 19.9273 \times S/(1000 - (1.005109 \times S)) \tag{8.5}$$

where

pKa − logKa (acid dissociation constant of NH_4^+ ion),
T temperature in °C,
F ionic strength,
S salinity.

Removing ammonia nitrogen from wastewater is commonly performed through biological nitrification and denitrification [135, 136]. However, landfill leachates have a high concentration of ammoniacal nitrogen in their composition, making it appropriate to use the air stripping technique as a pretreatment for the removal of this pollutant in values that are not toxic to microorganisms and inhibit the biological nitrification process [11, 126, 137].

8.3.2 Alkalinity

Alkalinity is a parameter present at high values in landfill leachate [13]. Alkalinity in landfills is associated with the degradation of organic matter. It is related to the concentration of ions present in the medium that will react to neutralize hydrogen ions,it is the buffer capacity of the effluent to neutralize acids [58].

The main components related to alkalinity are bicarbonates (HCO_3^-), carbonates (CO_3^{2-}) and hydroxides (OH^-), as presented in Eqs. 8.6, 8.7 and 8.8 [138].

$$CO_{2(g)} + H_2O_{(l)} \Leftrightarrow H_2CO_{3(aq)} \tag{8.6}$$

$$H_2CO_{3(aq)} \Leftrightarrow HCO^-_{3(aq)} + H^+_{(aq)} \quad K_{1(25^\circ C)} = 10^{-6.35} \tag{8.7}$$

$$HCO^-_{3\,(aq)} \Leftrightarrow CO^{2-}_{3\ \ (aq)} + H^+_{(aq)} \quad K_{2(25^\circ C)} = 10^{-10.33} \tag{8.8}$$

Alkalinity is interrelated with pH values and CO_2 content. At room temperature, for pH values above 9.4, alkalinity is found in the forms of hydroxides and carbonates; at pH values between 8.3 and 9.4, the predominance of carbonates and bicarbonates forms occurs; and at pH values between 8.3 and 4.5, only bicarbonate is related to the alkalinity of the effluent [58].

According to Qasin and Chiang [139], inorganic components such as bicarbonates, ammonia and organic bases are responsible for the alkalinity of the leachate. Alkalinity, as well as ammonia nitrogen, is a pollutant that contributes to toxicity in landfill leachate [13].

Clément et al. [140] studied leachates from 22 landfills receiving different types of non-hazardous urban and industrial waste and concluded that alkalinity, expressed as bicarbonate, was a determining factor for the toxicity of the leachate observed in the microcrustacean *Daphnia magna.*

Ward et al. [141] evaluated leachate's physicochemical and ecotoxicological composition from six different landfills in Florida, USA. The study used acute tests with the organisms *Ceriodaphnia dubia* and *Vibrio fischeri* and chronic tests with the alga *Pseudokirchneriella subcapitata.* The authors concluded that the leading role of alkalinity in leachate toxicity is related to its buffering capacity and the effect on the concentration of un-ionized ammonia.

8.3.3 Organic Matter

Organic matter can be understood as a complex system of several substances of diverse nature that is in continuous transformation due to the action of chemical, physical, and biological factors inherent to the soil and climate [142]. The material resulting from this transformation can be divided into two large groups: non-humic substances—NHS (amino acids, proteins, fatty acids, among other organic compounds) and humic substances—SH.

In landfills, high dissolved organic carbon content is generated in the natural process of waste degradation [143]. The biodegradability of this organic material varies according to the stabilization phase and the landfill's age, expressed in the BOD_5/COD ratio. Values between 0.3 and 0.6 indicate biodegradable leachate from

Table 8.1 Chemical properties of humic substances

Parameters	Fulvic acid	Humic acid	Humin
Massa Molar (Da)	640–5000	10,000–100,000	> 100,000
C (%)	42–47	51–62	> 62
O (%)	45–50	31–36	< 30
N (%)	2.0–4.1	3.6–5.5	> 5
Cation exchange capacity (mol/kg)	~ 14	~ 5	< 5

Source Adapted from [148]

a young landfill. BOD_5/COD ratios below 0.2 represent low biodegradable leachate from old landfills.

Most of the organic material from an old landfill that is in the methanogenic phase is composed of molecules of high molar mass (> 10,000 Da) and low biodegradability, called humic substances [143–145]. These substances present themselves as one of the main problems faced in treatment plants due to the low efficiency obtained with conventional treatment techniques for the degradation of organic matter [16, 146], such as biological treatment processes.

The term "humic substances" refers to the three main high molar mass organic fractions such as humic acid (HA), fulvic acid (FA) and humin (HU) [147]. The chemical properties of these fractions are presented in Table 8.1.

Humic substances in the aquatic environment can cause different damages to organisms, ranging from interference with photosynthesis due to blocking sunlight irradiation to the production of stress related to alterations in ion exchange mechanisms at the cell membrane level [149]. However, the presence of humic substances may benefit organisms due to reducing the bioavailability of various pollutants, including xenobiotic compounds, aromatic hydrocarbons and heavy metals [150].

Organic material, generally expressed in the form of COD, is reported in the literature as a toxic component present in landfill leachate by different authors: [11, 47, 140, 151] among others.

Studies by Oliveiro-Verbel et al. [152] suggested that the increase in COD of landfill leachate from Cartagena-Colombia is inversely correlated with the LC50 of *Artemia franciscana,* i.e., the higher the concentration of COD in the leachate, the lower the LC50, which indicates greater toxicity.

8.3.4 Salinity (Chloride)

The dissolution of salts in the leachate contributes to the increase of chloride ion (Cl^-) concentration in the water body. According to Mor et al. [120], high chloride ion concentrations in natural waters indicate pollution, originating from sources such as domestic sewage, leachate, and fertilizers, among others.

Table 8.2 Lethal chloride ion concentration (Cl^-) to organisms in aqueous solution

Test organism	Exposure time (h)	EC50 or LC50 or IC (mg/L)	References
Daphnia similis	48	1300	[154]
Daphnia magna	48	3630	[155]
Ceriodaphnia dubia	48	1160	[154]
Danio rerio	48	4800	[156]
Pimephales promelas	96	4079	[155]

The chloride ion is a conservative contaminant, unaffected by the chemical and biological transformations occurring within the landfill. Therefore, they constitute a significant threat to the aquatic ecosystem since they remain in the contaminated environment in high concentrations for extended periods [153].

Table 8.2 presents the contribution of chloride ion to the ecotoxicity of some freshwater test organisms. According to Table 8.2, it was possible to observe that microcrustacean species have a greater sensitivity to chloride ion than the fish species analyzed.

According to Krull and Barros [157], the choice of test organisms in ecotoxicological tests should be appropriate, i.e., freshwater organisms should not be exposed to brackish or saline water, as their sensitivity will be increased due to the high level of chloride present in the water. This condition is also valid for marine organisms, which should not be used for the evaluation of freshwater samples because the addition of chloride alters the physicochemical composition of the sample, a fact that is much discussed in the literature related to the use of the bioluminescent bacterium *Vibrio fischeri* [39].

8.4 Ecotoxicological Effects of Brazilian Leachates

This item illustrates the research results that evaluate the pollutant potential causing acute ecotoxicity in leachates from municipal solid waste landfills in the State of Rio de Janeiro (Brazil). Leachates from two landfills with different modes of operation and ages were evaluated in this study.

The acute ecotoxicity of the leachates and the sensitivity of aquatic organisms of different trophic levels were investigated, such as decomposers (*Vibrio fischeri* and activated sludge microorganisms), primary consumers (*Daphnia similis* and *Artemia* sp.) and secondary consumer (*Danio rerio*). In addition, the contribution of the parameters of ammonia nitrogen, alkalinity, chloride and humic substances of the leachates in the acute ecotoxicity to the test organisms evaluated was determined.

8.4.1 Characterization of Samples

The leachate samples were obtained from landfills with different ages and modes of operation, these being the Centro de Tratamento de Resíduos (Waste Treatment Center, in English)—CTR Santa Rosa, also known as the Seropédica Landfill and the Gericinó Controlled Landfill, both located in the State of Rio de Janeiro. The Seropédica landfill is an operating landfill that receives about 10,000 tonnes of municipal solid waste per day from the Metropolitan Region of the City of Rio de Janeiro. In contrast, the Gericinó controlled landfill at the time of the study was closed but operated for 27 years, partly as a dump and partly as a controlled landfill. More details of the sites evaluated can be found in [158].

Leachate samples were collected at the Seropédica landfill (n = 6) and the Gericinó controlled landfill (n = 5) from 2017 to 2020. The collections were made at different times in each landfill, and single samples were obtained at the point of leachate accumulation. The physicochemical characterization analyses were performed in triplicate, according to the methodology described in Standard Methods for the Examination of Water and Wastewater [159]. The concentration of humic substances was determined by the spectrophotometric method described by [147] and adapted by [146].

Table 8.3 illustrates the characterization of the evaluated leachates. Much of the concentration of organic matter in the leachates studied results from the presence of recalcitrant compounds or those that are difficult to degrade, as can be seen from the high average concentrations of humic substances measured in the leachates of the Seropédica (1935 mg/L) and Gericinó (636 mg/L) landfills. These results corroborate the studies of Amaral et al. [160], who found that the concentration of humic substances present in landfill leachate in Minas Gerais State (Brazil) corresponded to a fraction of about 50% of the total COD concentration. Moreover, Lima et al. [146] observed a similar proportion of humic substances (40.66%) concerning COD of the leachate from the Gericinó landfill.

Another indication that corroborates the low biodegradability evidenced in the leachates of the landfills studied is the absorbance at 254 nm value. This parameter is related to aromatic organic compounds, such as humic substances [161, 162]. A high value of absorbance 254 nm was observed in the leachates of both landfills, evidenced mainly in the leachates of the Seropédica landfill (31.5), as expected since it presented in its composition a high concentration of humic substances compared to the leachates of Gericinó.

Regarding total ammoniacal nitrogen, average concentrations of 1376 mg/L (866–1576) were observed in the Gericinó leachate and 1919 mg/L (1278–2489) in the Seropédica leachate. Similar results were obtained by Costa et al. [8], who found high concentrations of TAN (2104–2231 mg/L) in leachate samples from the Seropédica landfill (n = 6) during the years 2013 and 2014.

High concentrations of ammonia nitrogen are found in leachates as a product of the degradation of waste protein and can constitute 0.5% of the dry mass of the waste [163]. According to Kjeldsen et al. [4], ammonia nitrogen concentrations can

Table 8.3 Physicochemical characterization of raw leachates from Seropédica (n = 6) and Gericinó (n = 5) landfills

Parameters	Values Average (minimum–maximum)	
	Seropédica landfill	Gericinó landfill
Absorbance at 254 nm	31.5 (25–37)	15 (11–20)
Total Alkalinity (mg $CaCO_3$/L)	11,311 (8622–13,166)	3852 (1830–4927)
Total Inorganic Carbon (IC) (mg/L)	2.598 (1699–3032)	1128 (1003–1197)
Total Organic Carbon (TOC) (mg/L)	1740 (1156–2059)	743 (612–852)
Chloride (mgCl/L)	3389 (1322–4681)	2037 (1549–2599)
Conductivity (mS/cm)	26.9 (23.0–31.7)	14.6 (10.0–17.8)
True color (mgPtCo/L)	4896 (3075–6220)	3164 (2457–4710)
Chemical Oxygen Demand (COD) (mg/L)	4712 (3267–5546)	2122 (1850–2383)
Total Ammonia Nitrogen (TAN) (mg/L)	1919 (1278–2489)	1376 (866–1576)
pH	8.2 (8.0–8.3)	8.1 (7.8–8.4)
Salinity	18 (15–20)	9 (6–11)
Total Dissolved Solids (TDS) (mg/L)	16,881 (12,629–20,585)	4534 (3461–5865)
Total Suspended Solids (TSS) (mg/L)	1428 (417–4689)	324 (222–484)
Volatile Suspended Solids (VSS) (mg/L)	479 (324–840)	132 (102–180)
Total Solids (TS) (mg/L)	18.787 (13.370–25.562)	4.213 (3.888–6.365)
Humic Substances (HS) (mg/L)	1.935 (1.329–2.191)	636 (586–688)
Turbidity (NTU)	274 (77–688)	6.5 (3.1–9.2)

remain high, as can chloride and alkalinity concentrations in leachates from mature landfills. A significant reduction in TAN is not observed over the years. Therefore, it is considered one of the main pollutants of leachate [13].

Alkalinity in the Seropédica landfill ranged from 8622 to 13,166 mg/L, similar to that found by Baettker et al. [164] (7133–14,855 mg/L) in leachate from the Caximba municipal landfill in Curitiba (Parana State, Brazil). In the case of the Gericinó landfill, which is closed, the mean value of this parameter was approximately three times lower than the value observed in Seropédica.

In addition, the mean chloride concentrations at Gericinó landfill (2037 mg/L) and Seropédica landfill (3389 mg/L) were within the expected range for Brazilian landfill leachate (500–5200 mg Cl^{-}/L), according to studies by [165]. Chloride is considered a conservative contaminant and does not attenuate significantly over the years, as it is not affected by the biochemical processes in the landfill [116].

The statistical evaluation between the compositions of the landfill leachates, using Fisher's Least Significant Difference (LSD) test, showed that only the chloride and pH parameters had statistically equal values for both leachates. Of course, this difference was expected, as the landfills have different modes of operation and ages, which are directly reflected in the composition of their leachates.

Table 8.4 illustrates other evaluated leachate components resulting from only one collection. Quantification of the elements Al, As, Ba, B, Cd, Ca, Pb, Co, Cu, Cr, Fe, Mg, Mn, Hg, Mo, Ni, K, Na and Zn was performed using EPA method 200.8 [166].

Except for boron in the Seropédica landfill, all other inorganic elements in the raw leachate from the Seropédica and Gericinó landfills were within the limits defined by the Rio de Janeiro State Legislation for effluent disposal Technical Standard 202/R.10 [167].

8.4.2 Experiments Performed

To evaluate the toxicity of the leachate, treatments were performed to selectively remove the target pollutants: ammonia nitrogen, alkalinity and humic substances. The tests used for the removal of pollutants are detailed below. Figure 8.7 illustrates a scheme of the experiments performed. Figure 8.8 presents the experimental outline of the tests carried out.

In air stripping, a 4.0-L beaker was used under agitation and aeration with an air compressor in a lab hood. For free ammonia removal, pH was adjusted with NaOH (1.0 mol/L) to pH = 11.0, and free ammonia was monitored until it reached NH_3 concentration $\leq$ 0.025 mg/L not to generate a toxic effect on the studied test organisms. For alkalinity removal, the pH of the sample was adjusted with HCl (37%) to pH = 4.0, and the IC was monitored.

In the membrane filtration experiments, a batch membrane module with dead-end flow was used, from the company PAM (Membranas Seletivas Ltda, Brazil), with a maximum capacity of 1.5 L and a permeation area of 63.59 cm^2.

Table 8.4 Characterization of inorganic elements of the raw leachates from the Seropédica landfill and the Gericinó controlled landfill

Parameters	Seropédica Leachate (mg/L)	Gericinó Leachate (mg/L)
Aluminun	0.780	0.287
Arsenic	0.144	0.025
Barium	0.329	0.978
Boron	5.962	2.090
Cadmium	0.001	< 0.0005
Calcium	9.962	61.79
Lead	0.051	0.001
Cobalt	0.050	0.069
Copper	0.038	0.008
Chromium	0.273	0.060
Iron	3.160	3.040
Magnesium	118.38	83.69
Manganese	0.228	0.330
Mercury	0.00009	ND
Molybdenum	0.021	< 0.0005
Nickel	0.185	0.144
Potassium	1.823	814.48
Sodium	2418	1188
Zinc	1.341	0.1

ND not detected

Filtration experiments were performed at natural sample pH and room temperature (25–27 °C). Nitrogen gas (N_2) was used to regulate the pressure in the filtration module, and a stirring plate was used for sample homogenization inside the filtration cell to reduce fouling.

Initially, the leachate samples were filtered in the vacuum system on a 0.45 μm glass fiber filter (Macherey–Nagel, Germany) for the removal of coarse solids in order to avoid fouling in the membrane system. In order to remove the recalcitrant organic matter from the leachate, ultrafiltration was performed with a 10 kDa membrane (UP010) and a pressure of 5 bar for the removal of the humic fraction and using the ultrafiltration permeate. Nanofiltration was performed with a 500 Da membrane (NP030) and a pressure of 20 bar to remove the fulvic fraction. The description of the membranes used is shown in Table 8.5.

According to Chian [168], Trebouet et al. [169], Zolfaghari et al. [170], in leachates from intermediate and mature landfills, humic acids constitute organic molecules with a molar mass greater than 10,000 Da, while fulvic acids constitute molecules with a molar mass between 500 and 1000 Da.

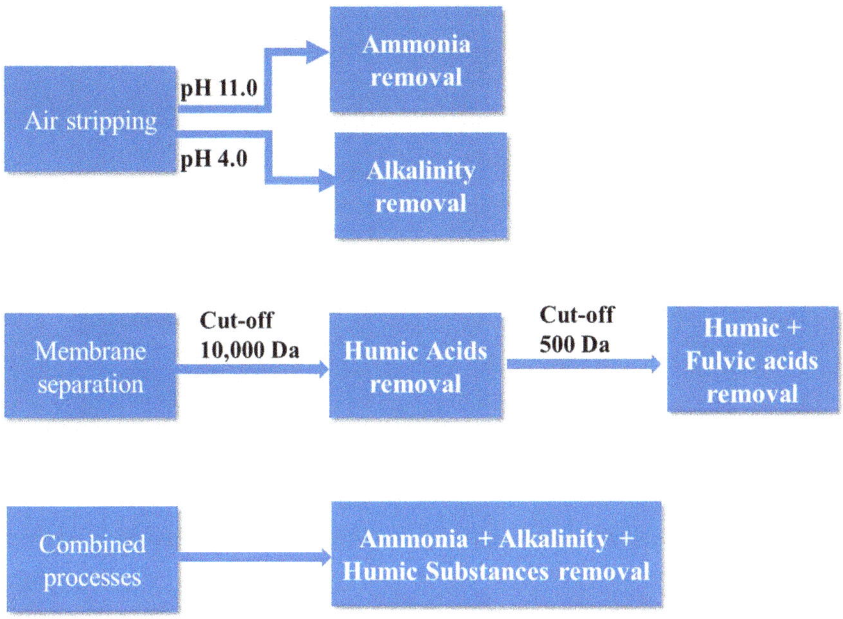

Fig. 8.7 Scheme of the processes carried out for selective removal of pollutants

In this study, UF and NF membranes can remove a significant ammonia nitrogen concentration from the leachate. Therefore, to evaluate only the toxic potential individually of humic substances and their fractions, the correction in ammonia nitrogen concentration with NH_4HCO_3 was performed for the initial concentration of the raw leachate in the UF and UF + NF permeates. According to studies by [126], the ammonia nitrogen in landfill leachate is in the form of carbonates and bicarbonates.

The combined removal of pollutants was conducted by matching the experiments: air entrainment at $pH = 11$ for ammonia nitrogen removal, followed by entrainment at $pH = 4.0$ for alkalinity removal and filtration with 10 kDa and 500 Da membranes.

Experiments with an aqueous solution containing NaCl were performed to reproduce the initial concentration of chloride present in the raw leachate samples and to evaluate the contribution of this component individually in the ecotoxicity in the test organisms studied.

In addition to the physicochemical parameters mentioned in 5.1, the polluting potential of the samples was evaluated for the toxicity of the test organisms and their respective methodologies:

- *Vibrio fischeri*: ABNT NBR 15411-3 [171], using lyophilized bacterial strains obtained by the company Umwelt (Environmental Biotechnology, Brazil) and the Microtox equipment (Modern Water, Model 500 Analyser).

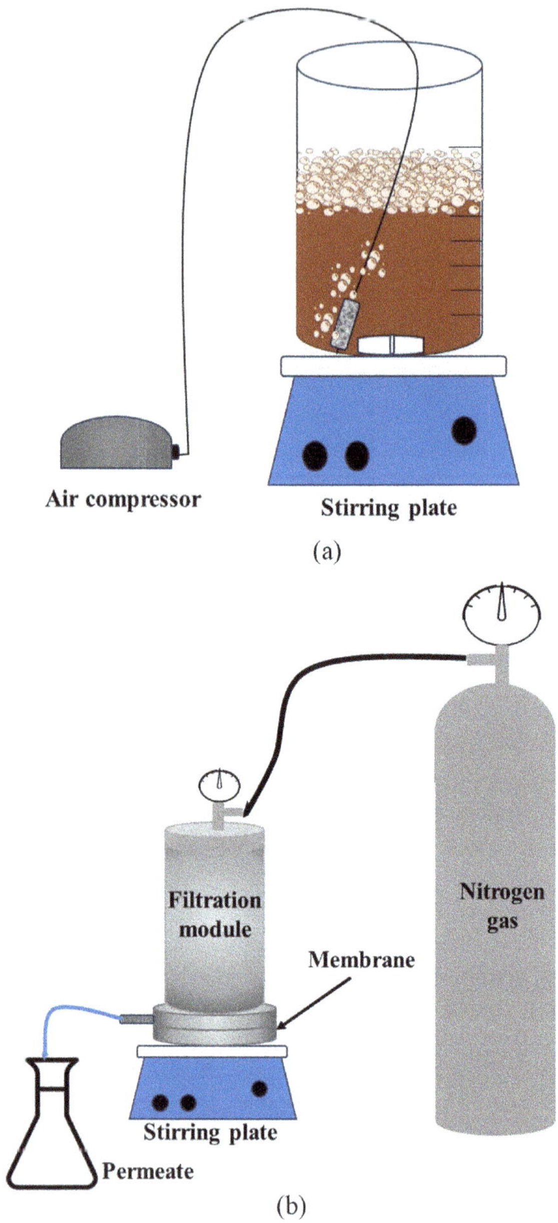

Fig. 8.8 Experimental scheme of **a** air stripping experiments **b** membrane filtration

Table 8.5 Specifications of the membranes used in the permeation experiments

Item	UP010	NP030
Manufacturer	NADIR®	NADIR®
Chemical composition	Polyethersulfone (PES)	Polyethersulfone (PES)
Operation pH	0–14	0–14
Permeability (L/m^2 h)	2.000[a]	40[b]
Rejection (%)	–	80–95[c]
Cut-off (Daltons)	10.000	500–600

[a] Conditions: Pure water (4 bar and 20 °C); [b] Conditions: Pure water (40 bar and 20 °C); [c] Na_2SO_4 rejection

- Activated Sludge Microorganism: The methodology described by Organisation for Economic Co-operation and Development [68] was used to perform the activated sludge respiration inhibition assay.
- *Daphnia similis*: The tests for evaluating acute ecotoxicity were carried out according to the methodology described in NBR 12713 [82].
- *Artemia salina*: The tests for evaluating acute ecotoxicity with the microcrustacean *Artemia* sp. were carried out according to the methodology described in NBR 16530 [172]. The organisms were obtained commercially as cysts by the company Bioartemia (Brazil).
- *Danio rerio*: The methodology described in NBR 15088 [173] was used. The use of the *Danio rerio* vertebrate specimen in this study was approved by the Ethics Committee on the Use of Animals in Research (UFRJ), process: 01200.001568/2013-87, November 1st, 2016.

In this study, the normality of the data was assessed by the Kolmogorov–Smirnov test and the comparison between the means of the landfill leachate parameters by Fisher's test (Least Significant Difference- LSD). Pearson's correlation performed the bivariate relationship, and the association between physicochemical and ecotoxicological parameters was evaluated by analysis of variance (ANOVA) and principal component analysis (PCA). Linear regression analysis was used to correlate the sensitivities of the test organisms. All these statistical tests were performed using STATISTICA software version 12, licensed by STAT SOFT.

The comparison between the results of the toxicity tests was performed according to the methodology proposed in USEPA [174]. This statistical test allowed the comparison of means between the EC50 or LC50 values, in addition to considering the confidence interval of the tests.

8.4.3 *Results and Discussion*

8.4.3.1 Physicochemical Parameters

Figure 8.9 shows that the TAN removal results achieved average efficiencies higher than 99% in the leachates from the studied landfills. The long experimental period of this study is due to the high average initial concentrations of the pollutant (1376–1919 mg TAN/ L), as well as the volume of sample used in the test (2 L), the temperature (25 °C ± 2 °C), the application of moderate air flow to control the formation of foam, and mainly, due to the need to reach final concentrations ($\leq$ 0.02 mg NH_3/L), as a way to avoid the toxicity of this pollutant to aquatic organisms. As a result of this extensive experiment period, it was possible to observe that organic matter, in the form of COD, showed average reductions of 13.30% in samples from the Gericinó landfill and 17.08% in leachate samples from the Seropédica landfill. In addition, reductions in alkalinity were also observed in the leachates from Seropédica (3.21%) and Gericinó (6.65%) after hauling. However, for both parameters, no statistically significant differences were found between the mean values of the raw leachates and after NAT removal (p-values > 0.05).

Campos et al. [126] evaluated the influence of pH, temperature and alkalinity parameters on treating TAN stripping in leachates from the Gramacho Metropolitan Landfill (RJ). The authors identified higher reduction efficiency of the pollutant in samples with pH 11 compared to the studied values of 8.3 and 9.0. Regarding temperature, high efficiencies were achieved at 60 °C, compared to tests performed at

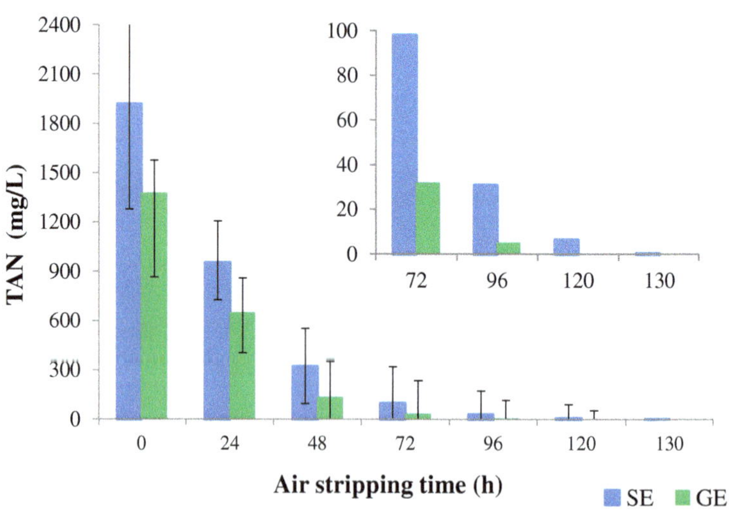

Fig. 8.9 Results of total ammonia nitrogen (TAN) during the stripping process. Highlighted: zoom of the graph for better visualization of the scale

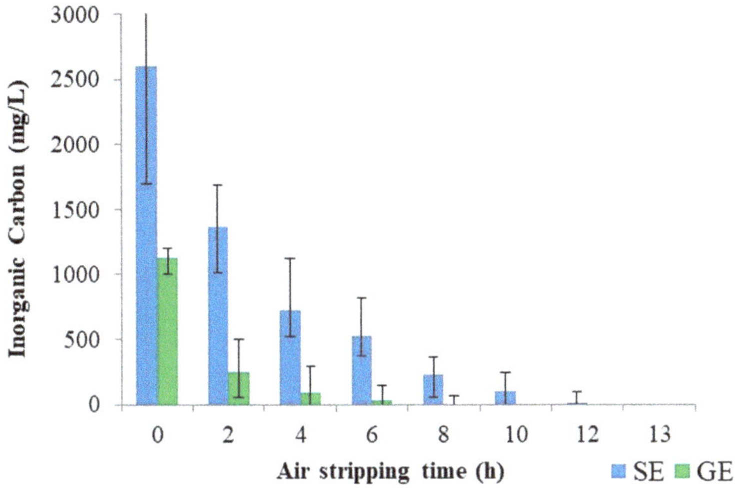

Fig. 8.10 Results of inorganic carbon during the stripping process

25 and 40 °C, without an alkalizing agent. Furthermore, finally, the authors identified the importance of alkalinity for ammonia reduction efficiency through entrainment.

As shown in Fig. 8.10, the removal of alkalinity by air stripping occurred during the average period of 8–13 h for the leachates from the Gericinó and Seropédica landfills, respectively. The 100% removal efficiency of inorganic carbon in the leachate samples from both landfills studied, that is, with the removal of inorganic carbon sources (carbonates and bicarbonates), it is understood that alkalinity was removed entirely from the leachates.

The alkalinity of the leachate is predominantly in the form of bicarbonates (which dissociate into CO_2 and H_2O) [126]. With the aid of aeration and continuous agitation of the sample, the CO_2 stripping process occurs, where it ends up being expelled from the reaction medium and consequently, alkalinity can be removed.

As a result of the alkalinity stripping, it was observed that organic matter, in the form of COD, showed a greater average reduction in leachates from the Gericinó landfill (11.15%) compared to the reduction of this same parameter in the Seropédica landfill (4.7%). Moreover, higher average reductions of ammonia nitrogen were also observed in the leachates of Gericinó (13.43%) about Seropédica (8.76%) after the entrainment period. However, for both parameters, no statistically significant differences were found between the mean values of the raw leachates and after alkalinity removal (p-values > 0.05).

Figures 8.11a–c illustrate the residual concentration in the permeates of the membrane processes used in terms of COD, TOC and Humic Substances, respectively. Permeate flux values in the ultrafiltration membrane (10 kDa) ranged from 73.29 to 81.46 and 80.51–89.34 L/m^2.h for leachates from Seropédica and Gericinó, respectively.

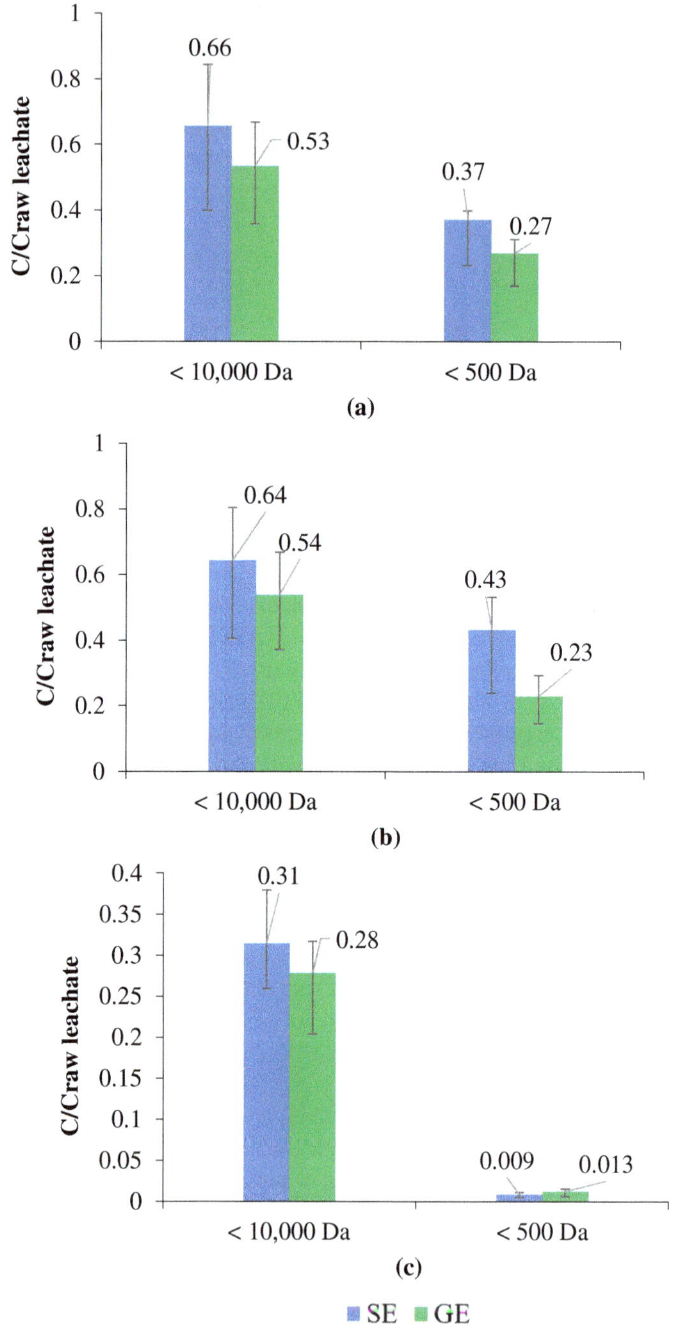

Fig. 8.11 Membrane fractionation results of residual concentration (concerning raw leachate—craw) of **a** COD, **b** TOC, **c** Humic substances

Figure 8.11a shows that the average COD reductions in ultrafiltration permeates were 34 and 47% in the leachates of Seropédica and Gericinó, respectively. Similar results were obtained by [175], who, by the fractionation of organic matter with XAD-8 resin, observed that the fraction of humic acids corresponded to about 38% of the COD of the leachate from the mature landfill of Changshengqiao, China.

Collado et al. [176] obtained a reduction of 18.5–49.6% of COD of leachate from a mature landfill after UF (150 kDa), resulting from the high molar mass organic fraction, the humic acids. TOC results followed a similar trend to COD.

The permeate from the ultrafiltration of the Seropédica leachate showed an average reduction efficiency of 69% of humic substances (in this case, the fraction of humic acids with a molar mass greater than 10 kDa) (Fig. 8.11c). In the leachate from the Gericinó landfill (a mature landfill), the average reduction efficiency of humic substances was even higher at 72%. These results corroborate the studies of Lima et al. [146], who performed the fractionation of recalcitrant organic matter with precipitation (pH < 1.0) and XAD-8 resin and found that the fraction of humic acids in leachate from Brazilian mature landfills corresponds to 60–70% of the total concentration of humic substances.

According to Kang et al. [145], humic acids constitute an important group of high molar mass organic matter (> 10 kDa) found in leachates from mature landfills. The authors analyzed leachates from landfills with different ages and observed that the highest proportion of humic acids (76%) was found in leachates from landfills older than 10 years. This fact corroborates the results obtained in this study, in which a fraction of about 70% of humic acids was found in the leachates of the Seropédica (10 years) and Gericinó (34 years) landfills.

This study observed reductions in ammonia nitrogen, alkalinity and chloride in the UF permeates from both landfills. In the Seropédica landfill, the average reductions were 41.06% ammonia nitrogen, 15.22% alkalinity, and 0.51% chloride. Similar values were observed in the permeates from Gericinó, which showed average reductions of 43.65% of TAN, 43.33% of alkalinity and 0.78% of chloride. By the Fisher LSD statistical test of correlation of means, it was verified that the TAN parameter presented statistical differences ($p \leq 0.05$) between the samples of the raw leachate and after UF, the reduction of this pollutant was corrected to the initial concentration of the raw leachate, in order not to compromise the individual evaluation of the toxicity of humic acids in the leachate samples.

Regarding the subsequent filtration through a 500 Da membrane, the values of the permeate flux in the nanofiltration membrane ranged from 16.51–18.94 L/m^2 h and 20.93–23.10 L/m^2 h for leachates from Seropédica and Gericinó, respectively.

The average values of COD reduction in the permeate < 500 Da of 63% for the Seropédica leachate and 73% for the Gericinó leachate. Permeates < 500 Da from both landfills showed an average reduction of more than 99% of humic substances, which evidences the efficiency of the applicability of this treatment for the reduction of recalcitrant organic matter from landfill leachate, as observed in the studies of [169, 177–180].

Just as performed in the UF permeate, the losses of pollutants from the UF + NF permeates were corrected for the initial concentration of the raw leachate not to compromise the individual evaluation of the toxicity of humic acids in the leachate samples studied. Table 8.6 shows the concentrations and percentages of reduction of inorganic elements in the leachates after permeation with the membranes.

The ultrafiltration (10,000 Da) followed by nanofiltration (500 Da) achieved high reduction efficiency of inorganic elements in both landfills studied, where reductions > 80% of aluminum, copper, lead and iron in the permeate of the Seropédica leachate and aluminum, barium, cobalt, chromium, iron, manganese and nickel in the permeate of the Gericinó leachate—corroborating with [170], who verified metal rejection efficiency by NF, such as aluminum and iron greater than 80%. Membrane filtration generated permeates with color reduction due to the removal of humic acids, as seen in Fig. 8.12.

The average true color reductions for permeate < 10 kDa were 60.5% for Seropédica leachate and 68.3% for Gericinó leachate. The true color reductions for permeate < 500 Da were average values of 85.3% for Seropédica leachate and 90.87% for Gericinó leachate. According to Wang et al. [181], organic compounds

Table 8.6 Concentrations and percentages of reduction of inorganic elements in the leachates after permeation with the membranes

Element	Seropédica Landfill		Gericinó Landfill	
	< 10 kDa mg/L (%)	< 500 Da mg/L (%)	< 10 kDa mg/L (%)	< 500 Da mg/L (%)
Al	0.120 (84.6)	0.011 (98.7)	< 0.0005 (> 99.8)	< 0.0005 (99.8)
As	0.124 (13.9)	0.09 (37.5)	0.018 (30.3)	0.015 (40.2)
Ba	0.178 (45.9)	0.128 (60.9)	0.108 (89.0)	0.080 (91.8)
B	4.909 (17.7)	4.787 (19.7)	1.90 (9.1)	0.860 (58.9)
Cd	0.0007 (57.4)	0.0007 (57.4)	< 0.0005 (0)	< 0.0005 (0)
Ca	5.557 (44.2)	4.899 (50.8)	19.47 (68.5)	17.95 (70.9)
Pb	0.003 (94.9)	0.003 (94.9)	< 0.0005 (> 61.5)	< 0.0005 (> 61.5)
Co	0.035 (30.1)	0.0136 (72.6)	0.028 (60.2)	0.004 (94.8)
Cu	0.021 (44.7)	0.001 (97.4)	< 0.005 (> 37.5)	< 0.005 (> 37.5)
Cr	0.145 (46.9)	0.060 (78.0)	0.022 (63.5)	0.004 (93.3)
Fe	1.15 (63.6)	0.61 (80.7)	0.318 (89.5)	< 0.005 (> 99.8)
Mg	87.16 (26.4)	82.11 (30.6)	75.7 (9.5)	62.06 (25.8)
Mn	0.101 (55.7)	0.087 (61.8)	0.017 (94.8)	0.014 (95.8)
Hg	0.00009 (0)	0.00009 (0)	ND (–)	ND (–)
Mo	0.012 (43.0)	0.009 (52.2)	< 0.0005 (0)	< 0.0005(0)
Ni	0.125 (32.3)	0.041 (77.9)	0.062 (56.9)	0.007 (95.1)
K	1429 (21.6)	1399.7 (23.2)	627 (23.1)	526.73 (35.3)
Na	1927 (20.3)	1869.5 (22.7)	927 (22.0)	775.52 (34.7)
Zn	0.611 (54.4)	0.599 (55.3)	0.74 (0)	< 0.05 (> 50.0)

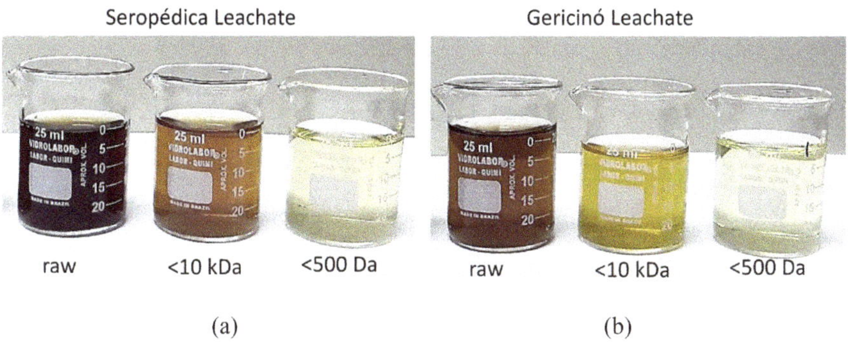

Fig. 8.12 Visual aspects of leachate samples. **a** Seropédica leachate; **b** Gericinó leachate. *Source* Authors' collection

with a molar mass greater than 1 kDa are the major contributors to the color of landfill leachate.

Figure 8.13 shows the performances of the combined processes, and Table 8.7 presents the concentrations and percentages of reduction of the inorganic elements in the leachates. High average reductions of ammonia nitrogen (99%), inorganic carbon (100%) and humic substances (> 98%) in the leachates of both landfills studied. In the Seropédica leachate, the average COD, TOC and chloride reductions were 75.92, 71.75 and 10.85%, respectively. In comparison, the reductions observed in Gericinó leachate reached average reductions of 85.29% COD, 86.40% TOC and 12.22% chloride.

The reductions of metals after the combination of treatments presented the following order: Al > Pb > Cu > Fe > Cr > Ni > Co > Ba > Mn > Cd > Zn > Ca > Mo > As > Mg > K > Na. In the Gericinó leachate, the reductions of metals were distinct if compared to the Seropédica leachate: Fe > Al > Mn > Ni > Co > Cr > Ba > Ca > B > As > Zn > Cu > K > Na > Mg. No reductions were verified in this leachate for the metals Cd, Pb and Mo.

Notably, the metals were quantified in their total fraction (complexed or not to organic matter). Therefore, the removal of these elements may have occurred together with the reduction of organic matter. Jensen et al. [182] evaluated heavy metals in groundwater polluted by leachate from the Vejen landfill in Denmark. The authors found that metals such as Cd, Ni, Cu and Pb were mainly associated with organic matter (dissolved organic carbon), while Ni and Zn were associated with inorganic fractions forming carbonate complexes. The authors also pointed out that heavy metals in leachates are mainly associated with organic molecules and small colloidal material (1–10 μm diameter).

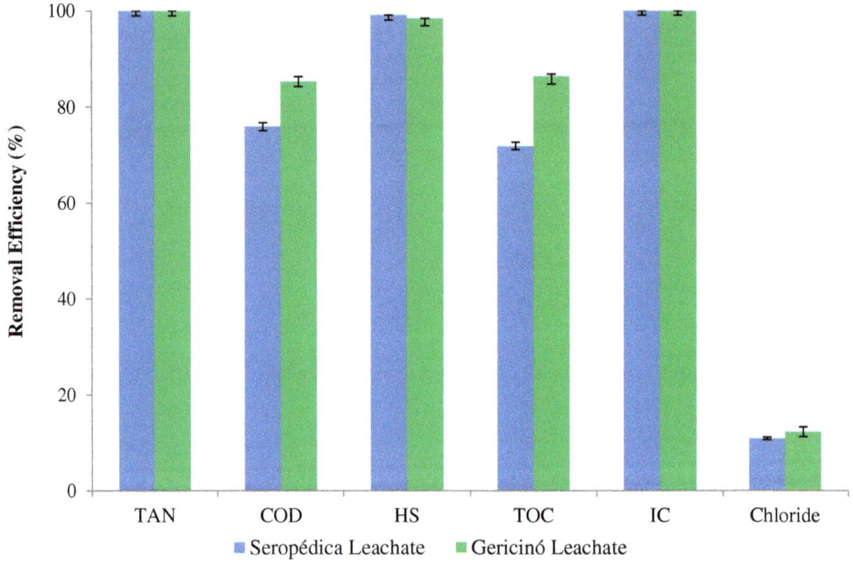

Fig. 8.13 Results of efficiency of removal of main pollutants in landfill leachates after combined treatment processes

Table 8.7 Characterization of the inorganic elements of the leachates from the Seropédica landfill and the Gericinó controlled landfill after removal of TAN, alkalinity and SH and (%) reduction concerning the raw leachate

Element	Seropédica Landfill mg/L (%)	Gericinó Landfill mg/L (%)
Al	< 0.005 (> 99.4)	< 0.005 (> 98.2)
As	0.0761 (47.2)	0.0122 (52.0)
Ba	0.1106 (66.4)	0.0986 (89.9)
B	4.31 (27.7)	0.82 (60.8)
Cd	0.0007 (61.7)	< 0.0005 (0)
Ca	4.838 (51.4)	17.25 (72.1)
Pb	0.0031 (93.9)	< 0.005 (0)
Co	0.0115 (76.8)	0.0033 (95.2)
Cu	< 0.005 (> 86.8)	< 0.005 (> 37.5)
Cr	0.06 (78.0)	0.0003 (95.0)
Fe	0.6 (81.0)	< 0.005 (> 99.8)
Mg	79.52 (32.8)	61.8 (26.2)
Mn	0.085 (62.7)	0.01 (97.0)
Hg	0.00009 (0)	ND
Mo	0.0107 (48.3)	0.003 (0)
Ni	0.0405 (78.1)	0.005 (96.5)
K	1360 (25.4)	518.62 (36.3)
Na	1790.4 (26.0)	772.51 (35.0)
Zn	0.52 (61.2)	< 0.05 (> 50)

8.4.3.2 Ecotoxicity Results

Figures 8.14a, b illustrate the results of EC50 and LC50 of the raw and treated samples with respective ranges. Leachate from the Seropédica landfill presented a higher toxic potential to the test organisms than the leachate from the Gericinó landfill. This fact is due to the physicochemical composition of the Seropédica leachate, which showed high concentrations of potential pollutants (ammonia nitrogen, alkalinity, humic substances and chloride) compared to the Gericinó landfill leachate, as previously presented.

The fish *D. rerio* and the microcrustacean *D. similis* showed greater sensitivity to the pollutants of the raw leachates of both landfills concerning the other organisms studied, confirmed by the statistical test of comparison of means proposed in USEPA [174]. As a response to this test, it is observed that the sensitivity of the organisms to the leachates of Seropédica presented the following order: *D. rerio* and *D. similis* > *V. fischeri* > *Artemia* sp. > Microorganisms from activated sludge. In the Gericinó leachate, the same order of sensitivity of the organisms was observed in the Seropédica leachate, except for the toxicity responses of the bacterium *V. fischeri* and the microcrustacean *Artemia* sp., which were statistically equal.

Still, concerning the sensitivity of organisms, the results of this study corroborate with [183], who found in leachate samples from a landfill in Slovenia that the test organisms D*anio rerio* ($LC50_{48h}$: 3.2–6.2%) and *Daphnia magna* ($EC50_{48h}$: 3.3–4.8%) showed similar sensitivities to raw leachate, and this similarity was not observed with the bacterium *Vibrio fischeri* ($EC50_{30min}$:10.8–18.2%) which showed a lower sensitivity to leachate. Moreover, the studies of [184] showed greater sensitivity of the microcrustacean *D. magna* ($EC50_{48h}$: 0.1–0.98%) to the pollutants of the raw leachate from experimental cells compared to the bacterium *V. fischeri* ($EC50_{15min}$: 2.62–12.14%).

According to Restrepo et al. [184], the microcrustacean *D. magna* is the most suitable organism for toxicity tests with complex effluents, such as landfill leachate, due to its high sensitivity. Therefore, it can be considered a model organism for this type of evaluation.

In this study, the microbial assays, the bacterium *V. fischeri* bacteria showed greater sensitivity to leachate samples from the Seropédica [$EC50_{30min}$: 3.90 (2.04–6.47)%] and Gericinó [$EC50_{30min}$: 32.16 (27.50–39.10)%], if compared to the activated sludge microorganisms that obtained average responses of [$EC50_{180min}$: 17.88 (13.73–20.37)%] for Seropédica leachates and of [$EC50_{180min}$: 47.95 (40.05–53.5)%] for Gericinó leachates. Our results corroborate with the results found by [185], who found greater tolerance of activated sludge microorganisms to leachate samples ($EC50_{180min}$: 46.6 ± 5.8%) and greater sensitivity of the bacterium *V. fischeri* ($EC50_{180min}$: 27.6 ± 5.3%). Gutiérrez et al. [50] compared the sensitivities of activated sludge microorganisms and *V. fischeri* bacteria to organic and inorganic pollutants. The authors concluded that the low sensitivity observed in activated sludge microorganisms is due to the heterogeneous composition of the biomass and its high adaptability.

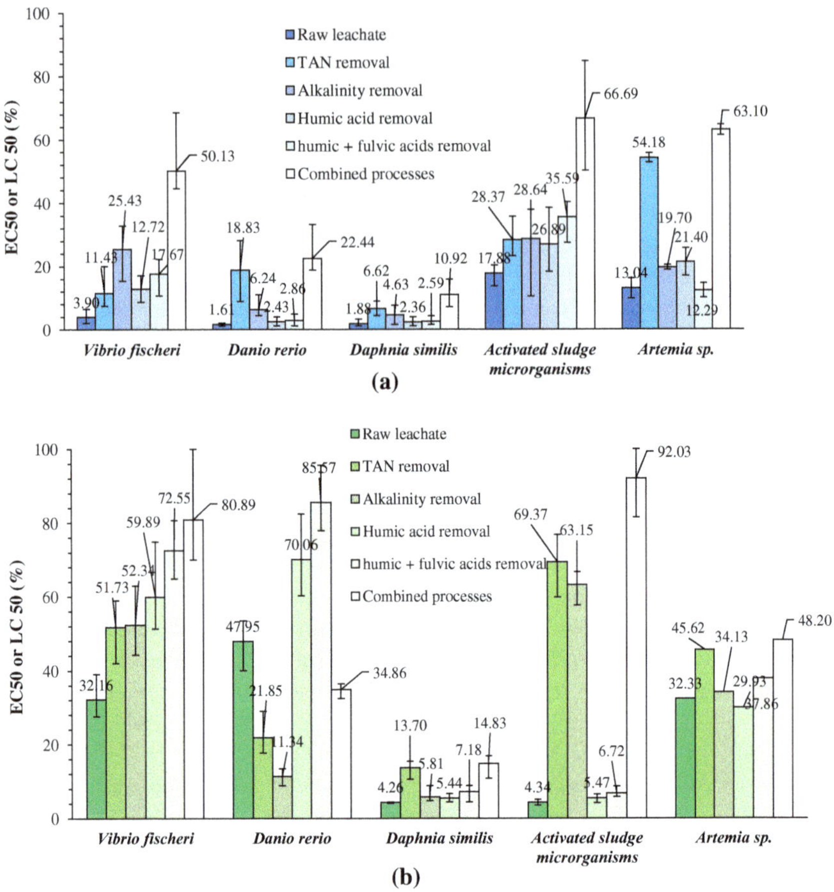

Fig. 8.14 Results of ecotoxicity before and after application of processes. N = 6 samples. **a** Seropédica Leachate; **b** Gericinó Leachate. *Note* Mean values: *TAN removal*: Seropedica Leachate: Initial concentration of TAN = 1919 mg/L and Final concentration of TAN = 0.25 mg/L. Gericinó Leachate: Initial concentration of TAN = 1376 mg/L and Final concentration of TAN = 0.22 mg/L); *Alkalinity removal*: Seropédica Leachate: IC concentration = 2598 mg/L and final IC concentration = 0 mg/L; Gericinó Leachate: Initial IC concentration = 1128 mg/L and final IC concentration = 0 mg/L; *Humic Substances removal*: Seropédica Leachate: Initial concentration of HS = 1935 mg/L and Final concentration of HS = 18 mg/L; Gericinó Leachate: Initial concentration of HS = 636 mg/L and Final concentration of HS = < 10 mg/L; *Combined Processes*: Seropédica Leachate: Initial concentrations of TAN = 1919 mg/L, IC = 2598 mg/L and HS = 1935 mg/L and Final concentrations of TAN = 0.25 mg/L, IC = 0 mg/L, HS = 18 mg/L) and Gericinó Leachate: Initial concentrations of TAN = 1376 mg/L, IC = 1128 mg/L and HS = 636 mg/L and final concentrations of TAN = 0.22 mg/L, IC = 0 mg/L and HS = < 10 mg/L

About the marine microcrustacean *Artemia* sp., the toxicity observed in the raw leachate of the Gericinó landfill [$LC50_{48h}$: 32.33 (28.88–35.96)%] was similar to the results obtained in the studies by [186] who verified toxicity to *A. salina* ($LC50_{48h}$: 28.4%) in the raw leachate.

Below are listed the main observations made of the results with the processes used:

- TAN: the microcrustacean *Artemia* sp. presented an $LC50_{48h}$ value: 54.18 (52.7–55.64)% higher than the other organisms studied in the leachates of Seropédica. In the Gericinó leachates, the microorganisms of the activated sludge ($EC50_{180min}$: 69.37 (59.82–76.78)%) were the organisms that obtained the best response with TAN removal. The statistical test of comparison of means [174] confirmed that the reduction in toxicity was statistically significant for all the organisms studied and observed in the leachate samples from both landfills.

According to the results obtained in this study and the studies of [11, 13, 47, 187, 188], it can be stated that ammonia nitrogen present in landfill leachate contributes significantly to the acute toxicity of aquatic organisms.

- Alkalinity: The removal of alkalinity favored the increase in the samples' EC50 (%) and LC50 (%) values, demonstrating that alkalinity contributes to the toxicity of the leachates of the landfills studied. In the Seropédica leachates, the bacterium *V. fischeri* ($EC50_{30min}$: 25.43 (15.4–32.1)%), activated sludge microorganisms ($EC50_{180min}$: 28.64 (10.47–37.90)%) and *Artemia* sp. ($EC50_{30min}$: 19.70 (18.93–20.46)%) showed similar responses with alkalinity removal, considering the confidence interval of the tests. In the Gericinó leachate, similar toxicity responses were observed for *V. fischeri* ($EC50_{30min}$: 52.34 (44.28–62.98)%) and activated sludge microorganisms ($EC50_{180min}$: 63.15 (57.69–66.63)%).

In removing alkalinity, statistically significant differences in toxicity reduction were observed for the test organisms in most samples tested. Studies individually correlating the alkalinity parameter with the toxicity of test organisms are scarce in the literature. Its presence can alter the bioavailability of metals, as observed in the studies of [189], because some metals can be complex with carbonate and bicarbonate ions, thereby reducing toxicity to organisms.

- Humic Substances: The removal of humic acids (i.e., molecules of molar mass > 10 kDa) favored the reduction of toxicity for the bacterium *V. fischeri* and the activated sludge microorganisms in the leachates of both landfills studied. With the joint removal of HA + FA (i.e., molecules higher than 500 Da), the toxicity reduction was even more expressive for these organisms. This reduction suggests that *V. fischeri* and activated sludge microorganisms are more sensitive to organic matter than *D. similis*, *Artemia* sp. and *D. rerio*.

The statistical test of comparison of means [174] confirmed that after the removal of humic acids and humic and fulvic acids, the reduction of toxicity was statistically significant for *V. fischeri* and activated sludge microorganisms in the leachates of both landfills, and this difference was not observed in all samples for *D. similis*, *Artemia* sp. and *D. rerio*.

After removing HA and HA + AF, the highest toxicity reductions were verified for *V. fischeri* and activated sludge microorganisms. Corroborating with the studies of Clément and Merlin [13] and [47, 141], who observed a high sensitivity for *V. fischeri* to organic compounds and low sensitivity to inorganics (metals) for leachate samples. According to He et al. [190], the bacterium *V. fischeri* shows high sensitivity to humic ($EC50_{15min} = 9.2$ mg/L) and fulvic acids ($EC50_{15min} = 130.8$ mg/L) present in landfill leachates.

Studies by Glover and Wood [191] showed that humic substances can bind to biological surfaces and alter membrane permeability, thus affecting the metabolism of organisms and the toxicity of some ions, such as metals. Tsiridis et al. [192] evaluated the bioavailability of some heavy metals in the presence of humic acids and found that the toxicity of the bacterium *V. fischeri* about metals (Zn^{2+}, Cu^{2+} and Pb^{2+}) is affected by humic acids. The authors observed that the presence of HA reduced the toxicity of Cu^{2+}, increased the toxicity of Pb^{2+} and did not alter the toxicity of Zn^{2+}.

The toxicity reduction after removing HA and HA + FA was low or non-existent for the organisms *D. similis* and *D. rerio*. This fact is explained because the high concentration of ammonia nitrogen found in the permeate is highly toxic to these organisms, in addition to the presence of remaining organic matter (< 500 Da), which comprises about 30% of the COD of the raw leachate.

Concerning the microcrustacean *Artemia* sp., no reductions in toxicity were observed with the removal of humic and fulvic acids from the Seropédica leachate. On the other hand, removing these pollutants resulted in a 50% reduction in toxicity in the Gericinó leachate. Noting that fulvic acids in the permeates of Gericinó leachates reached a lower average concentration (< 10 mg/L) compared to Seropédica (18 mg/L), and this may have influenced the toxicity of the organism.

Although it was observed that the removal of humic substances favored the reduction of toxicity, studies were scarce in the literature that evaluated the mechanisms of action of HS on test organisms.

- Combined processes: The potential pollutants from the leachate and greater reductions in toxicity were observed for the test organisms compared to the individual removals, as can be verified through the EC50(%) and LC50 (%) values. It is observed that the organisms *V. fischeri*, activated sludge microorganisms and *Artemia* sp. obtained the greatest reductions in toxicity with the combined removal of ammonia nitrogen, alkalinity and humic substances. On the other hand, *D. rerio* and *D. similis* showed greater sensitivities to landfill leachate samples, and even after removing potential pollutants, they presented the lowest EC50 and LC50 values among the others.

Figures 8.15a, b illustrate the results regarding toxicity units (100/EC or 100/LC) for the raw leachate and the samples submitted to the combined processes. *Daphnia similis* and *Danio rerio* are the most sensitive for the Seropédica leachate and *D. similis* and the activated sludge microorganisms for the Gericinó leachate. Table 8.8

illustrates the EC20 or EC50 or LC50 values of the inorganic elements evaluated for the test organisms *D. magna*, *Artemia* sp., *V. fischeri*, *D. rerio* and activated sludge microorganisms and the concentration of the elements in each landfill studied after the combined treatment.

Regarding salinity, the results of the toxicity tests with synthetic wastewater containing NaCl and the effect concentrations (EC10 and EC20) on the test organisms: *V. fischeri*, *Artemia* sp., activated sludge microorganisms, *D. rerio* and *D. similis* are presented in Table 8.9. The effect concentrations of chloride ion for freshwater organisms (*Danio rerio*, *Daphnia similis* and microorganisms in activated sludge) are lower than the concentrations of the leachates from Seropédica (1.075–4.369 g/L) and Gericinó (1.207–2.234 g/L) after the combined treatment. Therefore, the chloride ion concentration found in the leachates of the studied landfills contributes to the toxicity of the organisms *D. rerio*, D. *similis* and activated sludge microorganisms. In contrast, for the marine bacterium *V. fischeri*, the concentration of chloride ion present in the leachates was lower than salinity 20, requiring the adjustment of the sample's salinity for the organisms' survival (methodology requirement).

For *Artemia* sp., salinity values ≤ 1.12 generated 100% mortality, while values between 9.0 and 36.0 showed no toxic effects for the microcrustacean. For the combined processes, the low salinity contributed to *Artemia* sp. toxicity. Considering the raw leachate, the salinity present in the leachates studied (Seropédica = 18 and Gericinó = 10) was not a toxicity factor for this organism.

Daphnids and fishes are freshwater organisms, and because of this, they present elevated sensibility to the high salinity of this effluent, which can mask the effects of the other pollutants. In this specific case, using another salt-tolerant organism, such as artemia and *V. fischeri*, can be a valuable tool to provide information on the toxicity due to the other pollutants [204].

8.4.3.3 Statistical Correlation Between Physicochemical and Ecotoxicological Parameters

The statistical analysis of principal components showed the results listed below. The significance of the negative values is: the lower the physicochemical parameter, the higher the ecotoxicological parameter, i.e. the higher the EC50 or LC50 and the lower the ecotoxicity.[1]

- Leachate from Seropédica landfill: a strong correlation was observed between organic matter parameters (COD and HS) and toxicity of *V. fischeri* organisms (– 0.81 and 0.83) and activated sludge microorganisms (– 0.85 and – 0.86). Ammonia nitrogen showed significant correlations with the toxicity of *D. rerio* (– 0.87), *D. similis* (– 0.58) and *Artemia* sp. (– 0.49). Alkalinity showed weak correlations with the toxicity of the bacterium *V. fischeri* (– 0.33) and the fish *D. rerio* (– 0.26).

[1] In brackets: Pearson's correlation coefficient.

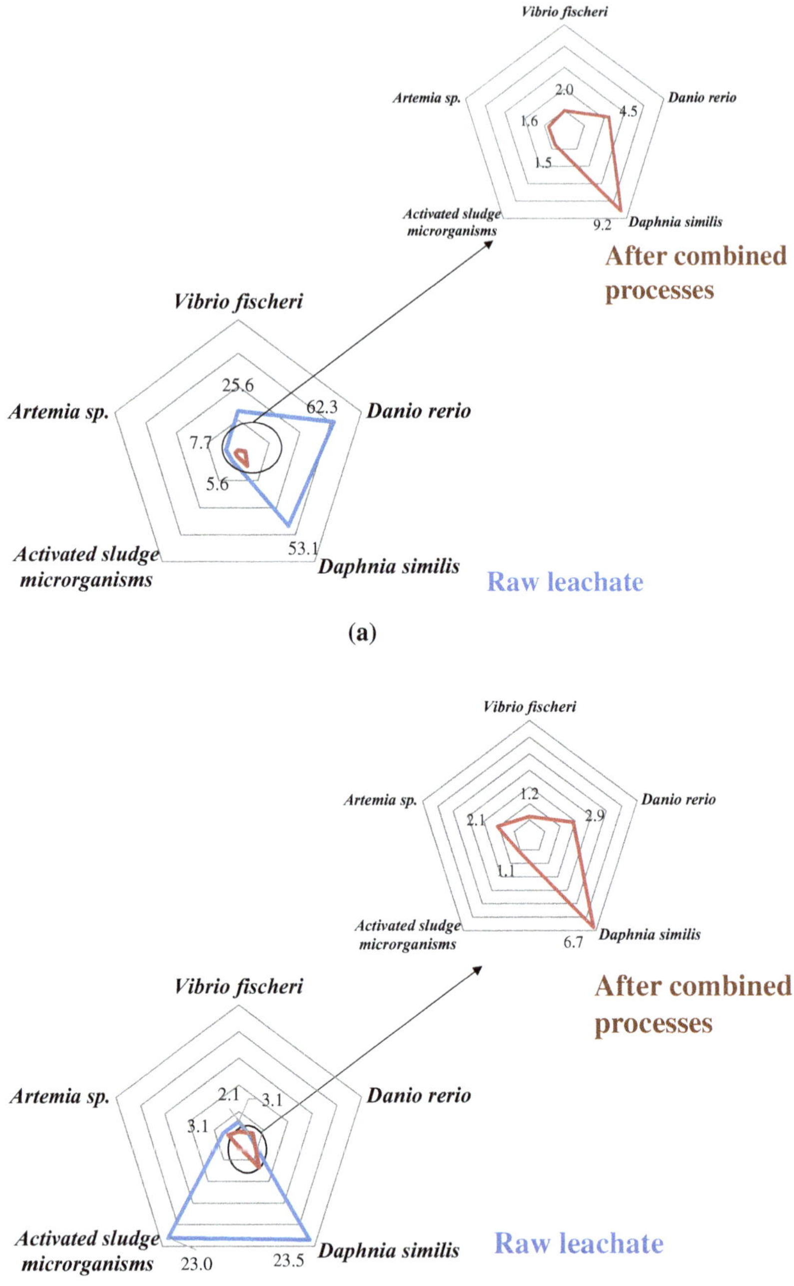

Fig. 8.15 Results of raw leachates and after combined processes expressed em TU (toxicity units). **a** Seropédica Landfill; **b** Gericinó Landfill

Table 8.8 Toxicity values of inorganic elements for *D. magna*, *Artemia* sp., *V. fischeri*, *D. rerio* and microorganisms from activated sludge and concentrations of inorganic elements of landfill leachate after combined treatment of pollutants

Element	EC50 or EC20 or LC50 (mg/L)					Concentration after processes combination (mg/L)	
	Daphnia magna	*Artemia* sp.	*Vibrio fischeri*	*Danio rerio*	Activated sludge microorganisms	Seropédica Leachate	Gericinó Leachate
Al	3.9[a]	–	–	–	–	< 0.005	< 0.005
As	2.4a	–	2.54[e]	–	300[k]	0.0761	0.0122
Ba	11.0[a]	–	–	–	–	0.1106	0.0986
B	141.0[b]	–	–	–	10.0[k]	4.31	0.82
Cd	3.6[a]	710.7[c]	9.96[f]	3.82[g]	0.27–5.55[i]	0.0007	< 0.0005
Ca	870[a]	–	–	–	–	4.838	17.25
Pb	8.7[a]	1.4 ± 0.015[d]	0.06[e]	0.059[g]	1.5–5.0[i]	0.0031	< 0.005
Co	0.710[a]	–	3.80[e]	32.9[g]	–	0.0115	0.0033
Cu	0.013[a]	19.5[c]	0.147[f]	0.212[g]	0.31–2.05[i]	< 0.005	< 0.005
Cr	0.130[a]	–	205[e]	45.5[h]	10.6–211[i]	0.06	0.0003
Fe	2.3[a]	13.9 ± 0.020[d]	–	–	–	0.6	< 0.005
Mg	140[a]	–	–	–	–	79.52	61.8
Mn	9.3[a]	51.8 ± 0.015[d]	–	–	–	0.085	0.01
Hg	0.065[a]	–	0.18[e]	0.23[h]	0.142[j]	0.00009	ND
Mo	1500[a]	–	–	–	–	0.0107	0.003
Ni	0.650[a]	11.2 ± 0.033[d]	–	13.12[g]	24.07[j]	0.0405	0.005
K	340[a]	–	–	–	–	1360	518.62
Na	1600[a]	–	–	–	–	1790.4	772.51
Zn	0.720[a]	12.3 ± 0.020[d]	0.46[e]	22.0[h]	0.17–84.7[i]	0.52	< 0.05

Note [a] Values of $EC50_{48h}$ mg/L (hardness = 72 mg $CaCO_3$/L and pH = 6.5–8.5). *Source* Okamoto et al. [193]; [b] Values of $EC50_{48h}$ mg/L (hardness = 170 mg $CaCO_3$/L and pH = 8.6 ± 0.1). *Source* [194], [c] $LC50_{48h}$ mg/L values (pH = 8.0 ± 0.5). *Source* [195], [d] $EC50_{48h}$ mg/L values (pH = 7.5 ± 8.4). *Source* [196], [e] $EC20_{15min}$ values mg/L (pH = 7.0). *Source* [197], [f] Values of $CE20_{15min}$ $EC20_{15min}$ mg/L (pH = 7.0). *Source* [198], [g] $LC50_{96h}$ mg/L values (hardness = 141 mg $CaCO_3$/L and pH = 7.0). *Source* [199], [h] $LC50_{96h}$ mg/L values (hardness = 100 mg $CaCO_3$/L and pH = not informed). *Source* Bertolleti [200]; [i] $EC50_{24h}$ mg/L values for ciliated protozoa (pH = 7.6 ± 0.2). *Source* [201], [j] $EC50_{30min}$ values (pH = 7.0). *Source* [202], [k] Maximum concentrations that do not interfere with biological nitrification of sludge (pH = 7.0). *Source* [203]

- Leachate from Gericinó landfill: A significant correlation between COD and HS parameters with the toxicity of *V. fischeri* organisms (− 0.85 and − 0.80) and activated sludge microorganisms (− 0.65 and − 0.59). Regarding the ammonia nitrogen parameter, as observed in the leachates of the Seropédica landfill, strong correlations were observed with the toxicity of the organisms *D. rerio* (− 0.76), *D. similis* (− 0.99) and *Artemia* sp. (− 0.84). In contrast, the alkalinity parameter

Table 8.9 Toxicity of chloride on test organisms

Test organisms	Cl^- (g/L)
Danio rerio (EC10)	0.9
Daphnia similis (EC10)	0.30
Microorganisms in activated sludge (EC20)	1.72
Vibrio fischeri (minimum threshold for survival)	29.0
Artemia sp. (minimum threshold for survival)	9.0

showed weak correlations with the toxicity of the bacterium *V. fischeri* (− 0.35) and the microorganisms of the activated sludge (− 0.27).

The correlations between the physicochemical and ecotoxicological data indicated that the parameters of COD, humic substances and ammonia nitrogen showed more significant influence on the toxicity of the leachates from the Seropédica and Gericinó landfills. The ecotoxicological results of the whole leachate probably resulted from interactions between organic matter, ammonia, and other pollutants in the leachate matrix [204].

This study also observed strong positive correlations concerning the sensitivities of *V. fischeri* bacteria and activated sludge microorganisms. However, studies are scarce in the literature on this correlation. According to Elnabarawy et al. [205], Ricco et al. [66], as the tests with activated sludge microorganisms use a non-pure culture, different from the bioluminescence inhibition test with *V. fischeri*, there are difficulties in reproducibility studies. With this, the respirometry test has become a frequent tool for monitoring the treatment processes with activated sludge and few studies are carried out to evaluate the toxicity of the effluent. Given this context, linear regression analysis was performed to determine the correlation between the toxicity responses obtained with the test organisms, as shown in Eq. (8.9).

$$EC50_{180min}(\%) = 15.358 + 0.96653 \times EC50_{30min}(\%) \tag{8.9}$$

where

$EC50_{180min}$ (%) = Activated Sludge microrganisms (%)
$EC50_{30min}$ (%) = *Vibrio fischeri* (%)

8.5 Essential Treatment Techniques for Ecotoxicity Removal

As evidenced in the study shown in the previous item, organic matter in the form of humic substances and ammonia nitrogen and chloride (salinity) are the main toxic pollutants in leachate, depending on the organisms evaluated. Therefore, technologies that remove these pollutants should be used to meet the characteristics of physicochemical parameters and toxicity for discharge into water bodies.

Considering the recovery of resources, stripping ammonia and recovery in the ammonium sulfate form is an interesting option. Some authors have evaluated the operational aspects of this process, including using temperature (40–60 °C) to convert all ammonia nitrogen into free ammonia [126].

Regarding humic substances, some authors agree that membrane processes are more effective [8, 206]. As seen in the study shown in the previous section, humic substances can be efficiently removed with membranes with small pore sizes (cut off ~ 500 Da), such as tight nanofiltration membranes and reverse osmosis. In the case of salinity removal, reverse osmosis is preferred to avoid the toxic effects of the permeate on the test organisms. Otherwise, tight nanofiltration may be suitable. A schematic of the process can be observed in Fig. 8.16.

The membrane process should operate in double steps for best performance, as described in [177]. In addition, the final permeate may have characteristics that can serve as reuse water for washing equipment and wetting roads in the landfill, among other uses. Humic substances can be recovered for agricultural use using loose nanofiltration [207, 208] (Fig. 8.17).

The first step is using a forward or reverse osmosis process or combining both for water recovery. [209] cite that the coupling of these technologies for water recovery is more energy efficient. The stream generated from this operation, if it has a high organic matter content, where HS will prevail, can be fractionated by tight UF or loose NF to recover the organic matter and have a subsequent use as fertilizer. The

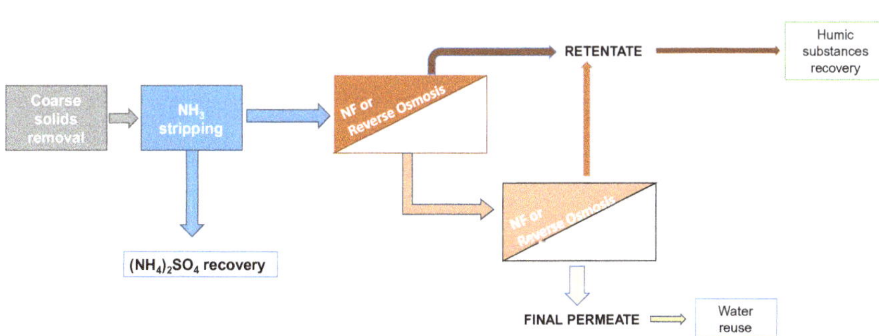

Fig. 8.16 Proposed landfill leachate treatment based on toxicity removal

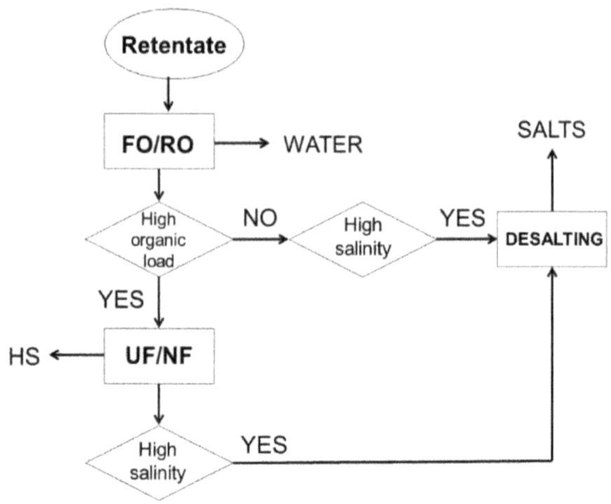

Fig. 8.17 Recovery train for the selection of resource recovery route. Adapted from [207]. *FO* forward osmosis, *NF* nanofiltration, *RO* reverse osmosis, *UF* ultrafiltration

salts in the saline streams can be recovered by evaporation/crystallization using the energy from the biogas generated in the landfill.

8.6 Conclusions

In this study, the importance of the choice of test organisms in the ecotoxicity assessment of landfill leachates was evidenced in at least two trophic levels of representativeness. Leachates generated in two different types of landfills in Rio de Janeiro State (Brazil) were evaluated.

The leachates from the Seropédica landfill showed greater toxic potential to the test organisms than those from the Gericinó controlled landfill. Concerning sensitivity, the fish *D. rerio* and the microcrustacean *D. similis* showed greater sensitivity to the pollutants of the raw leachates than the other organisms studied.

Organic matter (humic substances), ammonia nitrogen and chloride (salinity) were the main contributors to the leachate ecotoxicity. Therefore treatment train was proposed involving membrane separation and ammonia stripping. Resources recovery is possible in the proposed treatment system, including nitrogen, humic substances, water and salt recovery.

References

1. Kaza S, Yao L, Bhada-Tata P, Woerden FV (2018) What a waste 2.0. International bank for reconstruction and development. The World Bank. Washington, DC. Available on https://openknowledge.worldbank.org/server/api/core/bitstreams/92a50475-3878-5984-829e-0a09a6a9badc/content

2. Karak T, Bhagat RM, Bhattacharyya P (2012) Municipal solid waste generation, composition, and management: The world scenario. J Crit Rev Environ Sci Technol 42(15):1509–1630
3. Alfaia RGSM, Costa AM, Campos JC (2017) Municipal solid waste in Brazil. Waste Manag Res 35:1195–1209. https://doi.org/10.1177/0734242X17735375
4. Kjeldsen P, Barlaz MA, Rooker AP, Baun A, Ledin A, Christensen TH (2002) Present and long-term composition of MSW landfill leachate: a review. J Crit Rev Environ Sci Technol 32(4):297–336. https://doi.org/10.1080/10643380290813462
5. Matejczyk M, Plaza GA, Nałęcz-Jawecki G, Ulfig K, Markowska-Szcupak A (2011) Estimation of the environmental risk posed by landfills using chemical, microbiological and ecotoxicological testing of leachates. Chemosphere 82:1017–1023. https://doi.org/10.1016/j.chemosphere.2010.10.066
6. Andreotolla G, Cannas P (1992) Chemical and biological characteristics of landfill leachate. In: Christensen TH, Cossu R, Stegmann R (eds) Landfilling of waste: leachate. CRC Press, London. https://doi.org/10.1201/9781482288933
7. Christensen TH, Kjeldsen P, Bjerg PL, Jensen DL, Christensen JB, Baun A, Albrechtsen H-J, Heron G (2001) Biogeochemistry of landfill leachate plumes. Appl Geochem 16(7–8):659–718. https://doi.org/10.1016/S0883-2927(00)00082-2
8. Costa AM, Silva CAMC, Daflon SDA, Quintaes BR, Ferreira JA, Campos JC (2019) Evaluation of toxic potential of leachate originating from experimental landfill cells containing household waste and healthcare waste. Waste Manage Res 37(10):1003–1011. https://doi.org/10.1177/0734242X19843159
9. Lema JM, Mendez R, Blazquez R (1988) Characteristics of landfill leachates and alternatives for their treatment: a review. Water Air Soil Pollution 40(3):223–250. https://doi.org/10.1007/BF00163730
10. Ziyang L, Youcai Z, Tao Y, Yu S, Huili C, Nanwen Z, Renhua H (2009) Natural attenuation and characterization of contaminants composition in landfill leachate under different disposing ages. Sci Total Environ 407(10):3385–3391. https://doi.org/10.1016/j.scitotenv.2009.01.028
11. Marttinen SK, Kettunen RH, Sormunen KM, Soimasuo RM, Rintala JA (2002) Screening of phisical-chemical methods for removal organic material, nitrogen and toxicity from low strength landfill leachate. Chemosphere 46(6):851–858. https://doi.org/10.1016/S0045-6535(01)00150-3
12. Prestes JG, De Souza MRDP, Kandalski PK, Herrerias T, Machado C, De Arruda EM, Donatti L (2020) Biomarkers of oxidative stress and cell damage in freshwater bivalves Diplodon parodizi exposed to landfill leachate. Environ Sci Pollut Res 27(22):28384–28395. https://doi.org/10.1007/s11356-020-08721-4
13. Clément B, Merlin G (1995) The contribution of ammonia and alkalinity to landfill leachate toxicity to duckweed. Sci Total Environ 170(1–2):71–79. https://doi.org/10.1016/0048-9697(95)04563-G
14. Miao L, Yang G, Tao T, Peng Y (2019) Recent advances in nitrogen removal from landfill leachate using biological treatments—a review. J Environ Manage 235:178–185. https://doi.org/10.1016/j.jenvman.2019.01.057
15. Pablos MV, Martini F, Fernández C, Banbín MM, Herraez I, Miranda J, Martínez J, Carbonell G, San-Segundo L, García-Hortiguëla P, Tarazona JV (2011) Correlation between physicochemical and ecotoxycological approaches to estimate landfill leachate toxicity. Waste Manage 31(8):1841–1847. https://doi.org/10.1016/j.wasman.2011.03.022
16. Kurniawan TA, Lo W, Chan GYS (2006) Physico-chemical treatments for removal of recalcitrant contaminants from landfill leachate. J Hazard Mater 129(1–3):80–100. https://doi.org/10.1016/j.jhazmat.2005.08.010
17. Smith VH, Tilman GD, Nekola JC (1999) Eutrophication: impacts of excess nutrient imputs on freshwater, marine, and terrestrial ecossystem. Environ Pollut 100(1–3):179–196. https://doi.org/10.1016/S0269-7491(99)00091-3
18. Wang Z, Leung KMY (2015) Effects of unionised ammonia on tropical freshwater organisms: implications on temperature-to-tropic extrapolation and water quality guidelines. Environ Pollut 205:240–249. https://doi.org/10.1016/j.envpol.2015.05.045

19. Ghosh P, Thankur IS, Kaushik A (2017) Bioassays for toxicological risk assessment of landfill leachate: a review. Ecotoxicol Environ Saf 141:259–270. https://doi.org/10.1016/j.ecoenv.2017.03.023
20. Thomas DJL, Tyrrel SF, Smith R, Farrow S (2009) Bioassays for the evaluation of landfill leachate toxicity. J Toxicol Environ Health-Part B 12(1):83–105. https://doi.org/10.1080/10937400802545292
21. Truhaut R (1977) Ecotoxicology: objectives, principles and perspectives. Ecotoxicol Environ Saf 1(2):151–173. https://doi.org/10.1016/0147-6513(77)90033-1
22. Butler GC (1978) Principles of ecotoxicology. Wiley, UK
23. Boudou A, Ribeyre F (1997) Aquatic ecotoxicology: from the ecossystem to the cellular and molecular levels. Environ Health Perspect 105(1):21–35. https://doi.org/10.1289/ehp.97105s121
24. Costa CRC, Olivi P, Botta CMR, Espindola ELG (2008) Toxicity in aquatic environments: discussion and evaluation methods. Quim Nova 31(7):1820–1830. https://doi.org/10.1590/s0100-40422008000700038[InPortuguese]
25. Rand GM, Wells PG, Mccarty LS (1999) Introduction of aquatic toxicology. In: Rand MG (ed) Fundamentals of aquatic toxicology, 2nd edn. Taylor & Francis, London
26. Park J, Shin K, Lee H, Choi S, Kim G, Depuydt S, De Saeger J, Heynderickx PM, Wu D, Asselman J, Janssen C, Han T (2023) Evaluating ecotoxicological assays for comprehensive risk assessment of toxic metals present in industrial wastewaters in the Republic of Korea. Sci Total Environ 867:161536. https://doi.org/10.1016/j.scitotenv.2023.161536
27. Moiseenko TI (2008) Aquatic ecotoxicology: Theoretical principles and pratical application. Water Resour 35(5):530–541. https://doi.org/10.1134/S0097807808050047
28. Novais M, Penha AM, Catarino A, Martins I, Fialho S, Lima A, Morais M, Palma P (2023) The usefulness of ecotoxicological tools to improve the assessment of water bodies in a climate change reality. Sci Total Environ. https://doi.org/10.1016/j.scitotenv.2023.166392
29. Ghisi NC, Oliveira EC, Guiloski I, Lima SB, Assis HCS, Longhi SL, Prioli AJ (2017) Multivariate and integrative approach to analyze multiple biomarkers in ecotoxicology: a field study in Neotropical region. Sci Total Environ 609:1208–1218. https://doi.org/10.1016/j.scitotenv.2017.07.266
30. Lam PKS (2009) Use of biomarkers in environmental monitoring. Ocean Coast Manag 52(7):348–354. https://doi.org/10.1016/j.ocecoaman.2009.04.010
31. Newman MC (2010) Ecotoxicology: the history and presents directions. In: Jørgensen SE (ed) Ecotoxicology: a derivative of encyclopedia of ecology. Academic Press, Amsterdam. https://doi.org/10.1016/B978-008045405-4.00431-6
32. Adams WJ, Rowland CD (2002) Aquatic toxicology test methods. In: Hoffman DJ, Rattner BA, Burton Jr GA, Cairns Jr J (eds) Handbook of ecotoxicology, 2nd edn. CRC Press, Boca Raton. https://doi.org/10.1201/9781420032505
33. Birge WJ, Black JA, Westerman JA, Short TM, Taylor SB, Bruser DM, Wallingford ED (1985) Recommendations on numerical values for regulating iron and chloride concentrations for the purpose of protecting warm water species of aquatic life in the commonwealth of Kentucky. Technical Report. Lexington, KY. Available on https://hero.epa.gov/hero/index.cfm/search/exportris/reference_id/2079214
34. Di Giulio RT, Newman MC (2012) Environmental toxicology. In: Klaassen CD, Watkins III JB (eds) Fundamentals of ecotoxicology, 2nd edn. AMGH Editora Ltda, Porto Alegre, Brazil (In Portuguese)
35. Hassan SHA, Van Ginkel SW, Hussein MAM, Abskharon R, Oh S-E (2016) Toxicity assessment using different bioassays and microbial biosensors. Environ Int 92–93:106–118. https://doi.org/10.1016/j.envint.2016.03.003
36. Keddy CJ, Greene JC, Bonell MA (1995) Review of whole-organism bioassays: soil, freshwater sediment, and freshwater assessment in Canada. Ecotoxicol Environ Saf 30(3):221–251. https://doi.org/10.1006/eesa.1995.1027
37. Buikema AL Jr, Niederlehner BR, Cairns J Jr (1982) Biological monitoring. Part IV—toxicity testing. Water Res 16(3):239–262. https://doi.org/10.1016/0043-1354(82)90188-9

38. Chapman PM (2002) Integrating toxicology and ecology: putting "eco" into ecotoxicology. Mar Pollut Bull 44(1):7–15. https://doi.org/10.1016/S0025-326X(01)00253-3
39. Farré M, Barceló D (2003) Toxicity testing of wastewater and sewage sludge by biosensors, bioassays and chemical analysis. TrAC Trends Anal Chem 22(2):299–310. https://doi.org/10.1016/S0165-9936(03)00504-1
40. Macdonald DD, Ingersoll CG (2003) A guidance manual to support the assessment of contaminated sediments in freshwater, estuarine and marine ecosystems in British Columbia. British Columbia Ministry of Water, Land, and Air Protection, Victoria, CA. Available on https://www2.gov.bc.ca/assets/gov/environment/air-land-water/site-remediation/docs/technical-guidance/x19_v1.pdf
41. Erzinger GS, Schmoeller F, Pinto LH, Américo L, Hemmersbach R, Hauslage J, Häder D-P (2018) Bioluminescence systems in environmental biosensors. In: Häder DP, Erzinger GS (eds) Bioassays. Elsevier, pp 241–262. https://doi.org/10.1016/B978-0-12-811861-0.00012-7
42. Girotti S, Ferri EN, Fumo MG, Maiolini E (2008) Monitoring of environmental pollutants by bioluminescent bacteria. Anal Chim Acta 608(1):2–29. https://doi.org/10.1016/j.aca.2007.12.008
43. Abbas M, Adil M, Ehtisham-Ul Ul-Haque S, Munir B, Yameen M, Ghaffar A, Shar GA, Tahir MA, Iqbal M (2018) *Vibrio fischeri* bioluminescence inhibition assay for ecotoxicity assessment: a review. Sci Total Environ 626:1295–1309. https://doi.org/10.1016/j.scitotenv.2018.01.066
44. Bulich AA (1982) A practical and reliable method for monitoring the toxicity of aquatic samples. Process Biochem 17(2):45–47
45. Hernando MD, De Vettori S, Bueno MJM, Fernández-Alba AR (2007) Toxicity evaluation with Vibrio fischeri test of organic chemicals used in aquaculture. Chemosphere 68(4):724–730. https://doi.org/10.1016/j.chemosphere.2006.12.097
46. Norberg-King TJ, Embry MR, Belanger SE, Braunbeck T, Butler JD, Dorn PB, Farr B, Guiney PD, Hughes SA, Jeffries M, Jornel R, Lèonard M, Mcmaster M, Oris JT, Ryder K, Segner H, Senac T, Kraak GVD, Whale G, Wilson P (2018) An international perspective on the tools and concepts for effluent toxicity assessments in the context of animal alternatives: reduction in vertebrate use. Environ Toxicol Chem 37(11):2745–2757. https://doi.org/10.1002/etc.4259
47. Pivato A, Gaspari L (2006) Acute toxicity test of leachate from traditional and sustainable landfills using luminescent bacteria. Waste Manage 26:1148–1155. https://doi.org/10.1016/j.wasman.2005.10.008
48. Rosen G, Osorio-Robayo A, Riveira-Duarte I, Lapota D (2008) Comparison of bioluminescent dinoflagellate (QwikLite) and bacterial (Microtox) rapid bioassays for the detection of metals and ammonia toxicity. Arch Environ Contam Toxicol 54:606–611. https://doi.org/10.1007/s00244-007-9068-3
49. Dalzell DJB, Alte S, Aspichueta E, De La Sota A, Etxebarria J, Gutierrez M, Hoffmann CC, Sales D, Obst U, Christofi N (2002) A comparison of five rapid direct toxicity assessment methods to determine toxicity of pollutants to activated sludge. Chemosphere 47(5):535–545. https://doi.org/10.1016/s0045-6535(01)00331-9
50. Gutiérrez M, Etxebarria J, De Las Fuentes L (2002) Evaluation of wastewater toxicity: comparative study between Microtox® and activated sludge oxygen uptake inhibition. Water Res 36(4):919–924. https://doi.org/10.1016/s0043-1354(01)00299-8
51. Parvez S, Venkataraman C, Mukherji S (2006) A review on advantages of implementing luminescence inhibition test (Vibrio fischeri) for acute toxicity prediction of chemicals. Environ Int 32(2):265–268. https://doi.org/10.1016/j.envint.2005.08.022
52. Teodorovic I, Planojevic I, Knezevic P, Radak S, Nemet I (2009) Sensitivity of bacterial vs. acute *Daphnia magna* toxicity test to metals. Central Eur J Biol 4(4):482–492. https://doi.org/10.2478/s11535-009-0048-7
53. Ma XY, Wang XC, Ngo HH, Guo W, Wu MN, Wang N (2014) Bioassay based luminescent bacteria: interferences, improvements, and applications. Sci Total Environ 468–469:1–11. https://doi.org/10.1016/j.scitotenv.2013.08.028

54. Zou X, Xiao X, He Y, Hu L, Hu C, Huang X (2017) Hormetic effects of metal ions upon *V. fischeri* and the application of a new parameter for the quantitative assessment of hormesis. J Hazard Mater 322:454–460. https://doi.org/10.1016/j.jhazmat.2016.09.045
55. Aguilar MI, Lloréns M, Fernández-Garrido JM, Pérez-Marín AB, Ortuño JF, Meseguer VF (2020) Heavy metals effect on the heterotrophic activity of activated sludge. Int J Environ Sci Technol (IJEST) 17(5):3111–3118. https://doi.org/10.1007/s13762-020-02704-1
56. Rossetti S, Levantesi C, Tandoi V (2017) The microbiology of the activated sludge process. In: Rossetti S, Tandoi V, Wanner J (eds) Activated sludge separation problems: theory, control measures, practical experiences. IWA Publish, London, pp 21–52
57. Madoni P (1994) A sludge biotic index (SBI) for the evaluation of the biological performance of activated sludge plants based on the microfauna analysis. Water Res 28(1):67–75. https://doi.org/10.1016/0043-1354(94)90120-1
58. von Sperling M (2007) Basic principles of wastewater treatment. In: Biological wastewater treatment. IWA Publishing, London
59. Nielsen PH, Thomsen TR, Nielsen JL (2004) Bacterial composition of activated sludge-importance for floc and sludge properties. Water Sci Technol 49(10):51–58. https://doi.org/10.2166/wst.2004.0606
60. Bento AP, Sezerino PH, Philippi LS, Reginatto V, Lapolli FR (2005) Microfauna characterization in an activated sludge sewage treatment plant: a tool for process evaluation and control. Engenharia Sanitária Ambiental 10(4):329–338 (In Portuguese). https://doi.org/10.1590/S1413-41522005000400009
61. von Sperling M (1997) Principles of biological wastewater treatment: activated sludge, 2nd edn. UFMG, Belo Horizonte, Brazil (In Portuguese)
62. Seviour RJ, Nielsen PH (2010) Microbial ecology of activated sludge. Water intelligence online 9. https://doi.org/10.2166/9781780401645
63. Puigagut J, Salvadó H, García J (2005) Short-term harmful effects of ammonia nitrogen on activated sludge microfauna. Water Res 39(18):4397–4404. https://doi.org/10.1016/j.watres.2005.08.008
64. Spanjers H (1998) Respirometry in modeling and control of wastewater treatment. In: Barnett MW, Stenstrom MK, Andrews JF (ed) Dynamics and control of wastewater systems, vol 6, 2nd edn. Technomic Publish Company Inc., Lancaster
65. Andreottola G, de Oliveira EL, Foladori P, Dallago L, Peterlini R, Cadonna M (2005) Método respirométrico para o monitoramento de processos biológicos. Engenharia Sanitária Ambiental 10(1):14–23. https://doi.org/10.1590/S1413-41522005000100003
66. Ricco G, Tomei MC, Ramadori R, Laera G (2004) Toxicity assessment of common xenobiotic compounds on municipal activated sludge: comparison between respirometry and Microtox. Water Res 38(8):2103–2110. https://doi.org/10.1016/j.watres.2004.01.020
67. Kungolos A (2005) Evaluation of toxic properties of industrial wastewater using on-line respirometry. J Environ Sci Health 40(4):869–880. https://doi.org/10.1081/ese-200048292
68. OECD (2010) Test No. 209: activated sludge, respiration inhibition test (carbon and ammonium oxidation), OECD guidelines for the testing of chemicals, Section 8.2. OECD Publishing, Paris. https://doi.org/10.1787/9789264070080-en
69. Damasceno EP, Figuerêdo LP, Pimentel MF, Loureiro S, Costa-Lotufo LV (2017) Prediction of toxicity of zinc and nickel mixtures to Artemia sp. at various salinities: from additivity to antagonism. Ecotoxicol Environ Saf 142:322–329. https://doi.org/10.1016/j.ecoenv.2017.04.020
70. Koutsaftis A, Aoyama I (2007) Toxicity of four antifouling biocides and their mixtures on the brine shrimp Artemia salina. Sci Total Environ 387(1–3):166–174. https://doi.org/10.1016/j.scitotenv.2007.07.023
71. Lu J, Tian S, Lv X, Chen Z, Chen B, Zhu X, Cai Z (2018) TiO_2 nanoparticles in the marine environment: Impact on the toxicity of phenanthrene and Cd^{2+} to marine zooplankton *Artemia salina*. Sci Total Environ 615:375–380. https://doi.org/10.1016/j.scitotenv.2017.09.292
72. Zhu S, Luo F, Chen W, Zhu B, Wang G (2017) Toxicity evaluation of graphene oxide on cysts and three larval stages of Artemia salina. Sci Total Environ 595:101–109. https://doi.org/10.1016/j.scitotenv.2017.03.224

73. Libralato G, Prato E, Migliore L, Cicero AM, Manfra L (2016) A review of toxicity testing protocols and endpoints with Artemia spp. Ecol Ind 69:35–49. https://doi.org/10.1016/j.ecolind.2016.04.017
74. Nunes BS, Carvalho FD, Guilhermino LM, Van Stappen G (2006) Use of the genus artemia in ecotoxicity testing. Environ Pollut 144(2):453–462. https://doi.org/10.1016/j.envpol.2005.12.037
75. Abatzopoulos TJ, Kappas I, Bossier P, Sorgeloos P, Beardmore JA (2002) Genetic characterization of *Artemia tibetiana* (Crustacea: Anostraca). Biol J Lin Soc 75(3):333–344. https://doi.org/10.1046/j.1095-8312.2002.00023.x
76. Cong Y, Wang Y, Zhang M, Jin F, Mu J, Li Z, Wang J (2021) Lethal, behavioral, growth and developmental toxicities of alkyl-PAHs and non-alkyl PAHs to early-life stage of brine shrimp, Artemia parthenogenetica. Ecotoxicol Environ Saf 220:112302. https://doi.org/10.1016/j.ecoenv.2021.112302
77. Kalčíková G, Zagorc-Končan J, Gotvajn AŽ (2011) Evaluation of landfill leachate quality with battery of biotests. Acta Environ Univ Comenianae (Bratislava) 19:145–50. Available on https://fns.uniba.sk/fileadmin/prif/actaenvi/ActaEnvi_2011_Suppl/27_Kalcikova_et_al.pdf
78. Smirnov NN (2017) Physiology of the Cladocera, 2nd edn. Academic Press, London
79. Buratini SV, Bertoletti E, Zagatto PA (2004) Evaluation of Daphnia similis as a test specie in ecotoxicological assays. Bull Environ Contam Toxicol 73(5):878–882. https://doi.org/10.1007/s00128-004-0508-8
80. Freitas EC, Rocha O (2011) Acute toxicity test with the tropical cladocera *Pseudosida ramosa*: the importance of using native species as test organisms. Arch Environ Contam Toxicol 60(2):241–249. https://doi.org/10.1007/s00244-010-9541-2
81. Rodgher S, Espíndola ELG, Lombardi AT (2010) Suitability of Daphnia similis as na alternative organism ecotoxicological tests: implication for metal toxicity. Ecotoxicology 19:1027–1033. https://doi.org/10.1007/s10646-010-0484-1
82. ABNT (2022a) Brazilian Association of Technical Standards. NBR 12713 Aquatic ecotoxicology—acute toxicity—test method with *Daphnia* spp. (Crustacea, Cladocera), Rio de Janeiro (In Portuguese)
83. Ebert D (2005) Ecology, epidemiology and evolution of parasitism in Daphnia. Bethesda (MD), National Library of Medicine (US), National Center for Biotechnology Information. Available on https://www.ncbi.nlm.nih.gov/corehtml/pmc/homepages/bookshelf/pdf/daph_screenA4.pdf
84. Sharma C, Langer S, Noorani IA (2016) Population structure and some biological parameters of Daphnia similis, an important fish food organism. Int J Fish Aquat Stud 4(1):111–114. Available on https://www.fisheriesjournal.com/archives/2016/vol4issue1/PartB/3-2-86.pdf
85. Eads BD, Andrews J, Colbourne JK (2008) Ecological genomics in Daphnia: stress responses and environmental sex determination. Heredity 100(2):184–190. https://doi.org/10.1038/sj.hdy.6800999
86. Hebert PDN (1987) Genotypic chracteristics of the Cladocera. Hidrobiology 145(1):183–193. https://doi.org/10.1007/BF02530279
87. Ebert D, Duneau D, Hall MD, Luijckx P, Andras JP, Du Pasquier L, Ben-Ami F (2016) A population biology perspective on the stepwise infection process of the bacterial pathogen *Pasteuria ramosa* in Daphnia. Adv Parasitol 91:265–310. https://doi.org/10.1016/bs.apar.2015.10.001
88. Stollewerk A (2010) The water flea Daphnia—a 'new' model system for ecology and evolution? J Biol 9(21):2–4. https://doi.org/10.1186/jbiol212
89. Martins J, Teles LO, Vasconcelos V (2007) Assays with *Daphnia magna* and *Danio rerio* as alert systems in aquatic toxicology. Environ Int 33:414–425. https://doi.org/10.1016/j.envint.2006.12.006
90. Kato K, Okamura K, Hiki K, Kintsu H, Nohara K, Yamagishi T, Nakajima N, Watanabe H, Yamamoto H (2022) Potential differences in chitin synthesis ability cause different sensitivities to diflubenzuron among three strains of Daphnia magna. Aquat Toxicol 243:106071. https://doi.org/10.1016/j.aquatox.2021.106071

91. Lawrence C (2007) The husbandry of zebrafish (Danio rerio): a review. Aquaculture 269(1–4):1–20. https://doi.org/10.1016/j.aquaculture.2007.04.077
92. Spence R, Gerlach G, Lawrence C, Smith C (2007) The behaviour and ecology of the zebrafish, *Danio rerio*. Biol Rev 83(1):13–34. https://doi.org/10.1111/j.1469-185X.2007.00030.x
93. Singh AP, Nüsslein-Volhard C (2015) Zebrafish stripes as a model for vertebrate colour pattern formation. Current Biol 25(2):PR81–PR92. https://doi.org/10.1016/j.cub.2014.11.013
94. Pritchard VL, Lawrence J, Butlin RK, Krause J (2001) Shoal choice in zebrafish, *Danio rerio*: the influence of shoal size and activity. Anim Behav 62:1085–1088. https://doi.org/10.1006/anbe.2001.1858
95. Dutta SPS (1993) Food and feeding habits of Danio rerio (Ham. Buch.) inhabiting Gadigarh Stream, Jammu. J Freshw Biol 5(2):165–168
96. Lawrence C, Best J, James A, Maloney K (2012) The effects of feeding frequency on growth and reproduction in zebrafish (*Danio rerio*). Aquaculture 368:103–108. https://doi.org/10.1016/j.aquaculture.2012.09.022
97. Breder CM, Rosen DE (1966) Reproduction in fishes: modes of reproduction in fishes. The Natural History Press, New York
98. Wixon J (2000) Danio rerio, the zebrafish. Yeast 17:225–231. https://doi.org/10.1155/2000/784810
99. Westerfield M (1995) The Zebrafish book. A guide for the Laboratory use of Zebrafish (Danio rerio), 3 edn. University of Oregon Press Eugene, OR
100. Dammski AP, Müller BR, Gaya C, Regonato D (2011) Zebrafish. Animal husbandry manual, 1 edn. Curitiba, Brazil, Available on https://gia.org.br/portal/wp-content/uploads/2013/06/ZEBRAFISH.pdf.pdf
101. Detrich HW, Westerfield M, Zon LI (2009) Overview of the zebrafish system. In: Essential zebrafish methods: cell and developmental biology, Academic Press, Cambridge
102. Dai Y-J, Jia Y-F, Chen N, Bian W-P, Li Q-K, Ma Y-B, Chen Y-L, Pei D-S (2014) Zebrafish as a model system to study toxicology. Environ Toxicol Chem 33(1):11–17. https://doi.org/10.1002/etc.2406
103. Hill AJ, Teraoka H, Heideman W, Peterson RE (2005) Zebrafish as a model vertebrate for investigating chemical toxicity. Toxicol Sci 86(1):6–19. https://doi.org/10.1093/toxsci/kfi110
104. Scholz S, Fisher S, Gündel U, Küster E, Luckenbach T, Voelker D (2008) The zebrafish embryo model in environmental risk assessment—applications beyond acute toxicity testing. Environ Sci Pollut Res 15(5):394–404. https://doi.org/10.1007/s11356-008-0018-z
105. Wang X, Zhang J, Lu C, Liu Y, Yang X, Hou K, Du Z, Li B, Juhasz A, Zhu L (2023) Development toxicity and cytotoxicity of pyroxsulam on embryos and adults of zebrafish (*Danio rerio*). Environ Pollut 319:121040. https://doi.org/10.1016/j.envpol.2023.121040
106. Wang Z, Tan Y, Li Y, Duan J, Wu Q, Li R, Shi H, Wang M (2022) Comprehensive study of pydiflumetofen in *Danio rerio*: enantioselective insight into the toxic mechanism and fate. Environ Int 167:107406. https://doi.org/10.1016/j.envint.2022.107406
107. Mc Bean EA, Rovers FA, Farquhar GJ (1995) Solid waste landfill engineering and design. Prentice-Hall PTR, Englewood Cliffs
108. Farquhar GJ, Rovers FA (1973) Gas production during refuse decomposition. Water Air Soil Pollut 2(4):483–495. https://doi.org/10.1007/BF00585092
109. Barlaz MA, Ham RK (1993) Leachate and gas generation In: Daniel DE (ed) Geothechnical practice for waste disposal. Springer, New York. https://doi.org/10.1007/978-1-4615-3070-1
110. Christensen TH, Kjeldsen P (1989) Basic biochemical processes in landfills. In: Christensen TH, Cossu R, Stegmann R (eds) Sanitary landfilling: process, technology and environmental impact. Academic Press, London
111. O'Leary PR, Tchobanoglous G (2002) Landfilling. In: Tchobanoglous G, Kreith F (eds) Handbook of solid waste management, 2nd edn. United State of America, Mc-Graw Hill
112. Ehrig H-J (1983) Quality and quantity of sanitary landfill leachate. Waste Manage Res 1(1):53–68. https://doi.org/10.1016/0734-242X(83)90024-1
113. Aziz HA, Amr SA (2016) Control and treatment of landfill leachate for sanitary waste disposal. IGI Glob Hershey PA. https://doi.org/10.4018/978-1-4666-9610-5

114. Christensen TH, Kjeldsen P, Albrechtsen H-J, Heron G, Nielsen PH, Bjerg PL, Holm PE (1994) Attenuation of landfill leachate pollutants in aquifers. Crit Rev Environ Sci Technol 24(2):119–202. https://doi.org/10.1080/10643389409388463
115. Chen PH (1996) Assessment of leachates from sanitary landfills: impact of age, rainfall, and treatment. Environ Int 22(2):225–237. https://doi.org/10.1016/0160-4120(96)00008-6
116. El-Fadel M, Bou-Zeid E, Chahine W, Alayli B (2002) Temporal variation of leachate quality from pre-sorted and baled municipal solid waste with high organic and moisture content. Waste Manage 22(3):269–282. https://doi.org/10.1016/S0956-053X(01)00040-X
117. Kulikowska D, Klimiuk E (2008) The effect of landfill age on municipal leachate composition. Biores Technol 99(13):5981–5985. https://doi.org/10.1016/j.biortech.2007.10.015
118. Hepburn E, Madden C, Szabo D, Coggan TL, Clarke B, Currell M (2019) Contamination of groundwater with per- and polyfluoroalkyl substances (PFAS) from legacy landfills in an urban re-development precinct. Environ Pollut 248:101–113. https://doi.org/10.1016/j.envpol.2019.02.018
119. Lopes DL, Silva SMCP, Fernandes F, Texeira RS, Celligoi A, Dallántônia LH (2012) Geophysical techinique and groundwater monitoring to detect leachate contaminantion in the surrrounding area of a landfill-Londrina (PR-Brazil). J Environ Manage 113:481–487. https://doi.org/10.1016/j.jenvman.2012.05.028
120. Mor S, Ravindra K, Dahiya RP, Chandra A (2006) Leachate characterization and assessment of groundwater near municipal solid waste landfill site. Environ Monit Assess 118:435–456. https://doi.org/10.1007/s10661-006-1505-7
121. Salem Z, Hamouri K, Djemaa R, Allia K (2008) Evaluation of landfill leachate pollution and treatment. Dessalination 220(1–3):108–114. https://doi.org/10.1016/j.desal.2007.01.026
122. Teng C, Zhou K, Peng C, Chen W (2021) Characterization and treatment of landfill leachate: a review. Water Res 203:117525. https://doi.org/10.1016/j.watres.2021.117525
123. Christensen TH, Cossu R, Stegman R (2012) Sanitary landfilling: process and environmental impacts. Academic Press, London
124. UK Environmental Agency (2002) Ammonia fluxes in landfills. R& D Technical Report, p1–306/TR. Available on https://assets.publishing.service.gov.uk/government/uploads/system/uploads/attachment_data/file/290616/sp1-306-tr-e-e.pdf
125. Silva AC, Dezotti M, Sant'anna GL (2004) Treatment and detoxification of a sanitary landfill leachate. Chemosphere 55(2):207–214. https://doi.org/10.1016/j.chemosphere.2003.10.013
126. Campos JC, Moura D, Costa AP, Yokoyama L, Araujo FVF, Cammarota MC, Cardillo L (2013) Evaluation of pH, alkalinity and temperature during air stripping process for ammonia removal from landfill leachate. J Environ Sci Health A 48(9):1105–1113. https://doi.org/10.1080/10934529.2013.774658
127. Al-Zaidan AS (2017) The acute effects of un-ionized ammonia on zebrafish (Danio rerio). Fish Aquac J 8(3):1–11. https://doi.org/10.4172/2150-3508.1000212
128. Gonçalves AF, Páscoa I, Neves JV, Coimbra J, Vijayan MM, Rodrigues P, Wilson JM (2012) The inhibitory effect of environmental ammonia on Danio rerio LPS induced acute phase response. Dev Comp Immunol 36(2):279–288. https://doi.org/10.1016/j.dci.2011.04.008
129. Levit SM (2010) A literature review of effects of ammonia on fish. In: The Nature Conservancy. Center for Science in Public Participation. Bozeman, MT. Available on https://www.nature.org/content/dam/tnc/nature/en/documents/literature-review-ammonia.pdf
130. Randall DJ, Tsui TKN (2002) Ammonia toxicity in fish. Mar Pollut Bull 45:17–23. https://doi.org/10.1016/S0025-326X(02)00227-8
131. Metcalf & Eddy, Inc., Tchobanoglous G, Stensel H, Tsuchihashi R, Burton F (2014) Wastewater engineering: treatment and resource recovery, 5th edn. McGraw-Hill, New York
132. Constable M, Charlton M, Jensen F, Mcdonald K, Craig G, Taylor KW (2003) An ecological risk assessment of ammonia in the aquatic environment. Hum Ecol Risk Assess 9(2):527–548. https://doi.org/10.1080/713609921
133. Emerson K, Russo RC, Lund RE, Thurston RV (1975) Aqueous ammonia equilibrium calculations: effect of pH and temperature. J Fish Board Canada 32(12):2379–2383. https://doi.org/10.1139/f75-274

134. Whitfield M (1974) The hydrolysis of ammonium ions in sea water—a theoretical study. J Mar Biol Assoc UK 54(3):565–580. https://doi.org/10.1017/S002531540002275X
135. Ren Y, Ngo HH, Guo W, Wang D, Peng L, Ni BJ, Wei W, Liu Y (2020) New perspectives on microbial communities and biological nitrogen removal processes in wastewater treatment systems. Biores Technol 297:122491. https://doi.org/10.1016/j.biortech.2019.122491
136. Winkler MKH, Straka L (2019) New directions in biological nitrogen removal and recovery from wastewater. Curr Opin Biotechnol 57:50–55. https://doi.org/10.1016/j.copbio.2018.12.007
137. Cheung KC, Chu LM, Wong MH (1997) Ammonia stripping as a pretreatment for landfill leachate. Water Air Soil Pollut 94(1):209–221. https://doi.org/10.1023/A:1026413031434
138. Manahan SE (2009) Fundamentals of aquatic chemistry. In: Environmental chemistry, 9th edn. CRC Press, Boca Raton
139. Qasin SR, Chiang W (1994) Sanitary landfill leachate- Generation, control and treatment. CRC Press LLC, Raton
140. Clément B, Janssen RC, Le Dû-Delpierre A (1997) Estimation of the hazard of landfills through toxicity testing of leachates: 2. Comparison of physico-chemical characteristics of landfill leachates with their toxicity determined with a battery of tests. Chemosphere 35:2783–2796. https://doi.org/10.1016/S00456535(97)00332-9
141. Ward ML, Bitton G, Townsend T, Booth M (2002) Determine toxicity of leachates from Florida municipal solid waste landfills using a battery-of-tests approach. Environ Toxicol 17:258–266. https://doi.org/10.1002/tox.10053
142. Stevenson FJ (1994) Humus chemistry: genesis, composition, reactions, 2nd edn. Wiley, New York
143. Christensen JB, Jensen DL, Grøn C, Filip Z, Christensen TH (1998) Characterization of the dissolved organic carbon in landfill leachated-polluted groundwater. Water Res 32(1):125–135. https://doi.org/10.1016/S0043-1354(97)00202-9
144. Harmsen J (1983) Identification of organic compounds in leachate from a waste tip. Water Res 17(4):699–705. https://doi.org/10.1016/0043-1354(83)90239-7
145. Kang KH, Shin HS, Park H (2002) Characterization of humic substances present in landfill leachates with different landfill ages and its implications. Water Res 36(16):4023–4032. https://doi.org/10.1016/s0043-1354(02)00114-8
146. Lima LSMS, De Almeida R, Quintaes BR, Bila DM, Campos JC (2017) Evaluation of humic substances removal from leachates originating from solid waste landfills in Rio de Janeiro state, Brazil. J Environ Sci Health A 52(9):828–836. https://doi.org/10.1080/10934529.2017.1312182
147. Šir M, Podhola M, Patocka T, Honzajkova Z, Kocurek P, Kubal M, Kuras M (2012) The effect of humic acids on the reverse osmosis treatment of hazardous landfill leachate. J Hazard Mater 207–208:86–90. https://doi.org/10.1016/j.jhazmat.2011.08.079
148. Mcbride MB (1994) Environmental chemistry of soils. Oxford University Press Inc., New York
149. Steinberg CEW, Meinelt T, Timofeyev MA, Bittner M, Menzel R (2008) Humic substances. Part 2: interactions with organisms. Environ Sci Pollut Res 15(2):128–135. https://doi.org/10.1065/espr2007.07.434
150. Pflugmacher S, Tidwell LP, Steinberg CEW (2011) Dissolved humic substances can directly affect freshwater organisms. Acta Hydrochim Hydrobiol 29(1):34–40. https://doi.org/10.1002/1521-401X%28%2929%3A1%3C34%3A%3AAID-AHEH34%3E3.0.CO%3B2-8
151. Liu J, Gu Z, Wang X, Li Q (2022) The molecular differences of young and mature landfill leachates: molecular composition, chemical property, and structural characteristic. Chemosphere 287:132215. https://doi.org/10.1016/j.chemosphere.2021.132215
152. Oliveiro-Verbel J, Padilla-Bottet C, de la Rosa O (2008) Relationships between physicochemical parameters and the toxicity of leachates from a municipal solid waste landfill. Ecotoxicol Environ Saf 70:294–299. https://doi.org/10.1016/j.ecoenv.2007.05.016
153. Fatta D, Papadopoulos A, Loizidou M (1999) A study on the landfill leachate and its impacts on the groundwater quality of the greather area. Environ Geochem Health 21:175–190. https://doi.org/10.1023/A:1006613530137

154. Martins DVR (2008) Ecotoxicological evaluation of bleached eucalyptus pulp effluents during biological treatment. Master's Thesis in Civil Engineering, Universidade Federal de Viçosa, Minas Gerais, Brazil. Available on http://www.bibliotecaflorestal.ufv.br/handle/123456789/2768
155. Elphick JRF, Bergh KD, Bailey HC (2011) Chronic toxicity of chloride to freshwater species: effects of hardness and implications for water quality guidelines. Environ Toxicol Chem 30(1):239–246. https://doi.org/10.1002/etc.365
156. Castro AAS (2008) Ecotoxicological assessment of industrial effluents using Danio rerio Hamilton-Buchanan, 1982 (TELEOSTEI, CYPRINIDAE). Rio Grande do Norte. Master's Thesis in Aquatic Bioecology, Universidade Federal do Rio Grande do Norte, Brazil. Available on https://repositorio.ufrn.br/handle/123456789/12481 (In Portuguese)
157. Krull M, Barros F (2012) Key Issues in aquatic ecotoxicology in Brazil: a critical review. Ecotoxicol Environ Contam 7(2):57–66. https://doi.org/10.5132/jbse.2012.02.009
158. Costa AM (2021) Evaluation of the main pollutants present in municipal solid waste leachate using ecotoxicity tests. Doctoral Thesis, Graduate Program in Chemical and Biochemical Process Engineering—UFRJ, Rio de Janeiro, Brazil. Available on http://objdig.ufrj.br/61/teses/917161.pdf (In Portuguese)
159. APHA (2012) Standard methods for the examination of water and wastewater, 22nd edn. American Public Health, Washington DC
160. Amaral AL, Leal CS, Vaz AI, Vieira JC, Quinteiro AC, Costa ML, Castro LM (2018) Use of chemometric analyses to assess biological wastewater treatment plants by protozoa and metazoa monitoring. Environ Monit Assess 190(497):1–15. https://doi.org/10.1007/s10661-018-6882-1
161. Guvenc SY, Varank G (2021) Degradation of refractory organics in concentrated leachate by the Fenton process: central composite design for process optimization. Front Environ Sci Eng 15(1):1–16. https://doi.org/10.1007/s11783-020-1294-1
162. Zhang QQ, Tian BH, Zhang X, Ghulam A, Fang CR, He R (2013) Investigation on characteristics of leachate and concentrated leachate in three landfill leachate treatment plants. Waste Manage 33(11):2277–2286. https://doi.org/10.1016/j.wasman.2013.07.021
163. Jokela JPY, Kettunen RH, Sormunen KM, Rintala JA (2002) Biological nitrogen removal from municipal landfill leachate: low-cost nitrification in biofilters and laboratory scale in-situ denitrification. Water Res 36(16):4079–4087. https://doi.org/10.1016/S0043-1354(02)00129-X
164. Baettker EC, Kozak C, Knapik HG, Aisse MM (2020) Applicability of conventional and non-conventional parameters for municipal landfill leachate characterisation. Chemosphere 251:126414. https://doi.org/10.1016/j.chemosphere.2020.126414
165. Souto GB, Povinelli J (2007) Characteristics of landfill leachate in Brazil. In: 24th Brazilian Congress of Sanitary and Environmental Engineering, Belo Horizonte 2–7 September, Brazil (In Portuguese)
166. USEPA, United States Environmental Protection Agency (1994) Determination of trace elements in waters and wastes by inductively coupled plasma-mass spectrometry. Method 200.8. US Environmental Protection Agency, Criteria and Standards Division, Washington, DC. Available on https://www.epa.gov/sites/default/files/2015-06/documents/epa-200.8.pdf
167. INEA (1986) Rio de Janeiro State Environmental Institute NT-202.R-10-Criteria and standards for discharging liquid effluents. Available on http://www.inea.rj.gov.br/wp-content/uploads/2018/12/NT-202-R-10.pdf (In Portuguese)
168. Chian ESK, Dewalle FB (1977) Characterization of soluble organic matter in leachate. Environ Sci Technol 11(2):158–163. https://doi.org/10.1021/es60125a003
169. Trebouet D, Schlumpf JP, Jaouen P, Quemeneur F (2001) Stabilised landfill leachate treatment by combined physicochemical nanofiltration processes. Water Res 35(12):2935–2942. https://doi.org/10.1016/S0043-1354(01)00005-7
170. Zolfaghari M, Dia O, Klai N, Drogui P, Brar SK, Buelna G, Dubé R (2018) Removal of pollutants in different landfill leachate treatment processes on the basis of organic matter fractionation. J Environ Qual 47(2):297–305. https://doi.org/10.2134/jeq2017.09.0360

171. ABNT (2021a) Brazilian Association of Technical Standards. NBR 15411–3:2012 aquatic ecotoxicology. Determination of the inhibitory effect of aqueous samples on the bioluminescence emission of *Vibrio fischeri* (luminescent bacteria assay) part 3: Method using lyophilised bacteria (In Portuguese)
172. ABNT (2021b) Brazilian Association of Technical Standards. NBR 16530. Aquatic ecotoxicology—acute toxicity—test method with *Artemia* sp. (Crustacea, Brachiopoda). Rio de Janeiro (In Portuguese)
173. ABNT (2022b) Brazilian Association of Technical Standards. NBR 15088: aquatic ecotoxicology—acute toxicity. Test method with fish (Cyprinidae), Rio de Janeiro (In Portuguese)
174. USEPA, United States Environmental Protection Agency (1985) Methods for measuring the acute toxicity of effluents to freshwater and marine organisms, 3rd edn. EPA/600/4-85/013. Available on https://nepis.epa.gov/Exe/ZyPDF.cgi/20012VMD.PDF?Dockey=20012VMD.PDF
175. Liu Z, Wu W, Guo SP, Cheng J (2015) Characterization of dissolved organic matter in landfill leachate during the combined treatment process of air stripping, Fenton, SBR and coagulation. Waste Manage 41:111–118. https://doi.org/10.1016/j.wasman.2015.03.044
176. Collado S, Nunez D, Oulego P, Rieira FA, Diaz M (2020) Effect of landfill leachate ageing on ultrafiltration performance and membrane fouling behaviour. J Water Process Eng 36:101291. https://doi.org/10.1016/j.jwpe.2020.101291
177. de Almeida R, Souza Couto JM, Gouvea RM, Almeida OF, Bila DM, Quintaes BR, Campos JC (2020) Nanofiltration applied to the landfill leachate treatment and preliminary cost estimation. Waste Manag Res 38(10):1119–1128. https://doi.org/10.1177/0734242X20933333
178. Li G, Wang W, Du Q (2010) Applicability of nanofiltration for the advanced treatment of landfill leachate. J Appl Polym Sci 116(4):2343–2347. https://doi.org/10.1002/app.31769
179. Mohammad AW, Hilal N, Pei LY (2004) Treatment of landfill leachate wastewater by nanofiltration membrane. Int J Green Energy 1(2):251–263. https://doi.org/10.1081/GE-120038756
180. Shu Z, Lu Y, Huang J, Zhang W (2016) Treatment of compost leachate by the combination of coagulation and membrane process. Chin J Chem Eng 24(10):1369–1374. https://doi.org/10.1016/j.cjche.2016.05.022
181. Wang F, Smith DW, El-Din MG (2006) Aged raw landfill leachate: membrane fractionation, H_2O_2/UV treatment and molecular size distribution analysis. J Adv Oxid Technol 9(1):86–96. https://doi.org/10.1515/jaots-2006-0112
182. Jensen DL, Ledin A, Christensen TH (1999) Speciation of heavy metals in landfill-leachate polluted groundwater. Water Res 33(11):2642–2650. https://doi.org/10.1016/S0043-1354(98)00486-2
183. Jemec A, Tišler T, Žgajnar-Gotvajn A (2012) Assessment of landfill leachate toxicity reduction after biological treatment. Arch Environ Contamin Toxicol 62:210–221. https://doi.org/10.1007/s00244-011-9703-x
184. Restrepo BJ, Flohr L, Melegari SP, Da Costa CH, Fuzinatto CF, Castilhos AB Jr, Matias WG (2017) Correlation between acute toxicity for *Daphnia magna*, *Aliivibrio fischeri* and physicochemical variables of the leachate produced in landfill simulator reactors. Environ Technol 38(22):2898–2906. https://doi.org/10.1080/09593330.2017.1281352
185. Kalčíková G, Zupančič M, Levei EA, Miclean M, Englande AJ, Gotvajn AZ (2015) Application of multiple toxicity test in monitoring of landfill leachate treatment efficiency. Environ Monit Assess 187(489):1–14. https://doi.org/10.1007/s10661-015-4670-8
186. Thode Filho S, Jorge ENDLF, de Oliveira AS, da Costa Marques MR, Franco HA (2017) Study on the use of Artemia salina as bioindicator in the ecotoxicological evaluation of landfill leachate. Revista Eletrônica em Gestão Educação e Tecnologia Ambiental 21:24–31. Available on https://periodicos.ufsm.br/reget/article/view/29718/pdf
187. Byrne M, Oakes DJ, Pollak JK, Laginestra E (2008) Toxicity of landfill leachate to sea urchin development with a focus on ammonia. Cell Biol Toxicol 24(6):503–512. https://doi.org/10.1007/s10565-008-9099-1

188. Luo H, Zeng Y, Cheng Y, He D, Pan X (2020) Recent advances in municipal landfill leachate: a review focusing on its characteristics, treatment, and toxicity assessment. Sci Total Environ 703:135468. https://doi.org/10.1016/j.scitotenv.2019.135468
189. Linbo TL, Baldwin DH, McIntyre JK, Scholz NL (2009) Effects of water hardness, alkalinity, and dissolved organic carbon on the toxicity of copper to the lateral line of developing fish. Environ Toxicol Chem Int J 28(7):1455–1461. https://doi.org/10.1897/08-283.1
190. He R, Tian B-H, Zhang Q-Q, Zhang H-T (2015) Effect of Fenton oxidation on biodegradability, biotoxicity and dissolved organic matter distribution of concentrated landfill leachate derived from a membrane process. Waste Manage 38:232–239. https://doi.org/10.1016/j.wasman.2015.01.006
191. Glover CN, Wood CM (2005) Physiological characterisation of a pH-and calcium-dependent sodium uptake mechanism in the freshwater crustacean, *Daphnia magna*. J Exp Biol 208(5):951–959. https://doi.org/10.1242/jeb.01426
192. Tsiridis V, Petala M, Samaras P, Hadjispyrou S, Sakellaropoulos G, Kungolos A (2006) Interactive toxic effects of heavy metals and humic acids on Vibrio fischeri. Ecotoxicol Environ Saf 63(1):158–167. https://doi.org/10.1016/j.ecoenv.2005.04.005
193. Okamoto A, Yamamuro M, Tatarazako N (2015) Acute toxicity of 50 metals to Daphnia magna. J Appl Toxicol 35(7):824–830. https://doi.org/10.1002/jat.3078
194. Maier KJ, Knight AW (1991) The toxicity of waterborne boron to *Daphnia magna* and *Chironomus decorus* and the effects of water hardness and sulfate on boron toxicity. Arch Environ Contam Toxicol 20(2):282–287. https://doi.org/10.1007/bf01055917
195. Kokkali V, Katramados I, Newman JD (2011) Monitoring the effect of metal ions on the mobility of Artemia salina nauplii. Biosensors 1(2):36–45. https://doi.org/10.3390/bios1020036
196. Gajbhiye SN, Hirota R (1990). Toxicity of heavy metals to brine shrimp Artemia. J Indian Fish Assoc 20:43–50. http://hdl.handle.net/1834/31739
197. Fulladosa E, Murat JC, Martínez M, Villaescusa I (2005) Patterns of metals and arsenic poisoning in Vibrio fischeri bacteria. Chemosphere 60(1):43–48. https://doi.org/10.1016/j.chemosphere.2004.12.026
198. Wang W, Lampi MA, Huang XD, Gerhardt K, Dixon DG, Greeberg BM (2009) Assessment of mixture toxicity of copper, cadmium, and phenanthrenequinone to the marine bacterium *Vibrio fischeri*. Environ Toxicol Int J 24(2):166–177. https://doi.org/10.1002/tox.20411
199. Alsop D, Wood CM (2011) Metal uptake and acute toxicity in zebrafish: common mechanisms across multiple metals. Aquat Toxicol 105(3–4):385–393. https://doi.org/10.1016/j.aquatox.2011.07.010
200. Bertoletti E (2009) Sensitivity of some freshwater fish species used in Brazil. J Braz Soc Ecotoxicol 4(1–3):9–13 (In Portuguese). https://doi.org/10.5132/jbse.2009.01.002
201. Madoni P, Davoli D, Gorbi G, Vescovi L (1996) Toxic effect of heavy metals on the activated sludge protozoan community. Water Res 30(1):135–141. https://doi.org/10.1016/0043-1354(95)00124-4
202. Rahmani A, Asadi A, Fatehizadeh A, Rahmani AR, Zare MR (2019) Interactions of Cd, Cr, Pb, Ni, and Hg in their effects on activated sludge bacteria by using two analytical methods. Environ Monit Assess 191(3):124. https://doi.org/10.1007/s10661-019-7241-6
203. Yu Y-H, Wu L-G (1978) Toxicity of some non-metal ions to activated sludge. J Chin Inst Eng 1(1):97–102. https://doi.org/10.1080/02533839.1978.9676604
204. Gripa E, Daflon SDA, de Almeida R, Fonseca FV, Campos JC (2023) Landfill leachate treatment by high-pressure membranes and advanced oxidation techniques with a focus on ecotoxicity and by-products management: a review. Process Saf Environ Prot 173:747–764. https://doi.org/10.1016/j.psep.2023.03.074
205. Elnabarawy MT, Robideau RR, Beach SA (1988) Comparison of three rapid toxicity test procedures: microtox, polytox and activated sludge respiration inhibition. Toxicity Assess 3:361–370. https://doi.org/10.1002/tox.2540030402
206. Renou S, Givaudan JG, Poulain S, Dirassouyan F, Moulin P (2008) Landfill leachate treatment: review and opportunity. J Hazard Mater 150(3):468–493. https://doi.org/10.1016/j.jhazmat.2007.09.077

207. de Almeida R, Porto RF, Quintaes BR, Bila DM, Lavagnolo MC, Campos JC (2023) A review on membrane concentrate management from landfill leachate treatment plants: the relevance of resource recovery to close the leachate treatment loop. Waste Manage Res J Int Solid Wastes Public Cleans Assoc 41(2):264–284. https://doi.org/10.1177/0734242X221116212
208. Ye W, Liu H, Jiang M et al (2019) Sustainable management of landfill leachate concentrate through recovering humic substance as liquid fertilizer by loose nanofiltration. Water Res 157:555–563. https://doi.org/10.1016/j.watres.2019.02.060
209. Singh SK, Sharma C, Maiti A (2021) A comprehensive review of standalone and hybrid forward osmosis for water treatment: Membranes and recovery strategies of draw solutions. J Environ Chem Eng 9:105473. https://doi.org/10.1016/j.jece.2021.105473

Chapter 9
Characteristics and Pollution Potential of Leachate from Municipal Solid Waste Landfills

Churchill Ebinimitei Simon

Abstract Landfill leachate occurs as one of the by-products of waste deterioration. It is the mineralization of pollutants within the waste stream which gives rise to a very concentrated, dark and toxic chemical mix, depending on the type of waste disposed, age of dumpsite and climatic conditions of the site. Leachate is generally collected at the bottom of the landfills, in unlined landfills, leachate usually migrates through the underground strata and may travel down to the groundwater table. Some lateral migration of the leachate is also possible depending on the surface and subsurface geology. The migrating leachate can dissolve soluble chemicals, many with toxic constituents that may result in soil and water resources pollution. Hence proper leachate management procedures should be developed and incorporated into the design of landfills.

9.1 Introduction

Uncontrolled waste dumps in municipalities of developing nations/cities have in recent times given rise to various forms of nuisance such as rodents' proliferation, air pollution (odor), soil pollution (percolation/infiltration) as well as water pollution (overland flow) [1]. The open dumping of municipal solid waste in landfills is one of the oldest and most common disposal methods adopted in most of the countries, particularly developing ones [2] Landfill leachate is generated when rainwater percolates through the waste layers in a landfill, in which process organic and inorganic constituents of the waste get dissolved, transported and deposited at the bottom of the landfill by gravity [3] Some of the components of landfill leachate may be categorized as a water-based solution of four groups of contaminants dissolved organic matter, inorganic macro-components, heavy metals, and xenobiotic organic compounds [4]. The most important potential environmental concern associated with landfill is the formation of leachate and the subsequent contamination of soil and water resources

C. E. Simon (✉)
Department of Civil Engineering, Federal University Otuoke, Yenagoa, Bayelsa State, Nigeria
e-mail: simonec@fuotuoke.edu.ng

A. Anouzla and S. Souabi (eds.), *A Review of Landfill Leachate*, Springer Water,
https://doi.org/10.1007/978-3-031-55513-8_9

[5, 6]. A landfill leachate may be regarded therefore as an enormously contaminated solution comprising of both organic and inorganic components emanating from a direct deposition of solid waste in a landfill. Landfill sites therefore poses a direct threat to the soil, ground and surface water resources as a result of the leaching effects of contaminants from municipal solid wastes deposited on it. This effect is determined by the landfill's age, climatic conditions, and the sort of waste generated as a result of the residents' lifestyle [1]. Jagloo [7], said that there are three essential attributes that distinguish any source of water resources contamination, the degree of localization, the loading history, and the kind of contaminants emanating from them. According to Al Sabahi and Abdul Rahim [5], the primary problem associated with landfills is the formation of leachate and eventual contamination of water sources and soil because of contaminant migration.

A landfill is a point source of environmental pollution, it can also be a non-point source of surface water pollution. The loading history describes how the concentration of a contaminant or its rate of production varies as a function of time at the source. Leachate rates at a landfill site are controlled by seasonal factors or by a decline in source strength as components of the waste such as organics, biodegrade. The sorts of waste put in the landfill, the amount of precipitation in the area, and other site-specific circumstances all influence leachate composition [7].

Through interactions within the hydrologic cycle, a leachate contaminated environment can have an impact on soil, ground and surface water quality. In a hydrologic continuum, groundwater and surface water are connected by their transitional zone, and contamination of one frequently affects the other [8].

9.2 Landfill By-Products

Initially, the biochemical reactions in a landfill take place under aerobic condition (where oxygen is the terminal electron acceptor), producing carbon dioxide (CO_2) as the principal gas. As most of the available oxygen (O_2) is depleted, decomposition reactions continue under partial aerobic to mostly anaerobic conditions, where the principal landfill gases generated are CO_2, methane (CH_4), trace amounts of ammonia (NH_3), and hydrogen sulfide (H_2S) [9].

The two main by-products generated during the life of a landfill are leachate and landfill gases. The composition and amount of leachate and gas generated varies spacio-temporily depending on the nature of waste and design of the landfill. This utilization of leachate and landfill gases are dependent on the associated management system that is needed for proper functioning of a sanitary landfill.

9.3 Formation of Leachate Plume

Infiltration from precipitation is caused by gravity to move leachates through the landfill, to the bottom and sides, and through the underlying soil until it reaches the groundwater zone or aquifer. Some of the key processes of leachate formation are presented below, (see Fig. 9.1).

Immediately after solid waste materials are deposited in disposal sites, the process of stabilization begins. This process, which forms the leachate within the site, occurs mainly through physical, chemical and biological processes. These processes brings about four key stages which include hydrolysis of solid waste, biological degradation of organic waste, solubilization of soluble salts contained within the waste mass and the transportation of waste as colloids or particulate matter [11].

As leachate is transported to the sides and base of the landfill through gravity, a percentage of it spreads across the surface of nearby soils which in excess, negatively affects available farmlands. Some part of the remaining leachate migrate to receiving surface water sources through runoffs from landfills that causes overland flows that finally ends up polluting sources of surface water. The rest leachate gets through the subsurface by infiltration and percolation, then mixes with groundwater held in soil voids forming a plume of contaminated groundwater that moves along the groundwater's flow pattern.

The leachate toxins initially reach the unsaturated zone before being transferred to the saturated zone's groundwater table, resulting in groundwater pollution that can later affect other water sources due to the geologic formation.

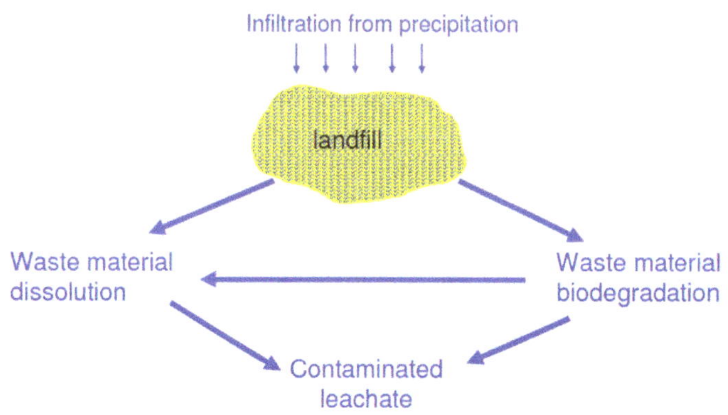

Fig. 9.1 Process of leachate formation [10]

9.4 Leachate Composition and Effects on the Environment

As a heavily contaminated “black chemical soup”, landfill leachate comprises both organic and inorganic components that originate directly from deposited solid waste materials. It is known generally to contain significantly more contaminant loads than raw sewage or many industrial wastes [12]. Below is a pictorial view of a typical landfill leachate collected in depression at the base of Yenagoa central waste dump site, Bayelsa State, Nigeria (Fig. 9.2).

Fig. 9.2 A typical leachate collected in a depression at a landfill in Bayelsa State, Nigeria

The relative quality of leachate varies widely depending on a series of complex but interrelated factors. Table 9.1 presents some of the organic and inorganic leachate constituents.

Leachate contains a host of toxic and carcinogenic chemicals, which may cause harm to both humans and environment as shown in Table 9.2. Furthermore, contaminated water from leachate can have a negative impact on domestic, industrial and agricultural activities that rely on it. Irrigating with contaminated water reduces soil productivity, brings about contaminated crops, and potentially moves hazardous pollutants up the food chain as animals and humans consume crops grown in contaminated water [7]. It can take years before groundwater pollution reveals itself and chemicals in the leachates often react synergistically and often in unexpected ways to affect the ecosystem but as for soil and subsurface water are concerned, leachate effects are immediate.

Table 9.1 Leachate parameters [12]

Constituent group	Parameters
Physical	pH, Redox Potential, Electrical Conductivity, Colour, Turbidity, Suspended Solids, Total Dissolved Solids, Volatile Suspended Solids, Volatile Dissolved Solids, Temperature, Odour
Organic	Chemical Oxygen Demand, Total Organic Carbon, Phenols, Volatile Acids, Tannins, Lignins, Organic Nitrogen, Ether Soluble, Methylene Blue Active Substances, Chlorinated Hydrocarbons, Other Organic Chemicals
Inorganic	Chloride, Sulphate, Phosphate, Nitrate-N, Nitrite, Ammonia-N, Sodium, Potassium, Calcium, Magnesium, Hardness, Heavy Metals Pb, Cu, Ni, Cr, Zn, Cd, Fe, Mn, Hg, Ba, Ag), Arsenic, Cyanide, Fluoride, Selenium
Biological	Biological Oxygen Demand, Coliform Bacteria (Total and Thermotolerant or Faecal), Faecal Streptococci, Standard Plate Count

Table 9.2 Health effects of landfill leachate (EPA 2003)

Contaminant	Potential health effects from exposure above the MCL
Arsenic	Skin damage; circulatory system problems; increased
Barium	Risk of cancer
Fluoride	Bone disease (fluorosis)
Mercury	Kidney damage
Nitrate	Methemoglobinemia (blue-baby syndrome)

* *Maximum Contaminant Level (MCL)*

9.5 Effects of Leachate on Water Quality (Surface Water and Ground Water)

According to Omole [13], the number one source of contamination of water resources, particularly ground water comes from leachates generated from wastes of industrial origin deposited in a landfill. Leachates from non-hazardous waste in a landfill can also contain organic compounds, chlorinated hydrocarbons and metals of various concentrations which could pose a danger to both surface and ground water resources.

The quantity of leachate produced in a landfill and its subsequent impact on the environment is essentially dependent on the metrological and hydrological factors governing the area in addition to the waste disposal methods. According to Chapman [14], the volume of leachate generated is consequently expected to be excessively high in humid regions with high rainfall or high runoff and shallow water table Previous researches have noted that most landfills have high levels of BOD, COD, Ammonia, chloride, sodium, Potassium, Hardness and Boron with respect to the time and age of the landfill because its conditions often vary from aerobic to anaerobic thus allowing different chemical processes to take place [15].

9.6 Effects of Leachate on Soil and Farmlands

Recently, leachate contamination and subsequent pollution of soils and farmlands has been on the table of global environmental discuss. Figure 9.3, shows the adverse effects of landfill leachate on nearby soil which affected an Okra farm beside an uncontrolled municipal waste dump site. The picture shows the early stages of leachate effects of crops wilting on the farmland occasioned by the spread of leachate across the parcel of land. Therefore, the part encircled in red was the most affected at the time of this evidence.

A closer view of the effects of landfill leachate pollution on a farmland is shown in Fig. 9.4. While Fig. 9.5 shows a researcher/farmer in a landfill leachate polluted Okra farm. The figure shows the advanced stages of leachate pollution and its wilting effects on the crops.

9.7 Health Effects of Environmental (Water Resource) Contamination

Water has remained the most abundant and most important resource of man, every life depends on it for various reasons. It has prided itself over the years as the source of life on planet earth without which nothing survives. Plants need it for their uptake of minerals and nutrients, animals need it for drinking and bathing while humans need

Fig. 9.3 A picture showing the early stages of leachate effect of an okra farm close to a dumpsite

Fig. 9.4 A closer look at the early stages of leachate effect of an okra farm close to a dumpsite

it for drinking, bathing, agriculture, industrial purposes, transportation, recreation, dust suppression etc.

Water covers about 71% of the earth surface yet many people throughout the earth have faced grave challenges of the lack of portable drinking water in which the Niger Delta is not an exemption.

Fig. 9.5 Advanced stages of landfill leachate pollution of an okra farm close to a dumpsite

Water needed by humans for consumption is by all means expected to meet the needs of potability, which means water must be odorless, colorless and tasteless. From a biological standpoint, water has many distinct properties that are critical for the proliferation of life. It helps to transport oxygen throughout the body, lubricate joints and helps to increase metabolic activities etc. The human body contains from 50 to 75% of water depending on the body size [16]. This means that an intake of polluted water will directly affect the human body system.

When foreign elements enter the water column, or when the minerals found in water exceed the acceptable amount, the water is said to be contaminated; water pollution describes the presence of materials in water that interfere unreasonably with one or more beneficial uses of water [17]. When this happens, some health concerns are raised, health hazards arising from waste management and disposal associated activities and their ability to pollute surface and ground water consequently need to be analyzed specifically for the conditions in a given setting.

Pollution takes place when pollutants are directly or indirectly introduced into the water bodies without adequate treatment to remove the harmful compounds. This has remained a source of concern to environmental stakeholders. According to Anne [16], the polluted ground water has an effect on the food chain, health and human environment. She stressed that both women and children are victims of pollution and that over 40,000 children die from disorders and other epidemic diseases everyday due to poor waste disposal. She pointed out that waste is an environmental disaster that also causes aesthetic decay. There are different sources of water pollution ranging from agricultural practices, releases from industries and waste management to say a few.

9.8 Impact of Landfill Leachate and Public Outcry

Migration of hazardous materials into nearby water resources is often a vital environmental concern in subject of landfill sites, which may represent a public health problem, especially when a site is located near aquifers supplying public drinking water. Landfill sites may be a source of airborne chemical infection via the off-site migration of gases and through particles and chemicals adsorbed in dust, especially during the period of active operation of the site [18]. Therefore, people residing within the vicinity of these sites are exposed to cases of landfill leachate interference with their everyday lives. In the early 1980s, pollutants from leaking underground storage tanks (USTs) were found to be affecting some drinking-water wells in the Santa Clara Valley area. Two of these wells were studied and discovered that Chlorinated solvents had leaked from an underground waste storage tank impacting negatively on the residents living near one of the contaminated wells who reported a cluster of adverse pregnancy outcomes, mainly spontaneous abortions and congenital heart defects [19]. A first investigation confirmed a significant excess of cardiac anomalies in the service area of the water company that operated the contaminated well compared to those among residents of an unexposed area. The excess was found within the potentially exposed time period and not in an unexposed time period after the well was closed.

More so, leakages from an industrial dump of chemical waste drums in New Jersey caused contamination of groundwater and well water with organic chemicals (including benzene, toluene, trichloroethylene, and lead) [20]. It was found that there were higher prevalence of respiratory diseases and seizures among the people living in a high-exposure area estimated on the basis of groundwater flow patterns. The people dwelling in these high-exposure areas used private drinking-water wells, eat homegrown food, and smoked more often than populations living in unexposed areas, and whilst these factors were adjusted, variations in health outcomes disappeared.

The Love Canal, New York State is another important issue to take into account when planning a landfill. During the 1930s and 1940s, large quantities of toxic materials (residues from pesticide production) were dumped at the landfill which was followed by the building of houses and a school on and around the landfill in the 1950s. By 1977 the site was leaking and chemicals were observed in neighborhood creeks, sewers, soil, and indoor air of houses. This caused one of the most well-known and publicized incidents of environmental pollution from landfill. Exposure of Love Canal citizens, although not well understood, may have occurred via inhalation of volatile chemicals in indoor air or via direct contact with soil on playgrounds or surface water [21]. The drinking water supply was not contaminated. Chemicals detected at Love Canal were primarily organic solvents, chlorinated hydrocarbons and acids, consisting of benzene, vinyl chloride, PCBs, dioxin, toluene, trichloroethylene, and tetrachloroethylene. Several researches had been carried out to detect whether Love Canal residents suffered adverse health effects. A cross-sectional study observed an increased incidences of seizures, learning problems, hyperactivity, eye irritation, skin rashes, abdominal pain, and incontinence in children living close to the Love Canal site compared to controls from other areas, as reported by the parents of the children, as noted in previous reviews [22].

Bringing it closer home, the Koko incident of June 1988 is yet another sad tale when issues of waste disposal, particularly hazardous waste are considered. Little wonder [23], stated that Nigeria responds to most environmental problems on an ad-hoc basis. According to him, the discovery of toxic waste dumps in Koko, a remote part of southern Nigeria in June 1988 and its attendant media and public outcry prompted the government to react swiftly. This informed the formation of the Federal Environmental Protection Agency 1988 (FEPA). It latter translated to the further formation of Harmful Waste (Special Criminal Provision) Act, 1988 to deal specifically with illegal dumping of harmful waste.

9.9 Different Methods of Landfill Leachate Treatment

Since leachate is made up of a complex mixture of organic and inorganic constituents, treatment methods will be according to the composition of landfill leachate. The removal of organic matter is the general premise of leachate treatment, including chemical oxygen demand (COD) removal, biological oxygen demand (BOD) removal and ammonia nitrogen removal [24]. Before the leachate is discharged into the natural water body, toxicity analysis should also be carried out by testing various organisms. At present, there are three kinds of convenient methods for leachate treatment: leachate transfer method, biodegradation method and physical and chemical method [25].

9.10 Case Study: Typical Landfill in Bayelsa State, Nigeria

9.10.1 Description of the Study Area

The study area was situated within the lower floodplain of the Niger Delta. The terrain is poorly drained with a gentle syncline to the Gulf of Guinea in a south-western direction [26]. It is characterized by sedimentary formations with a thickness of about 8000 m. It includes Akata Formation, Agbada Formation, Benin Formation,

Fig. 9.6 Map of Nigeria showing Bayelsa State

from bottom to top, which is Oligocene to Pleistocene in age. It consists predominantly of freshwater continental friable sands and gravel that are excellent aquifer properties, with occasional intercalation of shales [27]. The Niger delta has two basic hydrological regimes which are Coastal and Inland [2] (Fig. 9.6).

The landfill is specifically located off Edepie-Amassoma road, Etelebu in Yenagoa Local Government Area of Bayelsa State, and operated by Bayelsa State Environmental Sanitation Authority. As at the time of this study, the site which began operations from around 2008 had an average life span of about 11 years. The landfill also had a height of about 2 m and covers a total area of about thirty-six kilometers square (36 km^2). The hydrogeological condition of the landfill site was consistent with the regional hydrogeological setting of Port-Harcourt area as depicted by [28]. A pictorial view of the landfill is presented in Fig. 9.7.

The waste dump was unlined and serviced wastes generated from the Yenagoa municipality which include industrial, agricultural, institutional, commercial and domestic wastes, as well as sewage.

Fig. 9.7 An image showing a pictorial view of the waste dump

9.10.2 Study Objectives

The aim of the study was to access the impact of a waste dump site on the quality of surface water in a nearby stream within the study area. The specific objectives were to;

- Evaluate the characteristics of surface water from a nearby stream consequent upon waste disposal.
- Determine the variation in the physico-chemical and biological properties of the surface water sampled at Downstream point (SW 1), Central point (SW 2) and Upstream point (SW 3) of the stream.
- Compare the levels of contaminants from collected samples with standard acceptable limits, ie Nigerian Standard of Drinking Water Quality (NSDWQ) and World Health Organization (WHO).
- Ascertain the possible source of surface water pollution using correlational analysis.
- Discuss the potential adverse effects of ingesting or coming into daily contact with water from the stream.

Having regular contact with surface water from sources near municipal waste dump sites and possible ingestion of same, could result in endangering the individual as it has the potential to contain contaminants washed into it from the dumpsite.

9.11 Data and Methodology

9.11.1 Sample Collection

The study area was described using Geographic Information System (GIS). Every sampling point was picked with the help of a standard Global Positioning System (GPS). To determine the degree of surface water pollution, three surface water samples were taken around the dumpsite and labeled Downstream Point (SW 1), Central Point (SW 2) and Upstream Point (SW 3) respectively. Two samples were taken at the landfill's Northernmost and Southernmost ends, while the third sample was taken at the center of the nearby stream, which provided inhabitants with protein (fish) and other forms of livelihood (Table 9.3).

At each sampling location, surface water samples were taken with the help of clean 1.5L plastic bottles after initially rinsing the bottles with same water to be taken. Non-conservable parameters such as pH, temperature and electrical conductivity were determined in-situ. The pH of water samples was measured with a pH meter previously calibrated with buffer solutions. Conductivity was measured with a conductivity meter calibrated with potassium chloride solution. Temperature was measured with a thermometer [29].

After collection, the samples were immediately placed in iced coolers for transportation to the laboratory and stored in refrigerator. The water quality parameters dealt with were physical, chemical and biological characteristics [30] and were analyzed in accordance with standard laboratory methods [29]. The parameters were; pH, Electrical Conductivity (EC), Total Dissolved Solids (TDS), Chemical Oxygen Demand (COD), Biological Oxygen Demand (BOD), Total Hardness (TH), Ammonium ($NH4^+$), Sulphate ($SO_4{}^{2-}$), Nitrate ($NO_3{}^-$), Phosphate and Heavy Metals such as Cadmium (Cd), Chromium (Cr), Copper (Cu^{2+}), Lead (Pb), Zinc (Zn), Calcium (Ca^{2+}), Magnesium (Mg^{2+}), Sodium (Na^+), Iron (Fe^{2+}) and Potassium (K^+) ions, as well as Total coliform count.

In this study, a total of three surface water samples were taken for laboratory analysis. The three samples were analyzed for, Physico-chemical and microbiological characteristics. The variations of parameters from sampling points SW 1, SW 2 and SW 3 as compared with Nigerian Standard of Drinking Water Quality (NSDWQ) and World Health Organization (WHO) [31, 32] are shown in Table 9.2. The table also presents some descriptive statistics of the surface water parameters analyzed.

Table 9.3 Details of sampling sites

S/No.	Sampling locations	Coordinates		Description
		Latitude (N)	Longitude (E)	
1	Surface water (SW1)	4° 59′ 58.78788″	6° 20′ 11.51412″	Downstream point
2	Surface water (SW2)	4° 59′ 40.14″	6° 19′ 57.91″	Central point
3	Surface water (SW3)	4° 59′ 27.68064″	6° 19′ 43.92804″	Upstream point

9.12 Result and Discussions

The average concentration of parameters and their standard deviations as presented in Table 9.4 of the three samples analyzed showed that EC, TDS, TA, TH and Phosphate had higher average concentrations as compared to the NSDWQ and WHO standards. The average concentrations were 1533.33μS/cm, 766.67 mg/l, 339.53 mg/l, 260 mg/l and 2.95 mg/l respectively. Standard Deviation of these parameters were 824.58, 412.29, 175.97, 10 and 3.94. Similarly, parameters such as DOD, COD, Fe, K and Total plate count also indicated higher average concentrations as compared to the standard acceptable values. The average concentrations were 15.33 mg/l, 18 mg/l, 4.73 mg/l, 36.49 mg/l and 2,673,333.33 mg/l. While the Standard Deviations were 1.15, 6.56, 4.32, 30.83 and 202,319.88 respectively.

Figures 9.8 and 9.9 showed the percentages of the Physico-chemical and microbial characteristics of the surface water samples analyzed during the sampling period.

All the sampled parameters showed variations across the sampling points. It was observed from the chart above that the values for pH of the surface water samples ranged from 6.96 to 7.18, but were within the WHO recommended range. The EC which ranged from 852 to 2450 μS/cm and Total Dissolved Solid (TDS) with range from 426 to 125 mg/l showed similar trends. They both had higher percentages of values that were greater than the regulatory standards at the Downstream (SW 1) and the Central point (SW 2). Concentration of Sulphate (SO_4^{2-}), Nitrate (NO_3), Phosphate (PO_4^{3-}) and Ammonium (NH_4) also showed similar patterns. They all had the same concentrations at the Upstream point (SW 3) and Central point (SW 2), however, the values at the Downstream point (SW 1) were higher. When compared with NSDWQ and WHO standards, they fell within permissible limits except phosphate. Total Alkalinity, BOD, COD and Total Hardness values ranged between 193.5–534.9 mg/l, 14–16 mg/l, 12–25 mg/l and 250–270 mg/l respectively. When compared with the regulatory standards, it was observed that they were all above WHO standards except for the TA value at the Upstream point (SW 3). The higher BOD_5 and COD values indicate the presence of organic matter in water [4].

Summarily, the surface water analysis across the Downstream point (SW 1) showed that parameters such as pH, EC, TDS, TA, TH, Na, K, Fe, BOD, COD, Phosphate and Total Coliform Count, recorded 7.18, 2450 μS/cm, 1225 mg/l, 534.9 mg/l, 270 mg/l, 122.92 mg/l, 69.42 mg/l, 9.48 mg/l, 25 mg/l, 16 mg/l, 7.5 mg/l and 2.80 X 10^6 cfu/ml respectively. All these values were above NSDWQ and WHO recommended values for drinking water except pH. All the heavy metals in the surface water analyzed (Pb, Cu, Zn, Cd, and Cr) in this study were below instrument detectable limits (BDL) [4]. However, the microbial load within the surface water sampled at all points indicated a huge population much more than the acceptable NSDWQ and WHO standards for potability. The high concentration of parameters in the three surface water samples showed that the surface water was contaminated and poses a threat to those who access it for their needs.

Pearson's correlation was also used to obtain the common sources of contaminants within the samples of surface water analyzed. This was done to identify if the waste

Table 9.4 Summary of special variation and descriptive statistics of surface water parameters

Sample ID (Borehole Sampling points)	UNITS	NSDWQ (2015)	WHO GUIDELINES (2011)	SW 1	SW 2	SW 3	MIN	MAX	AVERAGE	Standard Deviation
pH		6.5–8.5	6.5–8.5	7.18	6.91	6.96	6.91	7.18	7.02	0.14
Temp	°C		25	22.3	21.3	21.6	21.3	22.3	21.73	0.51
EC	µS/cm	1000	1000	2450	1298	852	852	2450	1533.33	824.58
TDS	mg/l	NS	500	1225	649	426	426	1225	766.67	412.29
$SO^{2-}{}_4$	mg/l	100	500	20	24	24	20	24	22.67	2.31
NO_3	mg/l	50	50	0.4	0.1	0.1	0.1	0.4	0.2	0.17
$PO^{3-}{}_4$	mg/l	NS	0.1	7.5	0.67	0.67	0.67	7.5	2.95	3.94
NH_4	mg/l		NA	0.09	0.022	0.022	0.022	0.09	0.05	0.04
TA	mg/l	NS	200	534.9	290.2	193.5	193.5	534.9	339.53	175.97
BOD	mg/l		5	16	16	14	14	16	15.33	1.15
COD	mg/l		10	25	17	12	12	25	18	6.56
TH	mg/l	250	200	270	250	260	250	270	260	10
Pb	mg/l	0.01	0.01	< 0.001	< 0.001	< 0.001	0	0	#DIV/0!	NA
Cu	mg/l	1	2	< 0.001	< 0.001	< 0.001	0	0	#DIV/0!	NA
Zn	mg/l	3	3	< 0.001	< 0.001	< 0.001	0	0	#DIV/0!	NA
Fe	mg/l	0.3	0.3	9.48	3.67	1.03	1.03	9.48	4.73	4.33
Ca	mg/l	NS	75	14.06	8.15	11.8	8.15	14.06	11.34	2.98
Mg	mg/l	20	20	19.67	9.77	6.59	6.59	19.67	12.01	6.82
K	mg/l		20	69.42	31.72	8.32	8.32	69.42	36.49	30.83

(continued)

Table 9.4 (continued)

Sample ID (Borehole Sampling points)	UNITS	NSDWQ (2015)	WHO GUIDELINES (2011)	SW 1	SW 2	SW 3	MIN	MAX	AVERAGE	Standard Deviation
Na	mg/l	200	200	122.92	59.93	23.47	23.47	122.92	68.77	50.31
Cd	mg/l	0.003	0.003	< 0.001	< 0.001	< 0.001	0	0	#DIV/0!	NA
Cr	mg/l	0.05	0.05	< 0.001	< 0.001	< 0.001	0	0	#DIV/0!	NA
Total Plate Count	cfu/ml	10	0	2,800,000	2,440,000	2,780,000	2,440,000	2,800,000	2,673,333.33	202,319.88

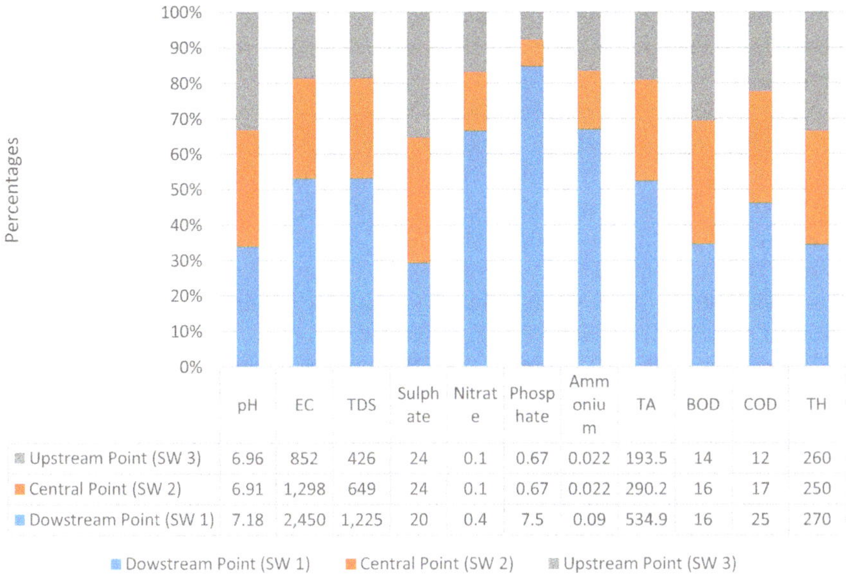

Fig. 9.8 Percentage composition of surface water parameters within the sampling points

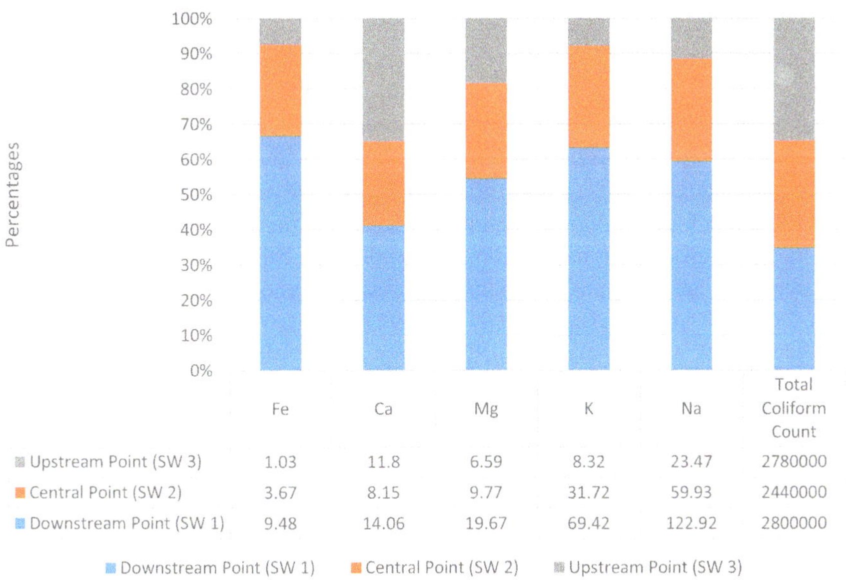

Fig. 9.9 Percentage composition of some metals and microbial load of surface water within the sampling points

dump was indeed responsible for the presence of these contaminants within the stream. The correlation matrices for 18 measured variables during sampling analysis are illustrated in Table 9.5.

A perfect positive correlation between TDS and EC (r = 1, $p \leq 0.01$), NH_4 and NO_3 (r = 1, $p \leq 0.001$) and PO_4^{3-} and NO_3 (r = 1, $p \leq 0.001$) was observed and this meant that they had exactly the same contributor which could be mud and putrescible wastes brought in by the infiltrating rain water and organics [4].

The strong positive correlation between EC and pH, TDS and pH, TA and EC, TA and TDS, COD and EC (r = 0.98, $p \leq 0.01$) signified that they had nearly the same contributors (the dissolved ions). A significant negative correlation was also observed between TA, COD, Fe, Mg, K, Na and SO_4 (r = – 0.9615, – 0.92447, – 0.92447, – 0.97246, – 0.92518, – 0.93204) respectively, indicating the opposing distribution of these pair variables.

9.13 Summary and Conclusions

When compared to samples from the Central Point (SW 2) and Upstream Point (SW 3) sampling locations, the Downstream Point (SW 1) samples had higher concentration of parameters. However, some metals and heavy metals were below equipment detectable limits for all sampled locations. It was confirmed therefore that SW 1, was the most contaminated among the three sampled surface water points, in this study. The microbial load within all the samples greatly exceeded the recommended WHO and NSDWQ standards for drinking water. The Pearson's correlation results also showed that some pollutants analyzed came from the same source.

Since the stream was the main resource that provided protein (fish) to nearby residents and other local farmers who came in contact with the water daily, its contamination may lead to a serious dislocation of the ecological balance. This may also result to a potential threat of bioaccumulation of pollutants as a result of fishers ingesting their catch and having dermal contact with the microbes infested water daily.

Therefore it was recommended that siting of landfills should be done after a proper study of the environment by policy makers since landfills had been confirmed to contribute contaminants into the environment. The surface water sources within landfills should also be properly monitored to avoid ingestion of contaminated water.

Table 9.5 Pearson's correlation for surface water analysis using microsoft excel

	pH	Temp.	EC	TDS	SO_2^{-4}	NO_3	PO_3^{-4}
pH	1						
Temp.	0.992603	1					
EC	0.900976	0.841638	1				
TDS	0.900976	0.841638	1	1			
SO_2^{-4}	− 0.98474	− 0.95632	− 0.96274	− 0.96274	1		
NO_3	0.984738	0.956325	0.962737	0.962737	− 1	1	
PO_3^{-4}	0.984738	0.956325	0.962737	0.962737	− 1	1	1
NH_4	0.984738	0.956325	0.962737	0.962737	− 1	1	1
Alkalinity	0.899014	0.839199	0.99999	0.99999	− 0.96151	0.96151	0.96151
BOD	0.341644	0.225018	0.715576	0.715576	− 0.5	0.5	0.5
COD	0.844011	0.772656	0.993129	0.993129	− 0.92447	0.924473	0.924473
Hardness	0.939829	0.974355	0.698535	0.698535	− 0.86603	0.866025	0.866025
Fe	0.884564	0.821398	0.999336	0.999336	− 0.92447	0.952241	0.952241
Ca	0.885313	0.935217	0.595898	0.595898	− 0.79087	0.790872	0.790872
Mg	0.91705	0.861854	0.999255	0.999255	− 0.97246	0.972457	0.972457
K	0.845005	0.773833	0.993344	0.993344	− 0.92518	0.92518	0.92518
Na	0.854757	0.785422	0.995306	0.995306	− 0.93204	0.932044	0.932044
Total plate count	0.680159	0.764125	0.294752	0.294752	− 0.54219	0.542194	0.542194

	NH_4	Alkalinity	BOD	COD	Hardness	Fe	Ca	Mg	K	Na	Total plate count
pH											
Temp.											
EC											

(continued)

Table 9.5 (continued)

TDS											
SO_2^{-4}											
NO_3											
PO_3^{-4}											
NH_4	1										
Alkalinity	0.96151	1									
BOD	0.5	0.718713	1								
COD	0.924473	0.993646	0.792406	1							
Hardness	0.866025	0.695307	0	0.609994	1						
Fe	0.952241	0.99949	0.740558	0.996734	0.671992	1					
Ca	0.790872	0.592278	– 0.13455	0.497825	0.990906	0.566236	1				
Mg	0.972457	0.999071	0.688083	0.987872	0.725632	0.99718	0.626448	1			
K	0.92518	0.993853	0.791272	0.999998	0.611464	0.996882	0.499434	0.988159	1		
Na	0.932044	0.995731	0.779821	0.999793	0.626002	0.998172	0.515381	0.990829	0.999829	1	
Total plate count	0.542194	0.290448	– 0.45658	0.1809	0.88968	0.259731	0.943025	0.331412	0.182725	0.200888	1

Acknowledgements I wish to acknowledge the efforts of Samuel Ihimikaiye of Federal University Otuoke, for providing some data used in this chapter.

References

1. Salami L, Fadayini MO, Madu C (2014) Assessment of a closed dumpsite and its impact on surface and groundwater integrity: a case of Oke Afa dumpsite, Lagos, Nigeria. IJRRAS 18(3)
2. Aderemi AO, Oriaku AV, Adewumi GA, Otitoloju AA (2011) Assessment of groundwater contamination by leachate near a municipal solid waste landfill. Afr J Environ Sci Technol 5(11):933–940
3. Churchill ES, Opololaoluwa OI (2022) Characterization of municipal solid waste in Yenagoa metropolis and its associated management problems, Nigeria. Int J Res Appl Sci Eng Technol (IJRASET):10(3):1460–1466. Available at www.ijraset.com
4. Maitia SK, Dea S, Hazrab T, Debsarkarb A, Duttab A (2016) Characterization of leachate and its impact on surface and groundwater quality of a closed dumpsite—a case study at Dhapa, Kolkata, India. Procedia Environ Sci 35:391–399
5. Al Sabahi E, Abdul Rahim S (2009) The characteristics of leachate and groundwater pollution at municipal solid waste landfill of Ibb City Yemen, Science Publications, Am J Environ Sci 5(3):256–266
6. Kjeldsen P, Barlaz B, Rooker PA, Baun A, Anna L, Christensen HT (2002) Present and long-term composition of MSW landfill leachate: a review. Crit Rev Environ Sci Technol 32(4):297–336
7. Jagloo K (2002) Groundwater risk analysis in the vicinity of a landfill, a case study in Mauritius. Royal Institute of Technology, Stockholm, pp 110
8. USEPA (United States Environmental Protection Agency) (2000) In: Proceedings of the ground-water/surface water interactions workshop, pp 102. Retrieved from: http://www.epa.gov/tio/download/remed/gwsw/gwsw_part1.pdf
9. Tchobanoglous G, Theisen H, Vigil S (1993) Integrated solid waste management: engineering principles and management issues. Irwin McGraw-Hill, Boston
10. Singh V, Mittal AK (2009) Toxicity analysis and public health aspects of municipal landfill leachate: a case study of Okhla Landfill, Delhi –8th world wide workshop for young environmental scientists WWW-YES 2009: Urban waters: resource or risks?
11. Kjeldsen P (2002) Present and long-term composition of MSW landfill leachate: a review. Crit Rev Environ Sci Technol 32:297–336
12. Yoshida M et al (2002) Characterisation of Leachate from Henchir El Yahoudia Close Landfill. Water Waste Environ Res 1(2):129–142
13. Omole DO, Longe EO (2008) An assessment of the impact of Abattior Effluents on River Illo, Ota, Ngeria. J Environ Sci Technol 1:56–64
14. Chapman D (1992) Water quality assessment: a guide to the use of Biota, sediments and water in the environmental monitoring. WHO, Geneva, pp 585
15. Taylor R, Allen A (2006) Waste disposal and landfill: potential hazards and information needs. J Environ Prot 6:5
16. Anne MH (2016) Water chemistry, how much of your body is water? Accessed on the 4 May 2023
17. Akhionbare SM (2009) The environment. MC Computer Press, 8 Owerri road, Nnwi, Anambara State, pp 131–140
18. Clark CS, Meyer CR, Gartside PS, Majeti VA, Specker B, Balistreri WF, Elia VJ (1982) An environmental health survey of drinking water contamination by leachate from a pesticide waste dump in Hardeman County, Tennessee. Arch Environ Health 37:9–18
19. Appe E (2003) State and local groundwater protection and cleanup programs, pp E-2

20. Najem GR, Terry S, Martin F (1994) Health effects of a superfund hazardous chemical waste. Am J Prev Med 10(3):151–155
21. Paigen B, Goldman LR, Magmant MM, Highland JH, Steegman AT (1987) Growth of children living near the hazardous waste site, Love Canal. Hum Biol 59:489–508
22. Matthew RF (2018) Environmental biology. Open Oregon Educational Resources, Revised Edition. Article 6, pp 195–196. Accessed from Environmental-Biology-1592954463_print.pdf
23. Ogbodo SG (2009) Environmental protection in Nigeria: two decades after the Koko incident, Annual survey of International and comparative law: Vol. 15(1):115. Article 2
24. Anqi T, Zhang Z, Suhua H, Xia L (2008) Review on landfill leachate treatment methods. IOP Conf Ser Earth Environ Sci 565:012038
25. Renou S, Givaudan JG (2008) Landfill leachate treatment: review and opportunity. J Hazard Mater 150:468–593
26. Tariwari CNA, Elijah IO, Cosboy EM (2015) The potability of groundwater in Bayelsa State, Central Niger Delta Nigeria: a review. J Environ Treatment Tech 3(2):127–135
27. Amangabara GT, Ejenma E (2012) Groundwater quality assessment of Yenagoa and Environs Bayelsa State, Nigeria between 2010 and 2011. Resour Environ 2(2):20–29
28. Koinyan AA, Nwankwoala HO, Eludoyin OS (2013) Water resources utilization in Yenagoa, Central Niger Delta. Int J Water Resour Environ Eng 5(4):177–186
29. American Public Health Association APHA (2015) Standard methods for the examination of water and wastewater, 22nd ed., Washington DC., pp 202
30. Rakh MS, Bhosle AB (2011) Adv Appl Sci Res 2(5):104–109
31. NSDWQ, Nigerian Standards for Drinking Water Quality (2015). ICS 13.060.20
32. World Health Organization Guidelines for drinking-water quality (2011) 4th ed. WHO Press Geneva, Switzerland, pp 541. Available at: www.who.int/water_sanitation_health/publications

Chapter 10
Assessing Leachate and Landfill Biogas Generation: Key to Sustainable Waste Solutions

Roukaya Bouyakhsass, Safaa Khattabi Rifi, Abdelaziz Madinzi, Khalid Digua, Abdelkader Anouzla, Younes Abrouki, Aysegul Pala, Tonni Agustiono Kurniawan, Hayat Laoukili, and Salah Souabi

Abstract Accurate landfill leachate and biogas production estimation is essential for effective waste management and environmental protection. In this study, we look at the range of techniques available for estimating landfill leachate and biogas production, considering various influencing factors such as waste composition, environmental conditions, and landfill design. Through the practical application of these estimation methods to the Oum Azza landfill, we discovered valuable information about the dynamics of leachate and biogas production at this specific landfill. In particular, our research identified the water balance method as the most accurate and suitable approach for estimating leachate quantities. In contrast, the LandGEM multiphase method was optimal for estimating biogas. The implications of our results go far beyond the scope of this study, as they highlight the profound importance of accurate estimation in the development of sustainable waste management strategies. The insights gained from this research pave the way for designing and implementing waste management practices that are both environmentally friendly and economically viable, thus contributing to a more sustainable and ecologically harmonious future.

Keywords Waste management · Leachate · Landfill gas · Estimations · Water balance method · LandGEM

R. Bouyakhsass · S. K. Rifi · A. Madinzi · K. Digua · A. Anouzla (✉) · H. Laoukili · S. Souabi
Laboratory of Process Engineering and Environment, Faculty of Science and Technology, University Hassan II, Mohammedia, Morocco
e-mail: aanouzla@gmail.com

Y. Abrouki
Mohammed V University in Rabat, Rabat, Morocco

A. Pala
Dokuz Eylul University, Faculty of Engineering, Environmental Engineering Department, Izmir, Turkey

T. A. Kurniawan
College of the Environment and Ecology, Xiamen University, Xiamen 361102, China

A. Anouzla and S. Souabi (eds.), *A Review of Landfill Leachate*, Springer Water,
https://doi.org/10.1007/978-3-031-55513-8_10

10.1 Introduction

The increase in the world's population has triggered an era of rapid urbanization and industrialization, accompanied by an unprecedented increase in waste production [10, 20, 21, 37]. This growing waste production represents a major challenge for municipalities, governments, and environmentalists worldwide. Waste management is not just a logistical concern but a multi-faceted problem with environmental, economic, and public health aspects [35].

Waste is generated in many sectors of society, from households to industry, commerce, public institutions, and agricultural activities. In fact, in Morocco, the amount of solid waste generated exceeded 6.8 million metric tonnes, representing 0.7 kg/day/inhabitant in urban areas and 0.3 kg/day/inhabitant in rural areas in 2013. Over 85% of solid waste is regularly collected in urban areas, totaling 5.5 million tonnes. Only 37% of the waste collected was disposed of in controlled landfills [12, 17].

Over the last few decades, waste treatment techniques have evolved to meet these challenges and minimize the negative impact of waste on our environment. Among the various waste management strategies, landfill remains a widely used method of waste disposal, particularly in developing countries in regions where alternatives are limited or costly to implement [4, 5, 18, 19, 30, 39]. However, landfilling is not without consequences for the environment.

Landfill sites store large quantities of municipal solid and industrial waste, creating an environment conducive to complex biological and chemical processes. The anaerobic decomposition of organic matter in landfill sites produces two essential by-products: landfill leachate and landfill biogas [1]. Leachate and biogas have significant environmental implications and must be managed effectively to mitigate their negative impact on soil, groundwater, and the atmosphere [16].

Leachate is a contaminated liquid from the percolation of rainwater or moisture through waste at landfill sites. It contains a variety of pollutants, including heavy metals, organic compounds, and microbial contaminants, making it a severe threat to local ecosystems and water resources [36]. Without proper containment and treatment, leachate can infiltrate the soil and groundwater, contaminating drinking water sources and aquatic ecosystems [25, 27, 29].

Landfill biogas also contains methane (CH_4), carbon dioxide (CO2), and traces of other gases. Methane, a potent greenhouse gas, is of particular concern because of its significant contribution to global warming when released into the atmosphere without being captured or used [31, 33]. In addition, uncontrolled landfill biogas emissions can cause odorous emissions, posing health risks and disturbing neighboring communities [13, 34, 38].

Given the environmental and health risks posed by landfill leachate and landfill biogas, it is imperative to accurately estimate the quantities involved in designing, sizing, and implementing effective treatment and recovery units [8]. This estimate is a fundamental step in sustainable waste management, as it enables decisions to be taken on infrastructure investment, environmental protection measures, and resource recovery initiatives. Accurate quantification of landfill leachate and biogas production enables authorities to assess the feasibility of various waste-to-energy projects, landfill gas capture systems, and leachate treatment processes, thereby minimizing their environmental footprint.

Several methods for estimating leachate flow are used, including the water balance model [2, 23], which considers rainfall, runoff, actual evaporation, and water contained in the waste. Leachate production ratios can also be used to estimate the leachate produced at landfills, such as the Ouled Berjal landfill ratio and the World Bank ratio [8].

In conjunction with this, quantifying biogas is essential for assessing its environmental impact and designing appropriate management systems. Estimation methods like the LandGem landfill gas emission model can be used. In Morocco, this model was used by [26] to estimate the production of energy from the methane produced by the waste of the inhabitants of Rabat. In addition, Saghir et al. [32] and Oukili et al. [28] used the LandGEM model to estimate the energy potential of the waste of the city of Oujda and the public landfill in the town of Mohammedia-Benslimane, respectively.

In this paper, we examine the critical aspects of estimating landfill leachate and biogas quantities in the context of landfilling as a waste management strategy. We explore the complex interplay of factors influencing leachate and biogas production, discuss the methods and models used for estimation, and highlight the importance of accurate quantification in developing sustainable waste management solutions. By shedding light on these critical aspects, we aim to contribute to the advancement of waste management practices that prioritize environmental protection and value for money at a time when waste-related challenges continue to grow.

10.2 Leachate Estimation Methods

Accurate assessment of the quantity and composition of leachate generated at landfills is essential to designing and operating effective leachate collection and treatment systems. Various factors, such as waste composition, climatic conditions, and landfill design, influence leachate generation, making it a multi-faceted challenge [23].

10.2.1 Water Balance Method

The water balance approach involves the comparison of water inputs and outputs within a specified system over a predetermined timeframe.

It is distinguished by its simplicity and minimal prerequisites in terms of input information [24]. It can be calculated using Eq. (10.1):

$$L = P - ET_a + E_W + R_{in} - R_{out} - I \tag{10.1}$$

where L is the amount of leachate produced (m^3),

P precipitations (quantity of precipitated water in the study area in m^3),

ET_a actual evaporation (eliminated amount of water by evaporation under natural conditions in m^3),

E_W the amount of water brought by the waste in m^3,

R_{in} runoff from outside to inside the waste cell in m^3,

R_{out} runoff from inside to outside the waste cell in m^3,

I the infiltration in m^3.

10.2.1.1 Leachate Production Ratios

By analyzing the current volume of waste deposited into a landfill and measuring the corresponding amount of leachate generated, we can calculate a ratio that expresses the relationship between leachate production and the quantity of waste in the landfill.

10.2.1.2 World Bank Ratio

Croitoru and Sarraf [11] used the World Bank ratio in a report to assess the costs associated with groundwater contamination. This ratio works on the assumption that for every ton of waste landfilled, it generates around 1.25 m^3 of leachate. To calculate this ratio, they first converted the weight of waste landfilled to volume using a density value of 0.4 tons per m^3. Next, it was assumed that the waste has a leachate level of 50% of the total volume.

10.2.1.3 Ouled Berjal Landfill Ratio

Using actual data on the waste deposited in a landfill and the corresponding leachate production, it becomes possible to derive a leachate production ratio, quantified in cubic meters (m^3) of leachate per ton of waste. A relevant example of this application can be seen at the Ouled Berjal public landfill, located within the city limits of Kenitra in Morocco. Covering an area of 20 ha, this landfill receives an average daily quantity of around 329 tons of waste. The leachate generated in this landfill is collected efficiently using a dedicated system equipped with a constant flow pump. In particular [7], in their comprehensive study focusing on the Ouled Berjal landfill, determined that the value of this leachate generation ratio varies around 0.26 m^3 per ton of waste.

10.3 Bigas Estimation Methods

Landfill biogas is primarily composed of methane (CH_4) and carbon dioxide (CO_2) [9]. It is a valuable resource that can be harnessed to mitigate greenhouse gas emissions and provide a renewable energy source. Estimating the quantity and composition of biogas produced within landfills is imperative for designing efficient gas collection systems and optimizing energy recovery processes.

10.3.1 LandGEM Model

This free, automated, user-friendly software program is accessible through the Microsoft Excel interface. It is created by the United States Environmental Protection Agency (US EPA), employing a first-order decomposition model to estimate methane production within landfills, outlined as follows:

$$Q_{CH_4} = \sum_{i=1}^{n} \sum_{j=0.1}^{1} k L_0 (\frac{Q_{Wi}}{10}) e^{-kt_{ij}} \quad (10.2)$$

With:

Q_{CH_4} represents the annual CH_4 generation in m^3/yr, k the methane generation constant in $year^{-1}$, L_0 the methane generation potential in m^3/ton, Q_{Wi} the amount of material buried in the year i in tons, t_{ij} the age of the j_{th} section of material buried in the i_{th} year (Decimal Years), i the increment of 1 year, j the increment of 0.1 years, n the number of years calculated (Calculation year—the initial year of acceptance of buried material).

10.3.2 Multiphase Model

The waste buried in landfill cells has diverse compositions, and each component behaves differently when subjected to storage conditions. To account for this complexity, multiphase models can be derived from the initial first-order models by treating each waste component individually with their respective methane generation constant and methane generation potentials. This section will focus on using the LandGEM model, where we will carefully examine each waste fraction separately, recognizing the unique characteristics and associated parameters.

10.4 Oum Azza Landfill Case Study

The Oum Azza landfill site, located in Rabat-Salé-Kénitra, Morocco, is geographically situated at latitude 33°52′11.686″ N and longitude 6°48′37.572″ W. This landfill was inaugurated in 2007, serving as a repository for household and similar waste as well as green waste and ordinary industrial waste from a consortium of municipalities including Rabat, Salé, Bouknadel, Témara, Skhirat, Harhoura, Ain Aouda, Mers Kheir, Al Manzeh, Ain Attig, Sebbah, S. Yahya Zaer, and Oum Azza. With a capacity of around 3 million inhabitants, this landfill treats an annual volume of approximately 850,000 tons of waste. The center extends over a large area, covering about 1.1 km^2, and has a sorting center dedicated to the recovery of recyclable materials. Non-recyclable waste is systematically deposited in cells to prevent soil and groundwater contamination. As a precautionary measure, it is systematically covered with a layer of soil to limit odors and the dispersion of light waste. Figure 10.1 shows the actual quantities and the projection of the waste received in the Oum Azza landfill.

The Oum Azza landfill registers an average waste production ratio of 0.95 kg per person daily. It's noteworthy that this figure surpasses the waste generation rate observed in other regions, such as the 0.67 kg per person per day reported by El Hamdouni et al. [15] in Khenifra city and the 0.8 kg per person per day documented in Mohammedia according to [14]. The highest values of this ratio tend to occur during July and August each year, primarily due to increased tourism activities and corresponding shifts in waste composition.

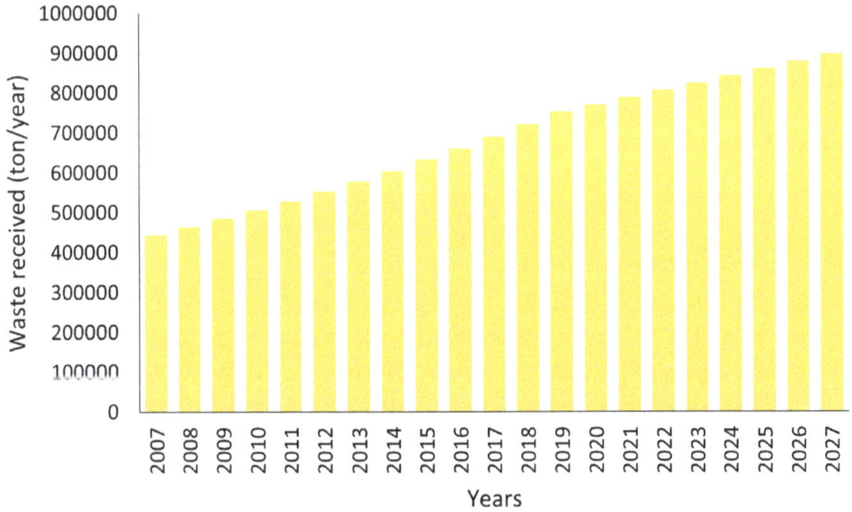

Fig. 10.1 Quantities and the projection of the waste received in the Oum Azza landfill

10.4.1 Leachate Estimation

The water balance method is valuable for estimating leachate production at the Oum Azza landfill. This model depends on the specific climatic conditions prevailing in the region. In the calculations presented, it has been assumed that the waste has a moisture content of 50%. However, this quantity is not fully released. According to [2], waste absorbs water until it reaches its field capacity. This parameter can be defined as the maximum amount of water that can be retained in the waste body against the force of gravity. As a result, excess water seeps out and moves towards the sides and bottom of the landfill. In this work, the waste's field capacity is supposed to be equal to 30%. These parameters played a crucial role in determining the leachate production rate. In addition, the calculations were based on a surface area of 5000 m^2 per month. On the other hand, the leachate production ratios are easy to use since they give leachate quantities that are directly proportional to the waste disposed of at the level of the landfill.

Figure 10.2 shows leachate estimation using the water balance method (a), the Ouled Berjal landfill ratio (b), and the World Bank ratio (c) for the year 2019.

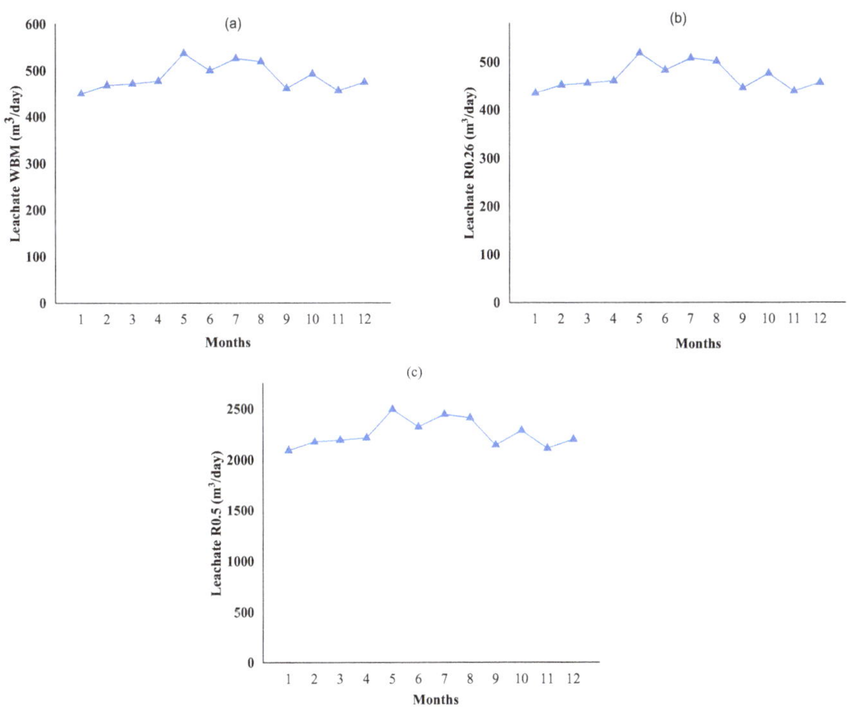

Fig. 10.2 Leachate estimations using the water balance method (**a**), the Ouled Berjal landfill ratio (**b**), and the World Bank ratio (**c**) for the year 2019

The water balance application at the Oum Azza landfill showed that the average leachate production equals 486.96 m^3/day. This method considers the actual evapotranspiration, the precipitation, and the water released by solid waste, since the Oum Azza landfill is surrounded by pits that stop precipitations from exiting the waste cells (Runoff water). It has geomembranes that prevent water from infiltrating the soil (Infiltration = 0).

From the water quantities entering the system, we can deduce that precipitation contributed in 2019 with a volume of 1325 m^3 and the water brought by the waste by 177,766 m^3.

When we apply the ratio for the Ouled Berjal landfill, the estimated average leachate production is 477 m^3 per day. In contrast, using the World Bank ratio gives significantly higher results, with an average production of 2295 m^3 per day. Significant differences exist between these estimation methods, with the Ouled Berjal landfill ratio producing lower results than those obtained using the World Bank ratio, with a substantial difference of 97%. It is important to note that the maximum quantities of leachate production, as estimated by both methods (Ouled Berjal ratio and World Bank ratio), occur during months 7 and 8. This observation highlights the direct proportionality between leachate production and the quantity of waste deposited in the landfill.

Using in situ measurements allows us to assess the accuracy of leachate generation ratios derived from estimation methods. Using direct measurements at the landfill site, we established a leachate production ratio of 0.25 m^3/ton of waste for the Oum Azza landfill. This ratio is 0.27 m3/ton for the water balance method, while for the Ouled Berjal landfill, it is 0.26 m^3/ton. In contrast, the World Bank ratio gives a much higher value of 1.25 m^3/ton. The ratio obtained using the World Bank method deviates considerably from actual measurements, with an average error of 394%. In contrast, the water balance method and the Ouled Berjal landfill ratio have much more minor errors, 6.65 and 2.85%, respectively. Although these two methods tend to overestimate the quantities of leachate, their results are very close to reality.

In general, landfill leachate volumes show considerable variability, mainly influenced by factors such as waste composition, water content, disposal practices, age of the landfill, and weather conditions. The water balance method has a significant advantage in considering these multi-faceted parameters, making it the most appropriate and accurate method for estimating leachate.

10.4.2 Biogas Estimation

Within landfill sites, the decomposition of organic matter in the waste, occurring in anaerobic conditions, produces biogas. This biogas primarily comprises methane varying around 60%, as noted by [22]. Only the biodegradable part of the waste will be considered to estimate the methane generated in the landfill, representing 80.15% of the total quantity of the received waste [6].

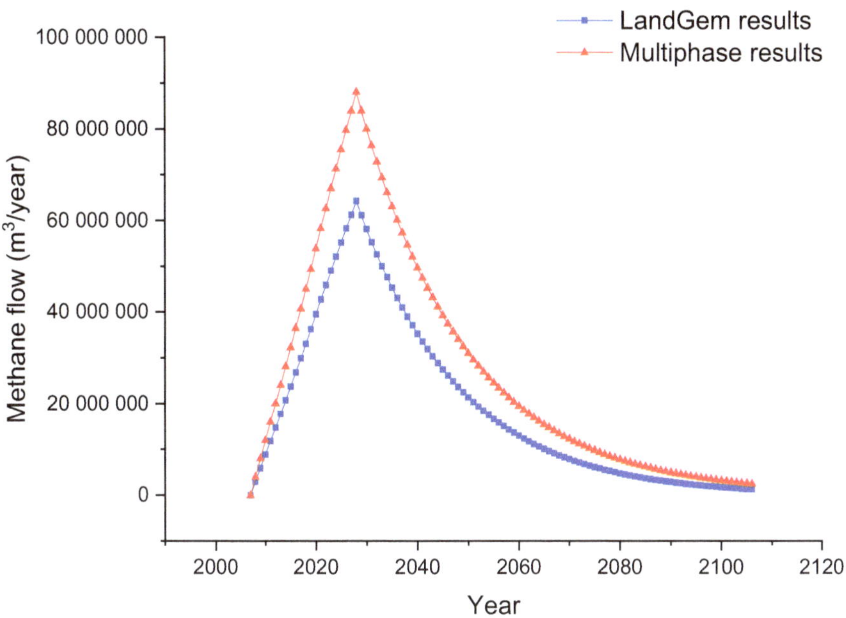

Fig. 10.3 The methane quantities estimated at the level of the Oum Azza landfill

The LandGEM model was used assuming k and L0 values equal to 0.05 $year^{-1}$ and 170 m^3/ton, respectively. On the other hand, for the multiphase model, we considered a rapidly biodegradable fraction representing 63.7% of the total amount of waste received and a moderately biodegradable fraction representing 16.45% of the total waste. These fractions, according to [32] have values that are equal to 262.5 m^3/ton and 0.05 yr^{-1} for L_0 and k for the rapidly biodegradable fraction and 170 m^3/ton and 0.03 yr^{-1} for the medium biodegradable fraction. Figure 10.3 shows the methane quantities estimated at the level of the Oum Azza landfill.

The application of the LandGem model predicts a maximum methane production estimate of 64,302,807.9 m^3 in the year 2028. In contrast, when employing the multiphase model, we obtained a higher maximum value of 88,175,052.08 m^3 for 2028.

Several other studies have also sought to estimate methane production levels within landfills. For instance, in the Kerman landfill in Iran, Amirmahani et al. [3] found that methane gas production ranged between 72–5015 m^3 per hour for landfills with waste production capacities ranging from 355,200 to 622,845 tons annually. Additionally, in a landfill in the municipality of Minas do Leão, in the southern region of Brazil, the peak methane generation was estimated at 107×10^5 m^3 using the LandGEM model.

While the results obtained through the second method exceed those of the first, they are considered closer to reality due to their consideration of each waste fraction, providing a more accurate depiction of methane production dynamics within landfills.

10.5 Conclusion

This work has illuminated the pivotal role of estimating landfill leachate and biogas production in shaping sustainable waste management practices. By examining a range of estimation techniques and their applicability, we have underscored the importance of tailoring these methods to specific landfill settings. Our case study of the Oum Azza landfill demonstrates the real-world application of these estimation techniques, highlighting their role in making informed decisions regarding waste treatment, environmental protection, and resource recovery. Accurate estimation of leachate and biogas production is essential for optimizing waste management strategies, mitigating environmental impact, and harnessing the potential of biogas as a renewable energy source. This research contributes to advancing environmentally responsible waste management, ensuring a sustainable future for communities worldwide.

References

1. Abuabdou SMA, Ahmad W, Aun NC, Bashir MJK (2020) A review of anaerobic membrane bioreactors (AnMBR) for the treatment of highly contaminated landfill leachate and biogas production: effectiveness, limitations and future perspectives. J Clean Prod 255:120215. https://doi.org/10.1016/j.jclepro.2020.120215
2. Abunama T, Othman F, Alslaibi T, Abualqumboz M (2017) Quantifying the generated and percolated leachate through a landfill's lining system in Gaza Strip, Palestine. Polish J Environ Stud 26(6):2455–2461. https://doi.org/10.15244/pjoes/73803
3. Amirmahani N, Sadeghi S, Yazdanpanah G, Tayebiyan A, Nasiri A, Malakootian M (2020) Estimating methane gas generation rate from Kerman City landfill using LandGEM software. Int J Environ Waste Manage 26(4):520. https://doi.org/10.1504/IJEWM.2020.110399
4. Anouzla A, Malika K, Azoulay I, Bencheikh, Ghizlane F, Jamal et al (2022) Multi-response optimization of coagulation-flocculation process for stabilized landfill leachate treatment using a coagulant based on an industrial effluent. Desalin Water Treatment 254:71–79. https://doi.org/10.5004/dwt.2022.28388
5. Anouzla A, Souabi S, Safi M, Younes Abrouki Y, Loukili H, Rhbal H (2014) Waste to treat waste of landfill leachates. Int J Environ Protect Policy 2(2):50. https://doi.org/10.11648/J.IJEPP.20140202.12
6. Bakraouy H (2018) Traitement des rejets de lixiviats des décharges publiques par voies physicochimique et biologique: Cas d'Ouled Berjal à Kénitra et d'Oum Azza à Rabat. Université Hassan II Faculté des Sciences et Techniques Mohammedia
7. Bakraouy H, Abouri M, Souabi S, Digua K, Yaacoubi A, Jada A (2016) Optimization of fresh landfill leachate treatment by a response surface methodology. J Colloid Sci Biotechnol 5(2):182–189. https://doi.org/10.1166/jcsb.2016.1142
8. Bouyakhsass R, Bouaouda S, Taleb A, Madinzi A, Baudu M, Souabi S (2023) Critical review for the quantification of leachate production: evaluation example for the "Oum Azza" public landfill in Rabat city, Morocco. Environ Q Manag 32(4):237–246. https://doi.org/10.1002/tqem.21981
9. Carriero G, Neri L, Famulari D, Di Lonardo S, Piscitelli D, Manco A, Esposito A, Chirico A, Facini O, Finardi S, Tinarelli G, Prandi R, Zaldei A, Vagnoli C, Toscano P, Magliulo V, Ciccioli P, Baraldi R (2018) Composition and emission of VOC from biogas produced by illegally managed waste landfills in Giugliano (Campania, Italy) and potential impact on the

local population. Sci Total Environ 640–641:377–386. https://doi.org/10.1016/j.scitotenv.2018.05.318
10. Chen DM-C, Bodirsky BL, Krueger T, Mishra A, Popp A (2020) The world's growing municipal solid waste: trends and impacts. Environ Res Lett 15(7):074021. https://doi.org/10.1088/1748-9326/ab8659
11. Croitoru L, Sarraf M (2017) Le Coût de la Dégradation de l'Environnement au Maroc (Discussion paper No 5 105633-M; Environment and Natural Resources Global Practice, p 147). World Bank
12. Dahchour A, Hajjaji SE (2020) Management of solid waste in Morocco. In Negm AM, Shareef N (éds) Waste management in MENA regions, pp 13–33. Springer International Publishing. https://doi.org/10.1007/978-3-030-18350-9_2
13. Duan Z, Scheutz C, Kjeldsen P (2021) Trace gas emissions from municipal solid waste landfills: a review. Waste Manage 119:39–62. https://doi.org/10.1016/j.wasman.2020.09.015
14. El Maguiri A, Laila I, Kissi B, Souabi S (2016) Towards recycling and waste reduction in Mohammedia, Morocco. Proc Instit Civil Eng Waste Resour Manage 169(2):101–107. https://doi.org/10.1680/jwarm.15.00005
15. Elhamdouni D, Arioua A, Karaoui I, Baaddi A, Ouhamchich KA (2019) Household solid waste sustainable management in the Khenifra region, Morocco. Arab J Geosci 12(23):744. https://doi.org/10.1007/s12517-019-4960-5
16. Ghosh A, Kumar S, Das J (2023) Impact of leachate and landfill gas on the ecosystem and health: research trends and the way forward towards sustainability. J Environ Manage 336:117708. https://doi.org/10.1016/j.jenvman.2023.117708
17. Gupta A, Paulraj R (2017) Leachate composition and toxicity assessment: an integrated approach correlating physicochemical parameters and toxicity of leachatesfrom MSW landfill in Delhi. Environ Technol 38(13–14):1599–1605
18. Gupta A, Rajamani P (2015) Toxicity assessment of municipal solid waste landfill leachate collected in different seasons from Okhala landfill site of Delhi. J Biomed Sci Eng 8(06):357
19. Gupta A, Rajamani P (2016) Evaluation of the leachate composition and contamination potential of municipal solid waste landfill sites in Delhi. Int J Environ Waste Manage 18(4):285–302
20. Gupta A, Gaharwar US, Verma A, Rajamani P (2022) Municipal solid waste landfill leachate induced cytotoxicity in root tips of Vicia faba: Environmental risk posed by non-engineered landfill. IJBB 59(11):1113–1125
21. Gupta P, Bhargava A, Kumari R, Lodhi L, Tiwari R, Gupta PK, Bunkar N, Samarth R, Mishra PK (2019) Impairment of mitochondrial-nuclear cross talk in lymphocytes exposed to landfill leachate. Environ Health Insights 13:1178630219839013
22. Hosseini SS, Yaghmaeian K, Yousefi N, Mahvi AH (2018) Estimation of landfill gas generation in a municipal solid waste disposal site by LandGEM mathematical model. Glob J Environ Sci Manage 4(4):493–506. https://doi.org/10.22034/gjesm.2018.04.009
23. Lekehal H, Bouchelta A, Benzougagh B, Mimich K (2016) Lixiviats Issus De La Décharge Publique De La Ville De Meknès: Calcul Et Résultat Du Bilan Hydrique. Eur Sci J ESJ 12(29):29. https://doi.org/10.19044/esj.2016.v12n29p242
24. Malmir T, Nomura H, Tojo Y (2015) Prediction of long-term leachate generation in a municipal solid waste landfill in an arid climate. 土木学会論文集g (環境) 71(7):III_247–III_258. https://doi.org/10.2208/jscejer.71.III_247
25. Mishra S, Tiwary D, Ohri A, Agnihotri AK (2019) Impact of municipal solid waste landfill leachate on groundwater quality in Varanasi, India. Groundw Sustain Dev 9:100230. https://doi.org/10.1016/j.gsd.2019.100230
26. Naimi Y, Saghir M, Cherqaoui A, Chatre B (2017) Energetic recovery of biomass in the region of Rabat, Morocco. Int J Hydr Energy 42(2):1396–1402. https://doi.org/10.1016/j.ijhydene.2016.07.055
27. Nurhasanah CMR, Riani E (2021) Micro- and mesoplastics release from the Indonesian municipal solid waste landfill leachate to the aquatic environment: case study in Galuga Landfill Area, Indonesia. Marine Pollut Bull 163:111986. https://doi.org/10.1016/j.marpolbul.2021.111986

28. Oukili AI, Mouloudi M, Chhiba M (2022) LandGEM biogas estimation, energy potential and carbon footprint assessments of a controlled landfill site. Case of the controlled landfill of Mohammedia-Benslimane, Morocco. J Ecol Eng 23(nr 3). https://doi.org/10.12911/22998993/145410
29. Parvin F, Tareq SM (2021) Impact of landfill leachate contamination on surface and groundwater of Bangladesh: a systematic review and possible public health risks assessment. Appl Water Sci 11(6):100. https://doi.org/10.1007/s13201-021-01431-3
30. Rajoo KS, Karam DS, Ismail A, Arifin A (2020) Evaluating the leachate contamination impact of landfills and open dumpsites from developing countries using the proposed Leachate Pollution Index for Developing Countries (LPIDC). Environ Nanotechnol Monitor Manage 14:100372. https://doi.org/10.1016/j.enmm.2020.100372
31. Rodrigo-Ilarri J, Rodrigo-Clavero M-E (2020) Mathematical modeling of the biogas production in MSW Landfills. Impact of the implementation of organic matter and food waste selective collection systems. Atmosphere 11(12), Article 12. https://doi.org/10.3390/atmos11121306
32. Saghir M, Naimi Y, Tahiri M (2018) First-order mathematical modeling of biogas production: application for the controlled landfill of fez. 2018 Renew Ener Power Syst Green Inclusive Econ (REPS-GIE) 1–6. https://doi.org/10.1109/REPSGIE.2018.8488783
33. Salvador R, Barros MV, Rosário JGDPD, Piekarski CM, da Luz LM, de Francisco AC (2019) Life cycle assessment of electricity from biogas: a systematic literature review. Environ Prog Sustainable Energy 38(4):13133. https://doi.org/10.1002/ep.13133
34. Shah AV, Srivastava VK, Mohanty SS, Varjani S (2021) Municipal solid waste as a sustainable resource for energy production: state-of-the-art review. J Environ Chem Eng 9(4):105717. https://doi.org/10.1016/j.jece.2021.105717
35. Sharma HB, Vanapalli KR, Samal B, Cheela VRS, Dubey BK, Bhattacharya J (2021) Circular economy approach in solid waste management system to achieve UN-SDGs: solutions for post-COVID recovery. Sci Total Environ 800:149605. https://doi.org/10.1016/j.scitotenv.2021.149605
36. Wijekoon P, Koliyabandara PA, Cooray AT, Lam SS, Athapattu BCL, Vithanage M (2022) Progress and prospects in mitigation of landfill leachate pollution : Risk, pollution potential, treatment and challenges. J Hazard Mater 421:126627. https://doi.org/10.1016/j.jhazmat.2021.126627
37. Yang H, Ma M, Thompson JR, Flower RJ (2018) Waste management, informal recycling, environmental pollution and public health. J Epidemiol Community Health 72(3):237–243. https://doi.org/10.1136/jech-2016-208597
38. Yao X-Z, Ma R-C, Li H-J, Wang C, Zhang C, Yin S-S, Wu D, He X-Y, Wang J, Zhan L-T, He R (2019) Assessment of the major odor contributors and health risks of volatile compounds in three disposal technologies for municipal solid waste. Waste Manage 91:128–138. https://doi.org/10.1016/j.wasman.2019.05.009
39. Zhang C, Xu T, Feng H, Chen S (2019) Greenhouse gas emissions from landfills: a review and bibliometric analysis. Sustainability 11(8), Article 8. https://doi.org/10.3390/su11082282

Chapter 11
Resources Recovery from Landfill Leachate: Current Status, Challenges, and Prospects

Jianchao Wang, Ao Shi, Dongbei Yue, Chunhui Wang, Yuan Liu, and Yanli Shi

Abstract Nowadays, the average generation of global municipal solid waste (MSW) is about 2.00 billion tons, and is anticipated to reach 3.40 billion tons by 2050 due to urbanization and increase in population. Landfill is the most popular method due to cost-effectiveness, and about 90% of global MSW is landfilled, producing massive landfill leachate. Landfill leachate includes various pollutants, such as organic compounds, heavy metals, ammonia, and inorganic salts, leading to serious pollution to the surrounding environment. In this work, the traditional methods for treating landfill leachate, and recovery technologies of nutrients (N and P), inorganic salts, bioenergy, and humic substances from landfill leachate are summarized, and the corresponding progress and bottlenecks of these technologies are identified. Among these resources recovery, nutrients (N and P) are extracted by struvite precipitation, adsorption, ammonium stripping, and membrane filtrations. The extraction of metals from landfill leachate is summarized. In terms of energy recovery, methane production, hydrogen production, and cultivating microalgae are included and discussed. In addition, humic substances as liquid fertilizer are recovered by using membrane technologies, adsorption, and other technologies are identified. Finally, the challenges, and prospects of traditional methods and resource recovery are discussed. Overall, this work can provide insights into resources recovery from landfill leachate, facilitating the sustainable development of landfill leachate disposal.

Keywords Landfill leachate · Resources recovery · Technologies · Prospects

J. Wang (✉) · A. Shi · C. Wang · Y. Liu · Y. Shi
School of Chemical and Environmental Engineering, China University of Mining and Technology (Beijing), Beijing 100083, China
e-mail: jcwang91@163.com

D. Yue
School of Environment, Tsinghua University, Beijing 100084, China

A. Anouzla and S. Souabi (eds.), *A Review of Landfill Leachate*, Springer Water,
https://doi.org/10.1007/978-3-031-55513-8_11

11.1 Introduction

The generation of municipal solid waste (MSW) has increased significantly over the past decades due to economic growth, improvement in living standards, urbanization, demographic growth, and changing life styles [1, 2]. Nowadays, the global generation of MSW is 2.01 billion tons, and it will reach 3.4 billion tons in 2025 [2]. In general, landfilling, incineration, and composting are widely-used methods for disposal of MSW. Among these methods, landfilling is the most prevalent method because of economic feasibility, strong stability, and low maintenance requirements. For example, in China, more than 80% of MSW is treated via landfilling [3]. During the stabilization process, MSW undergoes a series of biological and physicochemical processes, leading to the generation of highly polluted wastewater (landfill leachate, Fig. 11.1) [1, 4]. During the landfilling lifetime of MSW, it is estimated that one ton of MSW can produce 0.05–0.2 tons of landfill leachate [3]. In nature, landfill leachate is a mixture of the inherent water in MSW, water produced from the physical, chemical and biological processes, and percolated rainwater, which contains a large amount of highly hazardous organic compounds, ammonia, inorganic compounds, synthetic ligands, biological organisms, heavy metals, and the other emerging contaminants (microplastics, various surfactant, and trichloroethylene) [5, 6]. For example, the concentration of chemical oxygen demand (COD) in landfill leachate is as high as from 10,000 to 70,000 mg/L, and the concentration of ammoniacal nitrogen is approximately 1000–3000 mg N/L [7]. Such a high complexity leads to great challenges for efficient treatment of landfill leachate.

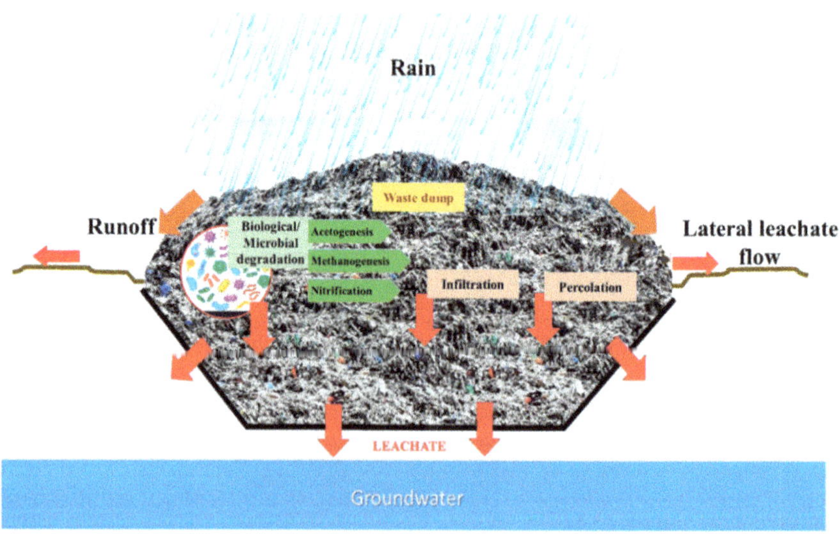

Fig. 11.1 The generation of landfill leachate during landfilling of MSW [1]

It has been proven that the organic compounds consist of total organic carbon (TOC), volatile fatty acids (VFA), humic acids (HAs), and fulvic acids (FAs) [8]. The inorganic compounds include Ca^{2+}, Mg^{2+}, Fe^{3+}, and Cl^{-}, and heavy metals (As, Hg, Cd, Cr, Cu, Mn, and Pb) [9]. The composition and water quality of landfill leachate are determined by various factors, including the type of waste, local climate and hydrogeology, operational conditions, and landfilling age. Among these factors, landfilling age is a crucial factor to determine the composition of landfill leachate. In general, according to landfilling age, landfill leachate can be classified into young and mature landfill leachate. In terms of young landfill leachate, the low molecular-weight hydrophilic organic matters are the main organic compounds with a high biodegradability index [biological oxygen demand (BOD_5)/COD], which are easy to be degraded by biology [10]. As to mature landfill leachate, the high molecular-weight HAs and FAs are the main organic compounds, and these compounds are refractory with a low biodegradability index. The pH value of young landfill leachate is low due to the presence of aliphatic acids, and mature landfill leachate has elevated pH value due to the decomposition of aliphatic acids. The concentration of heavy metals varies with landfilling age, and with landfilling age, heavy metals declined due to the decreased solubility. In addition, the elemental composition of young and mature landfill leachate is different; carbon, hydrogen, and nitrogen decrease, while oxygen and aromatic carbons increase as landfilling age increased [11].

Due to high-concentration and complicated pollutants in landfill leachate, the inadequate treatment of landfill leachate would cause serious environmental and ecological risks to the surrounding water and soil environment. For example, it has been reported that landfill leachate shows carcinogenic effects, acute toxicity, and genotoxicity to humans [12]. Klauck et al. [13] reported that at nearly all collection times, the raw landfill leachate possess high acute toxicity, cytotoxicity, and genotoxicity, and biological treatment has been demonstrated to be invalid for decreasing the genotoxicity of landfill leachate. Specially, the dissolved organic matter (DOM) of landfill leachate, including HAs and FAs, have been regarded as the emerging contaminants and can interfere with ultraviolet (UV) light transmittance [called UV-quenching substances (UVQS)] [14]. Moreover, UVQS can interact with organic and inorganic contaminants, affecting the fate of these contaminants in their transport behavior, stability and bioavailability. Besides, landfill leachate has been considered a significant source of emerging organic contaminants (EOCs), such as persistent organic pollutants, endocrine-disrupting chemicals, pharmaceutical and personal care products, antibiotic resistance genes, and disinfection by-products [5, 15]. Due to the ubiquity and potential hazard to human health, these EOCs attract more and more attention from academic researchers and policy makers. In particular, it has been reported that landfill leachate can contribute to microplastics (MPs) pollution, which has increasingly attracted more and more attention (as illustrated in Fig. 11.2) [16, 17]. MPs can adsorb and transport harmful chemicals such as heavy metals and persistent organic pollutants (POPs) due to hydrophobic nature, causing a range of negative health effects to humans because of physical contact or via the ingestion of food [18].

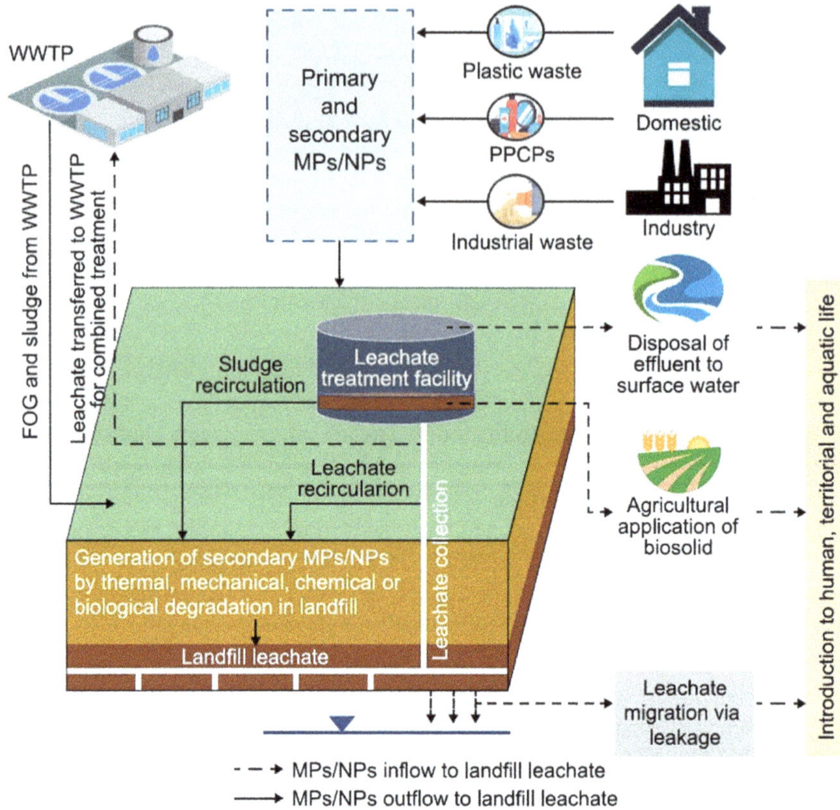

Fig. 11.2 The source and environmental pathways of MPs via landfill leachate; nanoscale plastics (NPs), pharmaceuticals and personal care products (PPCPs), fat, oil, and grease (FOG), and wastewater treatment plants (WWTPs) [17]

In the past decades, various techniques have been proposed for disposal of landfill leachate. Traditional techniques for disposal of landfill leachate include: combined treatment with domestic sewage (transferring), biodegradation, and chemical and physical methods. However, the presence of massive HAs and FAs can significantly interfere with UV disinfection in WWTPs, and thus at now the combined treatment of landfill leachate with WWTPs is limited in most countries [19, 20]. In general, biodegradation of landfill leachate consists of aerated lagoons, aerobic activated sludge, sequencing batch reactors, trickling filters, moving-bed biofilm reactor, and membrane biological reactor. Zhao et al. [4] reported that less than half of TOC in landfill leachate could be removed by biological treatment, and most of DOC (larger molecular weight fractions) in landfill leachate shows bio-refractory properties. Chemical and physical methods include oxidation (Fig. 11.3), adsorption, precipitation, coagulation/flocculation, sedimentation, air stripping, membrane, and froth flotation, which have been utilized to remove DOM, heavy metals, ammonia, and other emerging pollutants [21, 22]. For example, Amor et al. [23] found that using a combined physicochemical process (Fenton and solar photo-Fenton along with coagulation/flocculation pretreatment), the removal of COD, turbidity, and total polyphenols reached 65%, 80%, and 74%, respectively. Recently, various countries have adopted stricter regulations, particularly in China [24]. Membrane separation techniques including nanofiltration (NF) and reverse osmosis (RO) have been demonstrated highly effective for disposal of landfill leachate [25]. Nevertheless, about 30% of membrane concentrate cannot be treated, which is the greatest challenge in the field of landfill leachate treatment [2]. Importantly, the traditional methods fail to consider circular economy and carbon neutrality, and particularly cannot fully treat the pollutants of landfill leachate.

Owing to the climate change induced by carbon emission, many countries and regions have carbon neutrality goals, and particularly China will also achieve carbon neutrality before 2060. Carbon neutrality represents the balance of CO_2 emission and adsorption, namely a balance between emission footprint and voluntary compensation to realize an emission equilibrium [27]. This means that in the field of landfill leachate, the choosing of treatment methods and management policies needs to be from the view of carbon neutrality. As known, landfill leachate contains a large amount of metal elements, humic substances, nitrogen, phosphorus and some other valuable elements, and these resources can be utilized in agriculture as commercial synthetic fertilizers. For example, Liu et al. [28] utilized a novel loose nanofiltration-membrane distillation integrated process to effectively fractionate organics/salts, and subsequently recover salts and produce high-quality water (Fig. 11.3). Chang et al. [29] reported the high-efficiency recovery of nutrients from landfill leachate for microalgal lipids production. Huang et al. [30] reported that the antichlors of the Bi(III) containing oxides with quantum dots or two-dimensional structures showed high removal of Cl^- from landfill leachate, with removal rate of 61.8% and 66.1%, respectively. Wu et al. [31] found that calcium pretreatment can recover magnesium as struvite from landfill leachate, and under the condition [(Mg^+ $Ca_{residual}$): P molar ratio of 1:1.5 and pH 9.5], 98.6 ± 0.1% of magnesium could be recovered with a significantly lower residual PO_4^{3-}-P concentration. These reports just focused on resource recovery from landfill leachate [32, 33]. However, there are no comprehensive studies focused on the current status, challenges, and prospect of techniques towards recovery of resources from the view of carbon neutrality.

Therefore, the objective of the present paper is to summarize the current status of traditional techniques for disposal of landfill leachate, to focus on the state-of-the-art progress of resources recovery from landfill leachate in view of carbon neutrality, and finally to discuss the challenges and future research directions of treatment techniques of landfill leachate. Overall, this paper would provide valuable insights into the future sustainable and low-carbon disposal and management of landfill leachate.

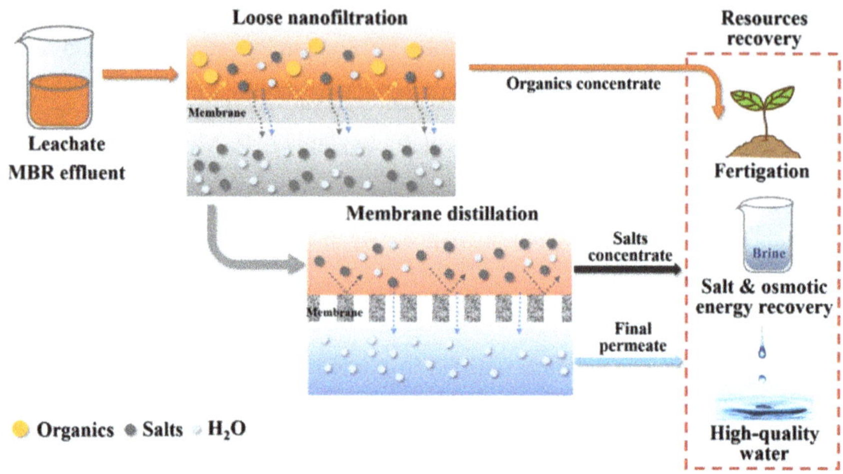

Fig. 11.3 The resource recovery from landfill leachate by using a novel loose nanofiltration-membrane distillation integrated process [28]

11.2 Traditional Treatment

11.2.1 Biological Treatment

Biological treatment of landfill leachate is often performed under aerobic or anaerobic conditions, while its performance was highly determined by the characteristics of landfill leachate. Zhao et al. [4]. found that landfill leachate exhibited refractory properties to biological treatment, and less that 50% of organic compounds in landfill leachate could be eliminated by the on-site biological treatment facilities. However, most BOD [these organic compounds with less molecular weight (less than 1 kDa)] of landfill leachate was eliminated by biological treatment. Song et al. [34] reported that under low dissolved oxygen condition, an anoxic rotating biological contactor (RBC) and four aeration tanks with gradient aeration were used to treat landfill leachate for 88 days. The results showed that the removal rate of COD, NH_4^+, and total nitrogen (TN) reached 85.65%, 99.92%, and 84.06%, respectively. It is apparent that biological treatment was more effective for treating the NH_4^+ of landfill leachate [35]. For young landfill leachate, it contains volatile fatty acids, which are easily biodegradable and are suitable for biological treatment [36]. However, the main organic compounds in mature landfill leachate are recalcitrant and are not suitable for biological treatment. Because of the ineffectiveness of biological treatment for mature landfill leachate, it has been widely considered as a pretreatment or posttreatment process, or the combined biological and chemical/physical methods. For instance, Bu et al. [37] reported that after sequencing batch reactor (SBR), aeration corrosive cell-Fenton and granular activated carbon (GAC) adsorption were utilized to treat landfill leachate. The results indicated that the COD, BOD_5, and DOC of landfill leachate were reduced by 97.2%, 99.1%, and 98.7%, respectively. Likewise, Zhao et al. [38]. reported that coagulation, NaClO oxidation, and Fe^{2+}/NaClO were effective for eliminating the COD, NH_4^+, and color degree of biologically treated landfill leachate, with the COD removal of 23.5%, 12.5%, and 53.5%, respectively. Thus, standalone biological treatment cannot meet the treatment requirements of landfill leachate.

11.2.2 Chemical Treatment

The chemical treatment of landfill leachate includes chemical oxidation (Fenton, ozone, photocatalysis, persulfate, and electrochemical treatment) and coagulation-flocculation, targeting organic pollutants in landfill leachate. Sruthi et al. [9] reported that using heterogeneous Fenton treatment (iron-manganese binary oxide loaded zeolite as catalysts), the COD removal of landfill leachate reached 87.5% under optimum conditions. Abu Amr and Aziz [40] found that using ozonation combined with Fenton, the COD, color, and NH_4^+ removal of landfill leachate were 65%, 98%, and 12%, respectively. In this regard, such advanced oxidation processes (AOPs) are not effective for NH_4^+ removal. Recently, since sulfate radical ($SO_4^{-}\cdot$) have higher redox potential, a wider pH range, and longer half-time that •OH, and thus $SO_4^{-}\cdot$ based AOPs (SR-AOPs) have been widely used to degrade organic pollutants [41]. Meanwhile, SR-AOPs has been used for treating landfill leachate. Karimipourfard et al. [42] reported the treatment of stabilized landfill leachate using persulfate activated by $CuFe_2O_4$ nano-catalyst. The results indicated that the COD, NH_4^+, and color removal reached 57%, 71%, and 63%, respectively. Ren et al. [43] found that using various coagulants, the COD, UV_{254}, total nitrogen, color, and turbidity of concentrated leachate reached 68.42%, 69.01%, 44.14%, 92.31%, and 87.44%, respectively. It can be seen that coagulation can eliminate more pollutants in landfill leachate than AOPs. In addition, Jiang et al. [44] utilized electrochemical oxidation combined with direct-current, monopulse, and double pulse to treat biologically treated leachate. The results showed that using monopulse, the COD, TOC, and total nitrogen removal reached as high as 80%, 30%, and 80%, respectively. In general, chemical treatment could achieve a high performance for treating landfill leachate, while it is limited by the high cost. In particular, chemical treatment cannot fully degrade the organic pollutants, and the residual small organics need to be further treated. In addition, in chemical treatment, the compounds in landfill leachate are considered as pollutants, and it does not recover these value-added compounds, meeting the requirements of circular economy and carbon neutrality.

11.2.3 Physical Treatment

Physical treatment of landfill leachate mainly includes adsorption, membrane separation, and ion exchange [45, 46]. As an effective and promising method, adsorption has been widely utilized to remove DOM, NH_4^+, and metals in landfill leachate. Zieliñska et al. [47] used powdered-activated-carbon (PAC) to adsorb the pollutants of real stabilized landfill leachate. The results indicated that under PAC dose of 3 g/L and time 30 min, the COD and color removal efficiency reached up to 73.1% and 94.4%, respectively. Mohan and Gandhimathi [48] applied coal fly ash as an adsorbent to eliminate the heavy metals from landfill leachate. The results showed that with coal fly ash of 2 g/L, the removal efficiencies of Cu, Mn, Pb, Zn and Cd were 39%, 28%, 74%, 42% and 71%, respectively. Modin et al. [49] tested the adsorption performance of activated carbon, bone meal, and iron fines for eliminating metals from landfill leachate. The results revealed that activated carbon showed a greater removal performance for Co, Cr, Cu, Fe, and Mn (more than 90%), while it was not effective for adsorption of Ni, Ca, Pb, Sr, and Zn. After adsorption by bone meal, over 80% of Cr, Fe, Hg, Mn and Sr, and about 20–80% of Al, Ca, Cu, Mo, Ni, Pb and Zn were removed. Iron fines showed removal performance for most metals (As, Ca, Co, Cr, Cu, Fe, Mg, Mn, Pb, Sr and Zn).

Similarly, ion exchange resin has been widely used for eliminating the COD, NH_4^+, UV_{254}, and color. Bashir et al. [50] studied the NH_4^+ removal of semi-aerobic landfill stabilized leachate from Malaysian using synthetic cation ion exchange resin and further optimize the process parameters by using central composite design (CCD) with response surface methodology (RSM). The results showed that under optimized conditions (24.6 cm^3 resin dosage), 6.0 min contact time, and 147.0 rpm shaking speed, the NH_4^+ removal reached 92%. Likewise, in another Bashir's study [51], CCD of RSM was used to optimize the COD, suspended solid (SS), and turbidity removal of landfill leachate by synthetic anion exchange resin. The results indicated that under 30.9 cm^3 anionic dosage, 90 min contact time, and 150 rpm shaking speed, the color, COD, SS, and turbidity removal was 91.5%, 70.3%, 93.1%, and 92.4%, respectively. In addition, ion exchange resins were also used to remove heavy metals from leachate. For example, Janin et al. [52] used various resins to eliminate copper, chromium, and arsenic from leachate. The results showed that the chelating resin Dowex M4195 possessed a greater copper selectivity, while the Amberlite IR120 resin exhibited a high removal performance for chromium. By combining these two resins in four successive columns, the copper removal and chromium removal reached 96% and 68%, respectively. Marañon et al. [45] used column experiments to test the removal of Zn and Cd from landfill leachate using various cationic exchange resins. It was found that Amberlite IRC 718 showed great removal efficiency for both these two metals, and the removal of Zn and Cd reached 93% and 50%, respectively.

Although some chemical and physical methods have been used in treatment of landfill leachate, these methods were effective for removal of DOC and showed a limited performance to remove inorganic salts and heavy metals. In this regard, membrane separation has been widely used to guarantee discharge levels. In general, various membrane separation techniques such as nanofiltration (NF), ultrafiltration (UF), and reverse osmosis (RO) have been adopted. Bohdziewicz and Kwarciak [53] used upflow anaerobic sludge blanker reactor (UASB) along with RO to treat landfill leachate. It was found that UASB showed a poor efficiency for removal of COD, BOD, NH_4^+, and Cl^{-1}, and thus RO was used as the post-treatment process. RO

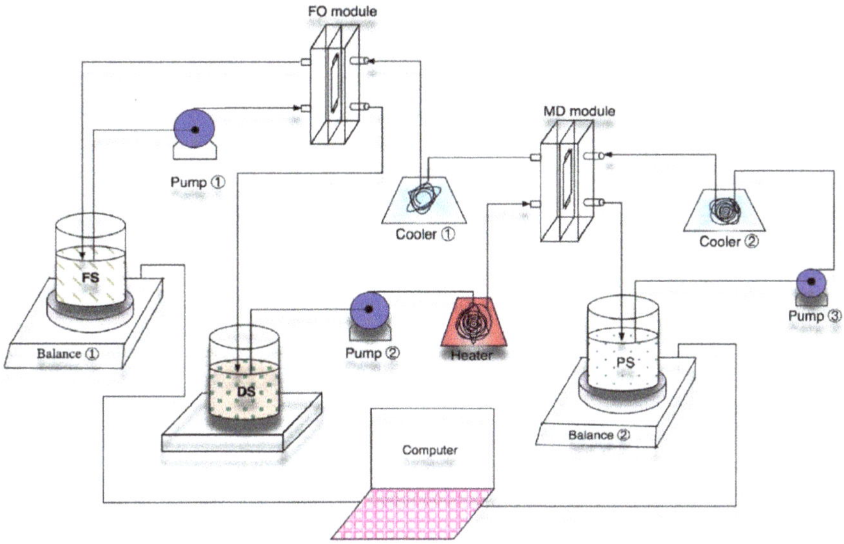

Fig. 11.4 The schematic of lab-scale FO-MD process for treating landfill leachate [55]

process achieved 95.4%, 90.2%, 85.4%, and 88.7% removal of COD, BOD, Cl^{-1}, and NH_4^+, respectively. Wang et al. [54] utilized anoxic/aerobic GAC, membrane bioreactor (MBR), NF, and RO to treat landfill leachate. The results indicated that more that 80% of COD and NH_4^+ can be removed by two pilot scale MBR, and NF exhibited a clearly greater color removal (93.75%) than AO-MBR (41.82%). Zhou et al. [55] utilized forward osmosis (FO) along with membrane distillation (MD) for treating landfill leachate (Fig. 11.4). After optimization using box-behnken design of RSM, the salt rejection rate was above 96%, the removal of TOC and total nitrogen was higher than 98%, and NH_4^+ and heavy metals (Hg, As, was Sb) showed a complete removal by such a FO-MD process. Fudala-Ksiazek et al. [56] evaluated the treatment efficiency of MBR/UF/NF process for landfill leachate. It was found that the effluent from the MBR/UF/NF process met the discharge requirements for wastewater entering WWTPs and for treated wastewater discharging into the receiver.

11.3 Resource Recovery

11.3.1 Nutrients Recovery

Nitrogen (N) and phosphorus (P) are the main nutrients in landfill leachate. N is a critical element for organisms due to its function in proteins and chlorophyll, which are essential for most living things and plants. P is critical to sustain the cellular growth of living organisms, since it is the main element in adenosine triphosphate, deoxyribonucleic acid, and lipids. Thus, these two nutrients are key components for fertilizer production [27]. Nevertheless, N and P are not easily accessible in the environment. Due to the strength of the covalent bond between the N atoms, the generation of N-based fertilizers from the atmosphere is energy-intensive. On the other hand, the generation of P-based fertilizer is highly dependent on mined phosphate rocks, and P total reserve will be depleted by 2100, resulting in threat to the supply of global food [57]. Given that N and P are abundant in landfill leachate, it is significant to recover N and P from landfill leachate, meeting the requirement of pollutants' treatment, circular economy, and carbon emission reduction in the field of landfill leachate.

In landfill leachate, the concentration of NH_4^+ is within the range of 200–15,000 mg/L. In general, struvite precipitation, ammonium stripping, nitrification, adsorption, membrane filtration have been used to recover N nutrients from landfill leachate. Struvite precipitation is based on forming magnesium ammonium phosphate, which is insoluble in aqueous solution and easily extracted from aqueous solution. In such a process, the recovery efficiency of NH_4^+ is mainly determined by the molar ratio of Mg: N: P and the pH value of landfill leachate. Huang et al. [58] studied the removal of ammonia from landfill leachate by using struvite precipitation with waste phosphoric acid as the phosphate source. It was surprising that the presence of Al^{3+} in the waste phosphoric acid had a clear effect on the struvite precipitation, and the removal of ammonic by using the waste phosphoric acid was comparable to that of the pure phosphate. Nevertheless, the H^+ in the waste phosphoric acid needs to be neutralized by adding a large amount of NaOH, and thus a low-cost magnesium source was proposed to be used as the substitution of alkali reagent in the struvite precipitation. The results showed that under the Mg: N: P molar ratio of 3:1:1, the removal of ammonia from landfill leachate was 85% and the residual phosphate was 56 mg/L.

The result indicated that the ammonia removal ratios were found to be 83%, with a remaining phosphate of 56 mg/L, by dosing the low-cost MgO in the Mg: N: P molar ratio of 3:1:1. Nevertheless, since landfill leachate has less magnesium and phosphate ammonium, struvite precipitation requires a large amount of chemical regents, including the external magnesium and phosphorus sources. In addition, the optimum pH value for P recovery is generally higher than 9.5, while the general pH of landfill leachate was about 6–8. Thus, the adjust of pH by adding a large amount of alkaline solution. Meanwhile, there are massive inorganic ions such as Ca^{2+}, K^+, Fe^{3+} and CO_3^{2-}, which could strongly interfere with the struvite precipitation, resulting in the impurity of the precipitation products.

In addition to precipitation, ammonium stripping is effective another method to recover ammonia from landfill leachate. Such a method is based on the principle of mass transfer, and a large amount of air is introduced into landfill leachate, forcing NH_3 from the aqueous solution to air stream. For example, Xiong et al. [59] utilized negative pressure steam-stripping technology to recover NH_3-N from mature landfill leachate. The results showed that before and after steam-stripping, the concentration of NH_3-N was 2941–3648 mg/L and 401–710 mg/L with an average removal rate of 82.03%, respectively. In particular, nitrogen was recovered as NH_4HCO_3, and the alkaline earth metals (Ca^{2+} and Mg^{2+}) were effectively eliminated by forming $CaCO_3$ and $MgCO_3$, reducing the potential of forming fouling during subsequent MBR system. In the study by Guo et al. [60], a high-gravity technology, namely a co-current-flow rotating packed bed in multi-stage cycle series mode, was used to enhance the ammonia stripping from landfill leachate. Furthermore, a machine learning approach was used to predict the removal performance of ammonia from landfill leachate. The results showed that under a pH value of 10 and ambient temperature, such a high-gravity field achieved the 82.73% removal of ammonia and the liquid volumetric mass transfer coefficient was $5.551 \times 10^{-4}\ s^{-1}$. Particularly, such a high-gravity technology exhibited apparent advantage, namely obviously reducing the consumption of alkaline chemicals and the economic costs of landfill leachate management. In comparison with struvite precipitation, the process of ammonia stripping can reduce the addition of phosphate and magnesium. In addition, ammonia stripping is generally used after anaerobic digestion, and thus the temperature of residual landfill leachate is about 30 °C, which is beneficial to perform ammonia stripping. Therefore, ammonia stripping shows promising prospects for ammonia recovery from landfill leachate (Fig. 11.5).

In addition to nitrogen, P is another main nutrient in landfill leachate, since it can sustain the cellular growth of living organisms and is a critical element in adenosine triphosphate, deoxyribonucleic acid, and lipids. Given that landfill leachate contains

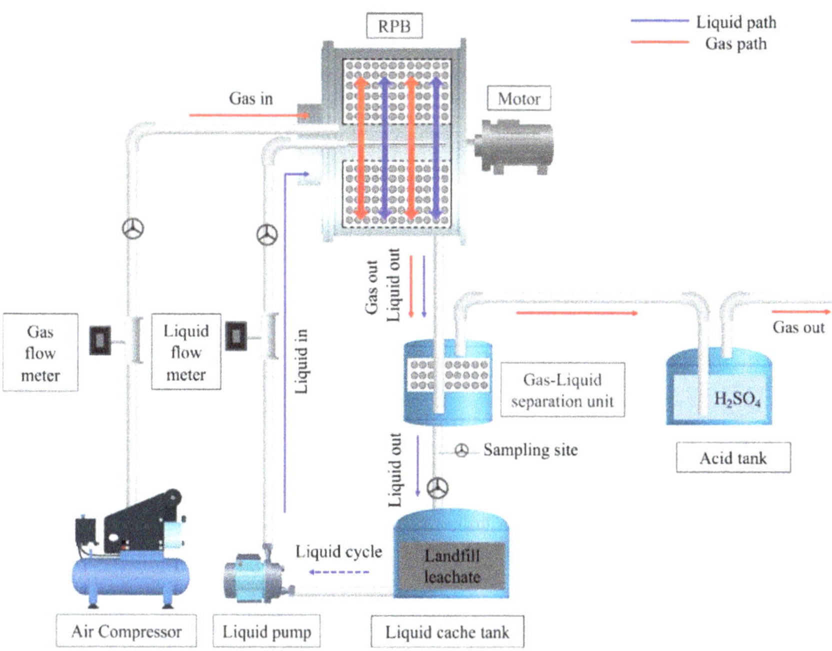

Fig. 11.5 The schematic diagram of a high-gravity system for the ammonia stripping from landfill leachate [60]

a large amount of P, precipitation, coagulation, adsorption, and membrane filtration have been proposed to recover P resource from landfill. Jaffer et al. [61] reported the application of struvite to recover the phosphorus from landfill leachate. It was found that as the increased dose of magnesium and the pH of the solutions were increased to 9.0, struvite crystallization occurred and achieved the highest removal (97%) of phosphorus. Furthermore, the presence of the molar ration of magnesium to phosphorus was found, and struvite formation occurred at such a ratio of least 1.05:1. If the molar ratio was below such a value, the formation of crystal was not struvite. Podder et al. [62] reported the coupling of struvite precipitation, aerobic granular sludge and anaerobic ammonia oxidation for recovering nutrients in landfill leachate, including phosphorus and NH_4^+. It was found that almost 77% recovery of phosphorus and 25% removal of NH_4^+ were recovered from landfill leachate via struvite precipitation under the optimized pH (9.0). Khanzada [63] reported that the application of microalgae can recover the phosphorus resources from landfill leachate. In this study, ultra-membrane treated landfill leachate was used as an example. Under different concentrations of phosphorus (15–100 mg/L) and nitrogen, the highest nitrogen removal was about 69.03% at concentration of 100 mg/L and P was completely removed from landfill leachate at all the tested phosphorus concentrations.

11.3.2 Energy Recovery

Landfill leachate contains a large amount of biodegradable organic compounds, which can be converted into bioenergy, such as methane, hydrogen and electricity. Mahmoud et al. [64] used a microbial electrolysis cell for pre-fermentation of poorly biodegradable landfill leachate to enhance current density. It was found that the current density, coulombic efficiency, coulombic recovery, and BOD_5 removal were 23 A/m^3 or 16 mA/m^2, 68%, 17.3%, and 83% with fermented leachate, and were 2.5 A/m^3 or 1.7 mA/m^2, 56%, 2.1%, and 5.6%, respectively. Similarly, Hassan et al. [65] used a microbial fuel cell to generate power and eliminate pollutants from landfill leachate. The results showed that in batch mode, the power generation and COD removal of 60% of young landfill leachate were 96.8 mW/m^2 and 90.0%, respectively, and in continuous mode these values were 75 mW/m^2 and 55.5% for 100% young landfill leachate. More recently, Chang et al. [66] proposed a novel technique by integrating microalgae photosynthesis and dark fermentation to achieve bioconversion of nutrients and energy from landfill leachate (Fig. 11.6). The results showed that the more intracellular proteins and carbohydrates were generated by the mixed microalgae, leading to higher energy yield of 10.80 kJ/L that 6.64 kJ/L than produced by the pure microalgae. In addition to power generation, landfill leachate was used to generate hydrogen. For example, Liu et al. [67] used the expanded granular sludge bed bioreactor along with granular sludge to recover the organic resources from landfill leachate by producing H_2. The results indicated that in the bioreactor, acid and ethanol were main metabolites and the fermentation of the bioreactor was ethanol-type. The maximum H_2 generation from landfill leachate was 2220 mL/(l

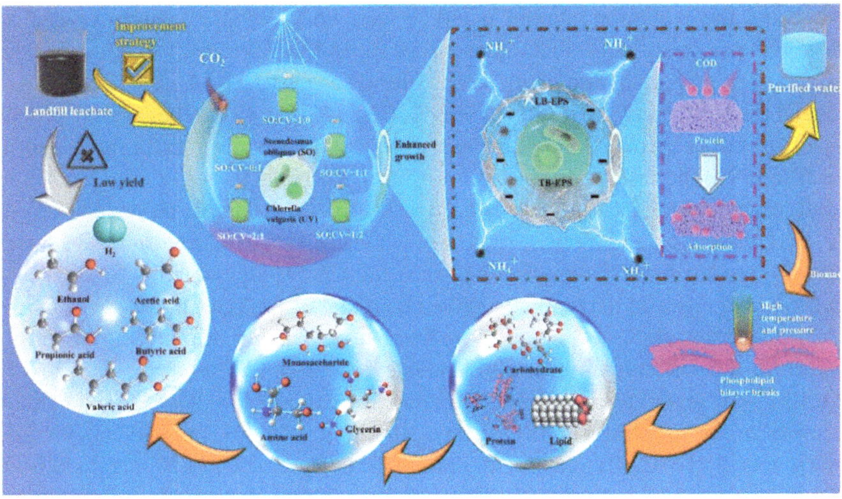

Fig. 11.6 The schematic of the energy conversion from landfill leachate to hydrogen by coupling microalgal photosynthesis and dark fermentation [66]

day), and the COD removal was 66.9%. Oz and Yarimtepe [68] reported that using pretreatment under low frequency ultrasound, the biogas production and methane yield from landfill leachate can be enhanced in anaerobic batch. The result indicated that the ratio of soluble COD to total COD was increased from 47% in raw landfill leachate to 63% after 45 min sonication at 600 W/L. In addition, using the ultrasonically pretreated landfill leachate, 40% more biogas was generated in anaerobic batch reactor.

11.3.3 Metal Recovery

There are many metals in landfill leachate, such as Zn, Pb, Cd, Ag, Cu, and Cr, and these metals possess both resources and pollutant properties [69]. Recovering or removing metals from landfill leachate can reduce the hazardous effects on microorganisms, enhancing the treatment performance of biodegradation. For example, Fernandez et al. [45] reported the application of cationic exchange resins, chelating resin, and adsorbent resin for the removal of Cd and Zn from the leachate. It was found that after treating by the chelating resin (Amberlite IRC 718), the removal of Cd and Zn showed the best performance, and reached 93% and 50% for Zn and Cd, respectively. Janin et al. [52] tested the extraction performance of ion exchange resins and precipitation for As, Cu, and Cr from the leachate. Under the initial concentration of As 447–651 mg/L, Cu 474–453 mg/L, and Cr 335–622 mg/L, the chelating resin Dowex M4195 showed a greater selectivity for Cu than other resins, and the Amberlite IR120 resin showed a greater adsorption performance for Cr. By combined these two resins, the extraction rate of Cu and Cr were 96% and 68%, respectively; after elution, the recovery of Cu and Cr reached 94% and 81%, respectively. In addition, coagulation and precipitation with ferric chloride and $Ca(OH)_2$ were used to extract As, and the removal of As was up to 99.9%. Besides resins, adsorbents such as activated carbon, bone meal, and iron fines were used to remove metals from landfill leachate [49]. The results showed that various adsorbents showed different affinity for different metals in landfill leachate. By using activated carbon, more than 90% of Co, Cr, Cu, Fe, Mn, and Ni were removed, while Ca, Pb, Sr, and Zn showed a lower removal performance. However, for bone meal, Cr, Fe, Hg, Mn, and Sr showed a much greater removal performance (80%) than that of Al, Ca, Cu, Mo, Ni, Pb, and Zn. In contrast to above adsorbents, iron fines showed a much lower removal performance for these metals from landfill leachate. Lee et al. [70] reported the recovery of critical raw materials from landfill leachate by recirculation. It was found that the recovery of the metal(loid)s was affected strongly by the pH, organic content, and type of metal(loid)s. After first recirculation, the concentration of Ni increased from 0.05 to 0.11 mg/L, showing prospects for recover metals.

11.3.4 HAs/FAs Recovery

Humic substances, including HAs and FAs, are ubiquitous in various landfill leachate, which are the main organic compounds in the old landfill leachate [71]. On the one hand, HAs and FAs are nonbiodegradable, and thus these compounds are the bottleneck problem for limiting the full treatment of landfill leachate, leading to the formation of massive membrane concentrate [72]. However, the membrane leachate concentrate fails to be used directly as a liquid organic fertilizer due to its low HAs and FAs content but high salinity. On the other hand, HAs and FAs are the main compounds of soil organic matter, and they are organic fertilizer for promoting plant growth and improving the structure of soil aggregates and fertilizer efficiency [3]. Meanwhile, the adding of HAs and FAs fertilizers can add carbon sinks to soils, meeting the requirements of carbon neutrality targets. Therefore, selective extraction of HAs and FAs is of important significance for resource recovery and treatment of landfill leachate. So far, the main methods for HAs and FAs extraction from landfill leachate include membrane filtration, chemical precipitation and adsorption. Among these methods, chemical precipitation is conducted under low pH ($\leq$ about 1.5), and thus the HAs and FAs fertilizers are not suitable for direct applications in soil.

Membrane filtration has been widely reported for recovering HAs and FAs from landfill leachate. For example, Ye et al. [73] reported that using persulfate as trigger, polydopamine-polyethylenimine complex was fabricated on the loose NF for recovery of humic substances from landfill leachate, leading to the decrease in the pore size of bio-inspired NF membrane (Fig. 11.7). Accordingly, the molecular weight cut-off was reduced from 640 to 298 Da after coating, and the rejection rate of humic substances increased from 95.1 to 98.9% for landfill leachate concentration; the concentration of humic substances was increased from 1779.4 to 17,247.1 mg/L

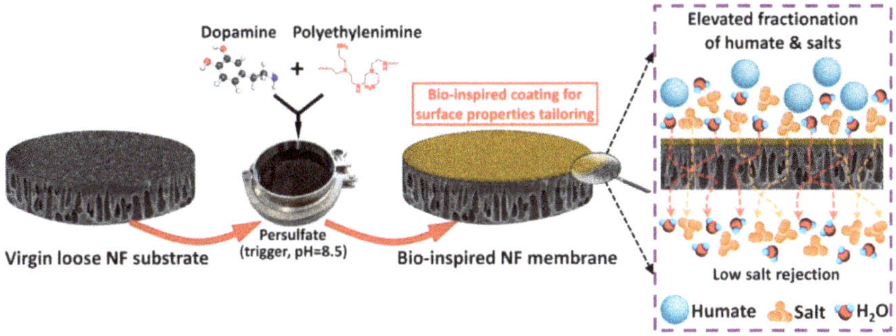

Fig. 11.7 The fabrication of bio-inspired NF membrane for recovery of humic substances from landfill leachate [73]

with 96.0% recovery. At the same time, the desalination efficiency was 99.5%, and thus the recovered humic substances showed a high purity, exhibiting great application potentials.

In another work, Ye et al. [74] utilized the loose NF membrane with molecular weight cut-off of 860 Da for recovery of humic substances from landfill leachate. Such a NF membrane possessed a permeation fluxes, and the transmissions for most inorganic ions were greater than 94.7%. The retaining humic substances reached up to 95.7%, showing a great promise for resource recovery from landfill leachate. During this process, cross-flow rates and temperature were demonstrated to have an obvious effect on the performance of recovering humic substances from landfill leachate. After filtration, the concentration of humic substances was increased from 1735 to 15,287 mg/L with a recovery of 91.2%, and meanwhile the desalination efficiency of this process was 85.7%. Furthermore, the practical application of the recovered humic substances was tested, and it can promote the growth of the green mungbean plants with no obvious phytotoxicity.

Iskander et al. [75] reported the application of chemical precipitation for recovery of HAs from the concentrated landfill leachate by FO. The results showed that the landfill leachate was concentrated by 3.2 times in 10 h using a 5-M NaCl draw solution via FO. Using this concentrated landfill leachate, chemical precipitation was performed under pH 2, and HAs recovery increased from 1.86 to 2.45 g/L. This was due to the O moieties in the molecular structures of HAs were replaced by other inorganics such as Cl, Na, and L, with higher molecular weights. At the same time, because the alkalinity causing species moved to the draw side driven by the concentration gradient, the amount of H_2SO_4 required in chemical precipitation (for HAs recovery) was reduced by 46.4%. Such a technique was beneficial to the subsequent treatment of landfill leachate. On the one hand, the HAs recovery reduced the production of sludge in Fenton treatment of landfill leachate by 30%. On the other hand, this technique also greatly reduced the amount of chemicals used in Fenton treatment, including H_2SO_4, $FeSO_4{\cdot}7H_2O$, and H_2O_2, in contrast to the general Fenton oxidation.

Xu et al. [76] utilized UF and evaporation combined process to promote the recycling of humic substances from mature landfill leachate. In mature landfill leachate used, FAs were the dominant organic compounds and accounted for 64% of DOC, and the concentration of humic substances was 441 mg/L. By using a two-stage UF, the concentrate of landfill leachate contained high-concentration humic substances (accounting for 91% of DOC) with low-concentration heavy metals. The concentrate of landfill leachate from UF was further concentrated via evaporation. The results showed that the contents of organic matter and dry-solids were 13 wt% and 25 wt%, respectively.

In another study, Xu et al. [77] reported the application of two-stage tight ultrafiltration (TUF) process to extract humic substances from MBR-treated landfill leachate, as water-soluble fertilizer. It was found that after concentration and desalting, the COD retention of MBR-treated landfill leachate increased by about 3%, while the rejections of metals showed a great performance. Divalent and monovalent ions showed 21–62 and 10–26% rejections. These results showed that metals can be effectively separated from MBR-treated landfill leachate by TUF process, generating humic substances fertilizers. Moreover, the generated humic substances fertilizers were reported to meet the requirements of Chinese nation professional standards for such water-soluble fertilizers.

Recently, Ye et al. [78] reported that using an electro-neutral NF membrane can achieve the effective recovery of humate from landfill leachate concentrate as fertilizers (Fig. 11.8). The results showed that by using such an electro-neutral NF membrane, the generated products contained 96.54% of humate and low salt concentration (3.47%). Under the pressure-driven concentration process, the concentration of humate in landfill leachate was concentrated from 1756 to 51,466 mg/L with a fold of 32.6, with a humate recovery of 90% and a desalination efficiency of 96.4%.

More recently, based on elucidating the selective adsorption mechanism and fabrication of novel adsorbents, we proposed an innovative approach for sustainable treatment of landfill leachate based on selective adsorption of HAs [8]. It is known that the main difference between HAs and other organics in landfill leachate was the greater aromaticity. Thus, based on such a property of HAs, it was speculated that the ideal adsorbent with π-π interaction would achieve the selective adsorption of HAs. Inspired by this speculation, the application of graphitic carbon nitride (g-CN) was evaluated for selective adsorption of HAs from landfill leachate. This is because that g-CN contains the delocalization electron structure, endowing it with strong π-π interaction with other aromatic compounds. In this study, ten types of landfill leachate samples including raw leachate and membrane concentrate were studied. The results showed that most samples, the removal of UV absorbance at 254 nm (UV_{254}) was 30–70%, and in some cases, the recovery rate and purity of the adsorbed humic-like components were 100%. These findings provide a promising direction for recovery of HAs from landfill leachate, as compared with the current membrane methods.

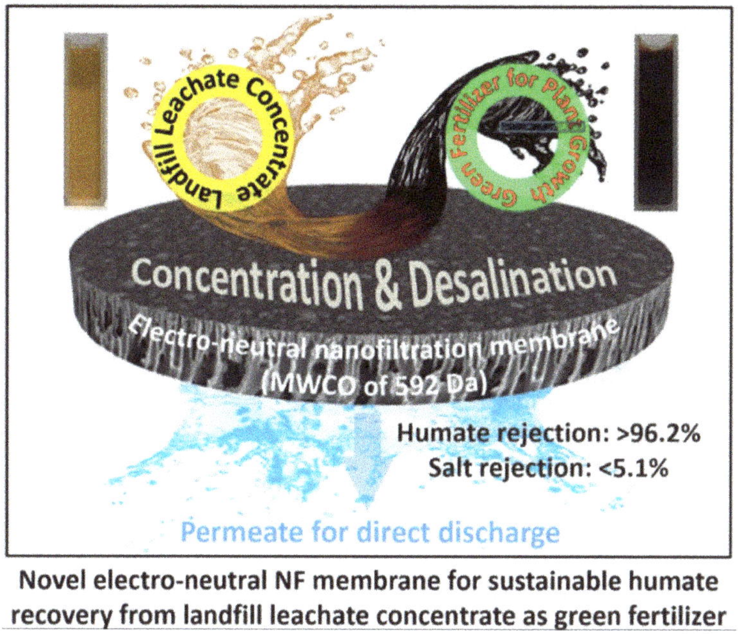

Fig. 11.8 The application of electro-neutral NF membrane for humate recovery from landfill leachate concentrate [78]

11.3.5 Other Resources Recovery

In addition, some studies reported the recovery of water, chloride ions, sulfur, volatile fatty acids, and other resources [79–82]. For example, Qin et al. [79] proposed a microbial electrolysis cell-forward osmosis system to recover water from landfill leachate along with nitrogen element (Fig. 11.9). It was found that using 2 M NH_4HCO_3 as the draw solution, such a process can achieve the 51% water recovery from the effluent of microbial electrolysis in 3.5 h operation. Similarly, Iskander et al. [80] used a microbial desalination cell along with FO to recover water from landfill leachate. The results showed FO can recover 51.5% water from the raw leachate, and the recovery of water increased to 83.5% from the concentrated leachate after desalination by the microbial desalination cell. Huang et al. [30] reported the application of Bi(III) containing oxides quantum dots/two-dimensional flakes for recovery of chlorite ions from the concentrated leachate (Fig. 11.10). The results showed that the maximum chlorite ions recovery rates were 61.8% and 66.1% under the optimum conditions. Zhang et al. [81] integrated partial nitrification, N-Anammox, and S-Anammox in a single system to recover valuable sulfur, and the recovery of sulfur reached 21.9%.

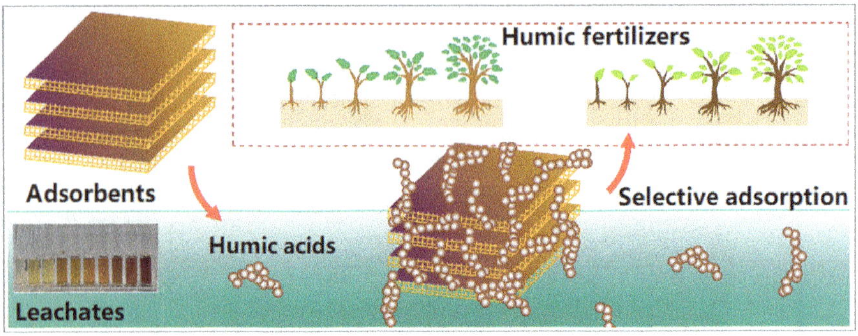

Fig. 11.9 The application of selective adsorption for recovery of HAs from landfill leachate by g-CN adsorbents [8]

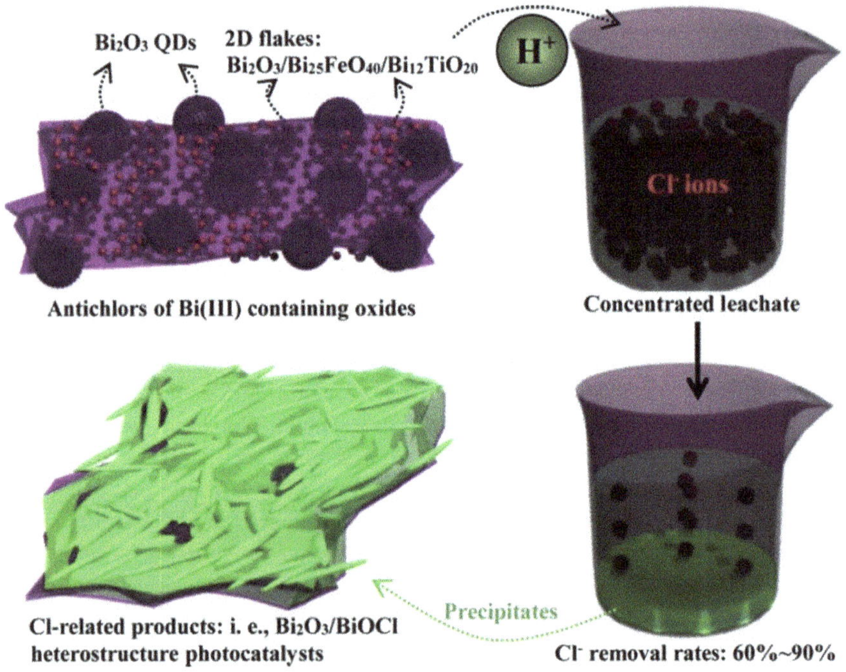

Fig. 11.10 Utilizing Bi(III) containing oxides quantum dots/two-dimensional flakes toward Cl^- recovery from leachate [30]

11.4 Challenge and Prospects of Landfill Leachate Treatment

11.4.1 Traditional Treatment

The environmental problems of landfill leachate have attracted more and more attention from researchers and policy-makers. However, landfill leachate is highly complicated in chemical composition, containing ammonia, metals, organic compounds and some emerging pollutants, and so far there is no one method or one integrated methods can fully treated the pollutants of landfill leachate. In terms of biological methods, including MBR, SBR, and up-flow anaerobic sludge blanket (UASB), these methods can remove the COD (the biodegradable fractions) of landfill leachate. Nevertheless, there is a large amount of refractory TOC that cannot be disposed by biological methods, particularly in old landfill leachate. In addition, there is a large amount of sludge produced in biological methods that need to be disposed. In this regard, physical and chemical methods, mainly including membrane and AOP, have been widely used with biological methods as pretreatment or posttreatment to meet the requirements of strict emission standard. Among these physical and chemical methods, membrane and AOP are the often used method. However, during membrane (NF, UF, and FO), there are some severe problems that cannot be addressed, and the generation of massive concentrate makes it challenging for full treatment of landfill leachate. In future, more effects need to be focused on the integrated treatment processes for improving the treatment efficiency of landfill leachate and the applicability for disposal of more pollutants in landfill leachate, particularly for some emerging pollutants. In addition, future studies need to address the bottleneck of treating landfill leachate, namely the membrane concentrate. Nevertheless, traditional treatment methods of landfill leachate do not consider the strategies of circular economy and carbon neutrality, and thus in near future, treatment of landfill leachate should consider both pollution elimination, resource recovery, and carbon emission reduction.

11.4.2 Resource Recovery

In contrast to the traditional methods, resource recovery from landfill leachate considers both the resources and pollutants properties. Thus, resource recovery possesses environmental, economic, and low-carbon profits. In general, landfill leachate contains nutrients (N, P), organic compounds, metals, humic substances, and other compounds that can be recovered. For example, ammonia can be recovered via struvite precipitation and ammonium stripping. Struvite precipitation needs a large amount chemicals and strict pH values, and by contrast ammonium stripping would be a promising method for ammonia recovery. Likewise, phosphorus was an important nutrient in landfill leachate, and can be effectively extracted via struvite precipitation. Phosphorus recovery can offset the treatment process costs of landfill leachate, and can make up the shortage of phosphorus resources. Nevertheless, the practical application of phosphorus recovered needs to be tested via field experiments. In addition to the nutrients, biodegradable organic compounds can be recovered as bioenergy, such as methane, hydrogen and electricity. These methods provide promising directions for resource utilization and pollutant treatment of landfill leachate. However, the environmental and economic profits and the credibility, engineering feasibility, and reliability of these methods should be further demonstrated. In terms of metals in landfill leachate, they can be extracted via resins and adsorption, while the efficiency and selectivity of these methods need to be substantially improved. This needs more efforts focused on the development of new methods for extracting metals from landfill leachate. As to the recovery of humic substances, given that humic substances are recovered as fertilizers, the toxic metals and other pollutants extracted along with humic substances need to be investigated, particularly focusing their toxicity in application in agricultural and forest areas. Moreover, more effective and selective methods need to be developed for extracted humic substances from landfill leachate, and selective adsorption would be a promising direction. Moreover, regardless of resources types and recovery methods, the carbon emission of these processes needs to be evaluated, as an indicator of technological advancement. In addition, there are few studies focused on the effect of emerging pollutants, such as perfluorooctane sulfonate, microplastics, and surfactants, on the resources recovery of landfill leachate. Hence, more studies should be conducted to develop more effective recovery methods to extract more available resources from landfill leachate, achieving reduction of both pollution and carbon emissions.

References

1. Wijekoon P, Koliyabandara PA, Cooray AT, Lam SS, Athapattu BCL, Vithanage M (2022) Progress and prospects in mitigation of landfill leachate pollution: risk, pollution potential, treatment and challenges. J Hazard Mater 421:126627
2. Keyikoglu R, Karatas O, Rezania H, Kobya M, Vatanpour V, Khataee A (2021) A review on treatment of membrane concentrates generated from landfill leachate treatment processes. Sep Purif Technol 259:118182
3. Iskander SM, Zhao R, Pathak A, Gupta A, Pruden A, Novak JT, He Z (2018) A review of landfill leachate induced ultraviolet quenching substances: Sources, characteristics, and treatment. Water Res 145:297–311
4. Zhao R, Novak JT, Goldsmith CD (2012) Evaluation of on-site biological treatment for landfill leachates and its impact: a size distribution study. Water Res 46:3837–3848
5. Siggins A, Thorn C, Healy MG, Abram F (2021) Simultaneous adsorption and biodegradation of trichloroethylene occurs in a biochar packed column treating contaminated landfill leachate. J Hazard Mater 403:123676
6. Qi C, Huang J, Wang B, Deng S, Wang Y, Gang Y (2018) Contaminants of emerging concern in landfill leachate in China: a review. Emerg Contam 4:1–10
7. Gu N, Liu J, Ye J, Chang N, Li YY (2019) Bioenergy, ammonia and humic substances recovery from municipal solid waste leachate: a review and process integration. Biores Technol 293:122159

8. Wang J, Wang C, Shi A, Shi Y, Yue D, Zhang L, Wang J, Wang H, Wang C, Cui D (2023) An innovative approach for landfill leachate treatment based on selective adsorption of humic acids with carbon nitride. Chem Eng J 461:142090
9. Sruthi T, Gandhimathi R, Ramesh ST, Nidheesh PV (2018) Stabilized landfill leachate treatment using heterogeneous Fenton and Electro-Fenton processes. Chemosphere 210:38–43
10. Ye Z, Zhang H, Yang L, Wu L, Qian Y, Geng J, Chen M (2016) Effect of a solar Fered-Fenton system using a recirculation reactor on biologically treated landfill leachate. J Hazard Mater 319:51–60
11. Teng C, Zhou K, Peng C, Chen W (2021) Characterization and treatment of landfill leachate: a review. Water Res 203:117525
12. Wdowczyk A, Szymańska-Pulikowska A (2021) Analysis of the possibility of conducting a comprehensive assessment of landfill leachate contamination using physicochemical indicators and toxicity test. Ecotox Environ Safe 221:112434
13. Klauck CR, Giacobbo A, de Oliveira EDL, da Silva LB, Rodrigues MAS (2017) Evaluation of acute toxicity, cytotoxicity and genotoxicity of landfill leachate treated by biological lagoon and advanced oxidation processes. J Environ Chem Eng 5:6188–6193
14. Deng Y, Jung C, Zhao R, Torrens K, Wu L (2018) Adsorption of UV-quenching substances (UVQS) from landfill leachate with activated carbon. Chem Eng J 350:739–746
15. Nika MC, Ntaiou K, Elytis K, Thomaidi VS, Gatidou G, Kalantzi OI, Thomaidis NS, Stasinakis AS (2020) Wide-scope target analysis of emerging contaminants in landfill leachates and risk assessment using risk quotient methodology. J Hazard Mater 394:122493
16. Singh S, Malyan SK, Maithani C, Kashyap S, Tyagi VK, Singh R, Malhotra S, Sharma M, Kumar A, Panday BK, Pandey RP (2023) Microplastics in landfill leachate: occurrence, health concerns, and removal strategies. J Environ Manage 342:118220
17. Kabir MS, Wang H, Luster-Teasley S, Zhang L, Zhao R (2023) Microplastics in landfill leachate: sources, detection, occurrence, and removal. Environ Sci Ecotechnol 16:100256
18. Su Y, Zhang Z, Wu D, Zhan L, Shi H, Xie B (2019) Occurrence of microplastics in landfill systems and their fate with landfill age. Water Res 164:114968
19. Pathak A, Pruden A, Novak JT (2018) Two-stage anaerobic membrane bioreactor (AnMBR) system to reduce UV absorbance in landfill leachates. Bioresour Technol 251:135–142
20. Jung C, Deng Y, Zhao R, Torrens K (2017) Chemical oxidation for mitigation of UV-quenching substances (UVQS) from municipal landfill leachate: Fenton process versus ozonation. Water Res 108:260–270
21. Reshadi MAM, Bazargan A, McKay G (2020) A review of the application of adsorbents for landfill leachate treatment: focus on magnetic adsorption. Sci Total Environ 731:138863
22. Han M, Duan X, Cao G, Zhu S, Ho S-H (2020) Graphitic nitride-catalyzed advanced oxidation processes (AOPs) for landfill leachate treatment: a mini review. Process Saf Environ 139:230–240
23. Amor C, De Torres-Socias E, Peres JA, Maldonado MI, Oller I, Malato S, Lucas MS (2015) Mature landfill leachate treatment by coagulation/flocculation combined with Fenton and solar Photo-Fenton processes. J Hazard Mater 286:261–268
24. Zhang L, Bai H, Zhang Y, Wang Y, Yue D (2021) Life cycle assessment of leachate treatment strategies. Environ Sci Technol 55:13264–13273
25. Chen W, Gu Z, Ran G, Li Q (2021) Application of membrane separation technology in the treatment of leachate in China: a review. Waste Manage 121:127–140
26. Li Y, Yi Q, Wang D, Wu Z, Wang Z (2022) Efficient treatment of landfill leachate using an electrochemical ceramic membrane filtration system: chlorine-mediated oxidation. Chem Eng J 450:138102
27. Kurniawan TA, Othman MHD, Liang X, Goh HH, Chew KW (2023) From liquid waste to mineral fertilizer: recovery, recycle and reuse of high-value macro-nutrients from landfill leachate to contribute to circular economy, food security, and carbon neutrality. Process Saf Environ 170:791–807
28. Liu G, Han M, Luo W, Liu Y, Fang J, Tian M, Wang L, Zhao S, Meng F (2023) Loose nanofiltration—membrane distillation integrated process for resource recovery from leachate MBR effluent: performance and membrane fouling. Desalination 545:116185

29. Chang H, Fu Q, Zhong N, Yang X, Quan X, Li S, Fu J, Xiao C (2019) Microalgal lipids production and nutrients recovery from landfill leachate using membrane photobioreactor. Bioresour Technol 277:18–26
30. Huang S, Li L, Zhu N, Lou Z, Liu W, Cheng J, Wang H, Luo P, Wang H (2020) Removal and recovery of chloride ions in concentrated leachate by Bi(III) containing oxides quantum dots/two-dimensional flakes. J Hazard Mater 382:121041
31. Wu S, Zou S, Liang G, Qian G, He Z (2018) Enhancing recovery of magnesium as struvite from landfill leachate by pretreatment of calcium with simultaneous reduction of liquid volume via forward osmosis. Sci Total Environ 610–611:137–146
32. Kurniawan TA, Singh D, Avtar R, Othman MHD, Hwang GH, Albadarin AB, Rezakazemi M, Setiadi T, Shirazian S (2021) Resource recovery from landfill leachate: an experimental investigation and perspectives. Chemosphere 274:129986
33. Chen G, Wu G, Li N, Lu X, Zhao J, He M, Yan B, Zhang H, Duan X, Wang S (2021) Landfill leachate treatment by persulphate related advanced oxidation technologies. J Hazard Mater 418:126355
34. Song J, Zhang W, Gao J, Hu X, Zhang C, He Q, Yang F, Wang H, Wang X, Zhan X (2020) A pilot-scale study on the treatment of landfill leachate by a composite biological system under low dissolved oxygen conditions: performance and microbial community. Bioresource Technol 296:122344
35. Saxena V, Kumar Padhi S, Kumar Dikshit P, Pattanaik L (2022) Recent developments in landfill leachate treatment: aerobic granular reactor and its future prospects. Environ. Nanotechnol Monit Manage 18:100689
36. Renou S, Givaudan JG, Poulain S, Dirassouyan F, Moulin P (2008) Landfill leachate treatment: review and opportunity. J Hazard Mater 150:468–493
37. Bu L, Wang K, Zhao Q-L, Wei L-L, Zhang J, Yang J-C (2010) Characterization of dissolved organic matter during landfill leachate treatment by sequencing batch reactor, aeration corrosive Cell-Fenton, and granular activated carbon in series. J Hazard Mater 179:1096–1105
38. Zhao X, Wei X, Xia P, Liu H, Qu J (2013) Removal and transformation characterization of refractory components from biologically treated landfill leachate by Fe^{2+}/NaClO and Fenton oxidation. Sep Purif Technol 116:107–113
39. Bove D, Merello S, Frumento D, Arni SA, Aliakbarian B, Converti A (2015) A critical review of biological processes and technologies for landfill leachate treatment. Chem Eng Technol 38:2115–2126
40. Abu Amr SS, Aziz HA (2012) New treatment of stabilized leachate by ozone/Fenton in the advanced oxidation process. Waste Manage 32:1693–1698
41. Soubh AM, Baghdadi M, Abdoli MA, Aminzadeh B (2018) Zero-valent iron nanofibers (ZVINFs) immobilized on the surface of reduced ultra-large graphene oxide (rULGO) as a persulfate activator for treatment of landfill leachate. J Environ Chem Eng 6:6568–6579
42. Karimipourfard D, Eslamloueyan R, Mehranbod N (2020) Heterogeneous degradation of stabilized landfill leachate using persulfate activation by $CuFe_2O_4$ nanocatalyst: an experimental investigation. J Environ Chem Eng 8:103426
43. Ren X, Xu X, Xiao Y, Chen W, Song K (2019) Effective removal by coagulation of contaminants in concentrated leachate from municipal solid waste incineration power plants. Sci Total Environ 685:392–400
44. Jiang Z, Cheng Z, Yan C, Zhang X, Tian Y, Zhang X, Quan X (2021) Simultaneous removal of nitrogen and refractory organics from a biologically treated leachate by pulse electrochemical oxidation in a multi-channel flow reactor. ACS Omega 6:25539–25550
45. Fernandez Y, Maranon E, Castrillon L, Vazquez I (2005) Removal of Cd and Zn from inorganic industrial waste leachate by ion exchange. J Hazard Mater 126:169–175
46. Oloibiri V, Ufomba I, Chys M, Audenaert WT, Demeestere K, Van Hulle SW (2015) A comparative study on the efficiency of ozonation and coagulation-flocculation as pretreatment to activated carbon adsorption of biologically stabilized landfill leachate. Waste Manage 43:335–342

47. Zielinska M, Kulikowska D, Stanczak M (2020) Adsorption-membrane process for treatment of stabilized municipal landfill leachate. Waste Manage 114:174–182
48. Mohan S, Gandhimathi R (2009) Removal of heavy metal ions from municipal solid waste leachate using coal fly ash as an adsorbent. J Hazard Mater 169:351–359
49. Modin H, Persson KM, Andersson A, van Praagh M (2011) Removal of metals from landfill leachate by sorption to activated carbon, bone meal and iron fines. J Hazard Mater 189:749–754
50. Bashir MJK, Aziz HA, Yusoff MS, Adlan MN (2010) Application of response surface methodology (RSM) for optimization of ammoniacal nitrogen removal from semi-aerobic landfill leachate using ion exchange resin. Desalination 254:154–161
51. Bashir MJ, Aziz HA, Yusoff MS, Aziz SQ, Mohajeri S (2010) Stabilized sanitary landfill leachate treatment using anionic resin: treatment optimization by response surface methodology. J Hazard Mater 182:115–122
52. Janin A, Blais JF, Mercier G, Drogui P (2009) Selective recovery of Cr and Cu in leachate from chromated copper arsenate treated wood using chelating and acidic ion exchange resins. J Hazard Mater 169:1099–1105
53. Bohdziewicz J, Kwarciak A (2008) The application of hybrid system UASB reactor-RO in landfill leachate treatment. Desalination 222:128–134
54. Wang G, Fan Z, Wu D, Qin L, Zhang G, Gao C, Meng Q (2014) Anoxic/aerobic granular active carbon assisted MBR integrated with nanofiltration and reverse osmosis for advanced treatment of municipal landfill leachate. Desalination 349:136–144
55. Zhou Y, Huang M, Deng Q, Cai T (2017) Combination and performance of forward osmosis and membrane distillation (FO-MD) for treatment of high salinity landfill leachate. Desalination 420:99–105
56. Fudala-Ksiazek S, Pierpaoli M, Luczkiewicz A (2018) Efficiency of landfill leachate treatment in a MBR/UF system combined with NF, with a special focus on phthalates and bisphenol A removal. Waste Manage 78:94–103
57. Ersahin ME, Cicekalan B, Cengiz AI, Zhang X, Ozgun H (2023) Nutrient recovery from municipal solid waste leachate in the scope of circular economy: recent developments and future perspectives. J Environ Manage 335:117518
58. Huang H, Xiao D, Zhang Q, Ding L (2014) Removal of ammonia from landfill leachate by struvite precipitation with the use of low-cost phosphate and magnesium sources. J Environ Manage 145:191–198
59. Xiong J, Zheng Z, Yang X, Dai X, Zhou T, He J, Luo X (2018) Recovery of NH_3-N from mature leachate via negative pressure steam-stripping pretreatment and its benefits on MBR systems: a pilot scale study. J Clean Prod 203:918–925
60. Guo S, Ao X, Ma X, Cheng S, Men C, Harada H, Saroj DP, Mang HP, Li Z, Zheng L (2023) Machine-learning-aided application of high-gravity technology to enhance ammonia recovery of fresh waste leachate. Water Res 235:119891
61. Jaffer Y, Clark TA, Pearce P, Parsons SA (2002) Potential phosphorus recovery by struvite formation. Water Res 36:1834–1842
62. Podder A, Reinhart D, Goel R (2020) Integrated leachate management approach incorporating nutrient recovery and removal. Waste Manage 102:420–431
63. Khanzada ZT (2020) Phosphorus removal from landfill leachate by microalgae. Biotechnol Rep 25:00419
64. Mahmoud M, Parameswaran P, Torres CI, Rittmann BE (2014) Fermentation pre-treatment of landfill leachate for enhanced electron recovery in a microbial electrolysis cell. Bioresour Technol 151:151–158
65. Hassan M, Wei H, Qiu H, Su Y, Jaafry SWH, Zhan L, Xie B (2018) Power generation and pollutants removal from landfill leachate in microbial fuel cell: variation and influence of anodic microbiomes. Bioresour Technol 247:434–442
66. Chang H, Feng H, Wang R, Zhang X, Wang J, Li C, Zhang Y, Li L, Ho S-H (2023) Enhanced energy recovery from landfill leachate by linking light and dark bio-reactions: underlying synergistic effects of dual microalgal interaction. Water Res 231:119578

67. Liu Q, Zhang X, Yu L, Zhao A, Tai J, Liu J, Qian G, Xu ZP (2011) Fermentative hydrogen production from fresh leachate in batch and continuous bioreactors. Bioresour Technol 102:5411–5417
68. Oz NA, Yarimtepe CC (2014) Ultrasound assisted biogas production from landfill leachate. Waste Manage 34:1165–1170
69. Upadhyay A, Singh R, Talwar P, Verma N, Ahire PD, Khatri H, Masakapalli SK, Pareek N, Kumar V, Kovalev AA, Zhuravleva EA, Litti YV, Vivekanand V (2023) Insights into sustainable resource and energy recovery from leachate towards emission mitigation for environmental management: A critical approach. J Environ Manage 343:118219
70. Lee H, Coulon F, Beriro DJ, Wagland ST (2023) Increasing recovery opportunities of metal(loid)s from municipal solid waste via landfill leachate recirculation. Waste Manage 158:116–124
71. Pap S, Stankovits GJ, Gyalai-Korpos M, Mako M, Erdelyi I, Sekulic MT (2021) Biochar application in organics and ultra-violet quenching substances removal from sludge dewatering leachate for algae production. J Environ Manage 298:113446
72. Liu XJ, Novak JT, He Z (2019) Removal of landfill leachate ultraviolet quenching substances by electricity induced humic acid precipitation and electrooxidation in a membrane electrochemical reactor. Sci Total Environ 689:571–579
73. Ye W, Liu R, Lin F, Ye K, Lin J, Zhao S, Van der Bruggen B (2020) Elevated nanofiltration performance via mussel-inspired co-deposition for sustainable resource extraction from landfill leachate concentrate. Chem Eng J 388:124200
74. Ye W, Liu H, Jiang M, Lin J, Ye K, Fang S, Xu Y, Zhao S, Van der Bruggen B, He Z (2019) Sustainable management of landfill leachate concentrate through recovering humic substance as liquid fertilizer by loose nanofiltration. Water Res 157:555–563
75. Iskander SM, Novak JT, He Z (2019) Reduction of reagent requirements and sludge generation in Fenton's oxidation of landfill leachate by synergistically incorporating forward osmosis and humic acid recovery. Water Res 151:310–317
76. Xu YD, Yue D-B, Zhu Y, Nie YF (2006) Fractionation of dissolved organic matter in mature landfill leachate and its recycling by ultrafiltration and evaporation combined processes. Chemosphere 64:903–911
77. Xu Y, Chen C, Li X, Lin J, Liao Y, Jin Z (2017) Recovery of humic substances from leachate nanofiltration concentrate by a two-stage process of tight ultrafiltration membrane. J Clean Prod 161:84–94
78. Ye W, Hong M, Huang X, Chen T, Gu A, Lin X, Li X, Chen X, Seo DH, Zhao S, Chen X, Van der Bruggen B, Xie M, Lin J (2023) Towards effective recovery of humate as green fertilizer from landfill leachate concentrate by electro-neutral nanofiltration membrane. Sci Total Environ 896:165335
79. Qin M, Molitor H, Brazil B, Novak JT, He Z (2016) Recovery of nitrogen and water from landfill leachate by a microbial electrolysis cell–forward osmosis system. Bioresour Technol 200:485–492
80. Iskander SM, Novak JT, He Z (2018) Enhancing forward osmosis water recovery from landfill leachate by desalinating brine and recovering ammonia in a microbial desalination cell. Bioresour Technol 255:76–82
81. Zhang F, Ren S, Liang H, Wang Z, Yan Y, Wang J, Peng Y (2023) Efficient nitrogen removal and partial elemental sulfur recovery by combined nitrite-anammox and sulfate-anammox: a novel strategy for treating mature landfill leachate. J Clean Prod 399:136553
82. Ciftcioglu-Gozuacik B, Sen G, Goze S, Ozbey-Unal B, Balcik C, Karagunduz A, Keskinler B (2023) Evaluation of volatile fatty acids and ammonia recovery approach from landfill leachate using pilot-scale mechanical vapor recompression. J Environ Manage 345:118720

Chapter 12
Leachate Discharge from a Public Landfill: Design and Sizing of a Treatment System

Salah Souabi, Abdelkader Anouzla, Nor-Edine Abriak, Aysegul Pala, and Tonni Agustiono Kurniawan

Abstract The production of solid waste, which is increasing on a global scale, presents considerable impacts linked to the environment. The storage of this waste in anaerobic conditions makes it possible to produce a large volume of leachate, which is harmful to the environment. This study concerns the sizing and design of a system for treating leachate (young and stabilized) by nitrification–denitrification, coagulation-flocculation followed by filtration on materials. The results showed that the optimal concentrations of $FeCl_3$ used as a coagulant vary between 6–8 and 4–6 for young and stabilized leachates for a pH of around 8.5. This work constitutes an interesting case study, making it possible to reduce leachate pollution produced by a public landfill using a simple technology. This work includes an interesting case study, making it possible to minimize leachate pollution produced by a public landfill using a simple technology. The results obtained in the laboratory on a pilot scale showed that the proposed steps are complementary and make it possible to reduce pollution effectively. However, using the treatment plan to discharge leachate stored in Basin 3 is more enjoyable, which presents a reduced pollutant load compared to the quality of raw leachate received by Basin 1. The sector's design based on the determined optimal conditions makes it possible to minimize pollution considerably.

S. Souabi · A. Anouzla (✉)
Laboratory of Process Engineering and Environment, Faculty of Science and Technology, University Hassan II, Mohammedia, Morocco
e-mail: aanouzla@gmail.com

N.-E. Abriak
IMT Lille Douai, GCE, 764 Bd Lahure, BP 10838, 59508 Douai, France

Université Lille Nord de France, LGCgE, Villeneuve d'Ascq, France

A. Pala
Faculty of Engineering, Environmental Engineering Department, Dokuz Eylul University, Izmir, Turkey

T. A. Kurniawan
College of Ecology and the Environment, Xiamen University, Xiamen 361102, Fujian, China

A. Anouzla and S. Souabi (eds.), *A Review of Landfill Leachate*, Springer Water,
https://doi.org/10.1007/978-3-031-55513-8_12

Furthermore, calculating the sector's costs remains moderate, which could encourage establishing the industry in development.

Keywords Leachate · Landfill · Design · Sizing treatment

12.1 Introduction

Waste management in general, and that of municipal solid waste in particular, is one of the main challenges facing our societies. With ever more significant and diversified consumption worldwide, waste production continues to increase in quantity, thus creating enormous risks for the environment and human health. This situation is much more worrying in developing countries. In addition, the lack of resources amplifies the difficulty of addressing this issue with an approach adapted to the regional context.

Municipal solid waste management is among the main challenges facing our societies. The combination of interrelated factors, from population growth, urban expansion, and socioeconomic and production activities development to lifestyles and consumption patterns, creates a growing waste deposit [3, 18]. In developing countries, solid wastes generated by the population and economic activities present severe problems and nuisances that can impact humans and environmental health.

The quantity of households and similar waste is continually changing in Morocco. It has increased from 6.3 to 7.4 million tons per year between 2007 and 2015 due to economic growth and urbanization [9, 13].

For most local Moroccan authorities, waste management has become a significant issue. The average collection rate for household and similar waste in the country varies between 80 and 100% in urban areas. Waste management is delegated to private companies, while most rural communities deal directly with their waste due to financial constraints.

Many controlled landfills are operational in Morocco, and others are being rehabilitated. All these discharges carried out are confronted with the problem of leachate management. Negative impact mitigation measures often remain limited, difficult to apply, and costly, such as evaporation, reverse osmosis or methanization [11, 14, 15].

Among the problems encountered caused by leachate, we note the salient alteration of the surface and groundwater quality (high load of organic, mineral, toxic matter, and bacteriological pollution). The need to find techniques adapted to treating these discharges with the initial conditions and the requirements of the receiving environment remains an obligation that cannot be circumvented.

The interest in landfills, whose establishment is uncontrolled and spontaneous, has grown considerably. Therefore, some priority actions have been given to control the impacts generated by the multiplication of these uncontrolled landfills. Indeed, uncontrolled waste dumping leads to visual effects, water resources, and air pollution.

Controlled landfills are defined as an installation or site meeting the regulatory technical characteristics and prescriptions where waste is deposited. These sites consider several elements, including realizing a leachate drainage system [17]. Issues related to leachate management and treatment in Morocco are technological and environmental. Indeed, the pressures on the environment, the quality of life of the populations, and the sanitary conditions are becoming increasingly alarming, leading researchers to find adequate solutions for good leachate management. In particular, the treatment by less expensive techniques in investment and operation [1, 4, 7]. Therefore, the stakes are high since it will be necessary to ensure adequate drainage and recovery of the leachate and its treatment for reuse or discharge into the receiving environment [7]. The genesis of leachate is the result of a highly complex process. Indeed, the waste is subjected to biodegradation processes resulting from numerous biological and physicochemical reactions. The water that infiltrates within the layers of the waste is the primary vector of this biodegradation. Thus, understanding the genesis of leachates involves knowing the nature of the landfilled waste and the operating mode of the waste storage facility (height of the waste, the surface area used, compaction, etc.) and the study of interactions between water and waste. The leachate composition depends on several parameters, including the age of the landfill, the type of waste, and the technique used for its burial [7]. El Kharmouz et al. [19] revealed high pollution levels from the public landfill of Oujda city. Leachate is, therefore, loaded with organic and mineral pollutants, which can be harmful (if infiltrated) to the quality of ground and surface water [7, 18] and can affect human health and hygiene [22]. The study by Gamar et al. [10] on the production of leachate from the wild landfill of El Hajeb showed an intense organic load exceeding Moroccan standards.

To control solid waste leachate pollution at the MohammediaBenslimane landfill, Kastali et al. [18] reported that the daily leachate flow rate calculation is essential for the treatment choice. Controlling the pollution generated by solid waste, particularly leachate, requires evaluating the flow rate produced in m^3/day. In addition, it makes it possible to estimate the pollutant load and the design and sizing of the treatment process selected.

Several methods for evaluating the leachate flow rate are used, particularly the water balance model [2, 6], which considers precipitation, runoff, evaporation, and wastewater. The last parameter is significant since waste in Morocco has very high moisture, between 60 and 75% [25]. A study in the landfill of Oujda City showed that the humidity of the waste received reached an average rate of 63.5% [5]. At the same time, Chiguer et al. [8] reported that this percentage is equal to 70% on average for the waste of Essaouira city. Therefore, evaluating landfill leachate flow is necessary to assess its environmental impact and choose a treatment technique. The factors influencing the leachate production process vary in time and space, making the assessment process more complex and complicated [2]. This paper aims to apply the water balance method to the OumAzza landfill and compare this method's results with leachate production ratios and the actual quantities produced on-site.

12.2 Material and Methods

12.2.1 *Presentation of the Study Area: City of Kenitra*

Kénitra, also called the capital of Gharb, is a city of 76 km^2 located 45 km north of Rabat. It is one of Morocco's most important industrial centers (the country's fourth industrial city).

It is considered one of the largest cities northwest of the Kingdom. Indeed, the city is located on the south shore of the Sebouwadi 12 km from the mouth of the Atlantic Ocean at Mehdya Beach.

The climate in this city is sub-humid to semi-arid, with an average rainfall of about 600 mm. According to the urban municipality of Kenitra, the population reached 2014 nearly 1,061,435 inhabitants.

The service of collection and cleaning of this waste to two professional operators: The company SEGEDEMA and the company TECMED. In comparison, the burial and treatment at the Kenitra ordinary landfill were entrusted to the company SOS.

The Ouled Berjal landfill is located northwest of Kenitra on the secondary road from Kenitra to Sidi Allal Tazi, on the left bank of the Oued Sebou estuary. It covers an area of 20 ha and receives an average of 329 tonnes of daily waste. The figure illustrates the Ouled Berjal landfill located in Kenitra (Fig. 12.1).

Fig. 12.1 Ouled Berjal landfill

The landfill has:

- Waste landfill area made up of a 4 ha compartment, which is subdivided into 4 cells;
- Leachate and rainwater drainage pipe;
- Three large basins to collect leachate and a reserve basin:
  - The first basin was built in 2010 with a capacity of 21945 m^3 or ($105 \times 55 \times 3.8$);
  - The second basin was built in 2012 with a capacity of 7363 m^3 or ($39 \times 59 \times 3.2$);
  - The third basin was built in 2013 with a capacity of 19,220 m^3 or ($88 \times 78 \times 2.8$);
  - The fourth reserve basin was built in 2014 with a capacity of 19,200 m^3 or ($100 \times 60 \times 3.2$).

Currently, the OuledBerjal landfill is encountering several difficulties. Indeed, there is an excellent risk of saturation of the landfill. In addition, the leachates are stored in basins and are not treated. This can lead to an overflow of leachate, altering the water table and the neighboring agricultural land.

12.2.2 Sampling Techniques

Sampling a liquid influent poses severe problems for controlling wastewater quality and its environmental impact.

For the OuledBerjal landfill, samples were taken from 3 storage basins (B1, B2, and B3) in three campaigns.

Leachate sampling is done in situ using a container directly from each basin.

The sampling technique consists of taking a sample of 5 L per hour for 8 h. The samples were used to diagnose the discharges and the physicochemical treatment tests by coagulation-flocculation followed by filtration. An automatic sampler was used to take samples from the storage basins.

12.2.3 Physico-Chemical Analysis

A pH meter labeled the "Accumet basic ab15" regulated the pH following the February 2001 NF T 90-008 standard (T 90-008).

Turbidity was analyzed using a HACH model 2100 N turbidimeter following the March 2000 edition of NF EN ISO 7027.

The total suspended matter (TSM) is measured according to NF T 90-105 January 1997 (T 90-105) protocol.

The chemical oxygen demand (COD) analysis was performed according to the NF T90-101 standard in February 2001 (T90-101). Using the manometric approach NF EN 1899 May 1998, the Biological Oxygen Demand (BOD) was calculated after five days (T90-103).

Nitrate concentrations were determined using the NF EN ISO 78-90 January 1997 (T 90-045) method.

The amount of ammoniacal nitrogen was tested according to the NF T 90-015 standard in January 1997.

NF T 90-023 from January 1997 determined the total phosphorus content. The Folin-Ciocalteu reagent-based colorimetric method was used to determine the concentration of phenolic compounds (Singleton and Rossi 1965).

12.2.4 Coagulation Flocculation Treatment

We used the Jar test technique, an experimental procedure that makes it possible to determine the optimal conditions for coagulation/flocculation: pH, coagulant, and Superfloc SD2065 flocculant. The leachate was treated by coagulation with chloride ferric and the flocculantSuperfloc SD2065. The optimal dose of $FeCl_3$ 40% was previously determined following Jartest tests.

The Superfloc SD2065 flocculant produced by the company is a cationic polyacrylamide as an emulsion. It is ideal for wastewater treatment and clarification in all industries. The characteristics of this flocculant are grouped in Table 12.1

The optimum concentrations determined for the coagulant and the flocculant are given in Table 12.2.

Table 12.1 Characteristics of superfloc SD2065 flocculant

Parameter	Value
Appearance	Opaque liquid
Density (g/cm^3)	1.01–1.05
Viscosity (cP) diluted à 0.50%	270
Freezing point (°C)	−18 °C
Large molecular	Weight

Table 12.2 Optimum concentrations determined for the coagulant and the flocculant

	Young leachates	Stabilizedleachates
FeI_3 g/l	6–8	4–6
Superfloc SD2065 mg/l	1–4	1–2
pH	8.6	8.5

12.2.5 Physico-Chemical Characteristics of Leachate Discharge

Table 12.3 illustrates the variation in the physicochemical parameters of leachate discharges (average values of three sampling companions) during three sampling companions (C1, C2 and C3) for three storage basins (B1, B2 and B3).

The physicochemical analysis of the leachate is an essential step before starting treatment. Indeed, the choice of the analysis technique is strongly linked to the composition. On the other hand, this analysis also makes it possible to determine the impact of the leachate on groundwater (in the event of infiltration), the environment neighboring the landfill, and human health.

The composition of the leachate depends on several parameters, including The age of the landfill, the type of waste buried, and the technique used for their burial.

Table 12.3 reports the average values of three sampling campaigns for the physicochemical parameters analyzed during the study.

The results show that the leachates stored in the different basins B1, B2 and B3 have a basic pH of 8 and 9. According to Kjeldsen et al. [20], with the aging of the landfill, the leachate becomes depleted in volatile organic compounds, increasing pH values up to 7 or more. Thus, the pH of the leachate is strongly influenced by natural dissolved materials in the environment.

Table 12.3 Variation in physicochemical parameters of leachate discharges

Parameters/basins	B1	B2	B3
pH	8.43	8.65	8.7
Conductivité (ms/cm)	14.2	14.3	14.4
Turbidité NTU	78.8	73.7	72.8
254 nm (FD = 200)	0.14	0.146	0.17
TSS g/l	0.38	0.35	0.293
COD mg/l	4519	4415.9	4372
BOD5 mg/l	992.5	895	899.3
Phenol mg/l	1126.6	11,139.5	1109.2
Surfactant mg/l	23.3	37.5	49.2
NH_4^+ mg/l	152.6	98	110.5
TKN mg/l	341	291	272
Organic nitrogen mg/l	188.4	193	161.5
NO_3^- mg/l	349.4	328	282.7
NH_4^+/NO_3	0.5	0.13	0.02
Phosphorus mg/l	24.1	14.2	16.7
COD/BOD5	3.95	4.6	5
COD/Pt	276	322	258.4

B1: basin1, B2: Basin 2, B3: Basin 3

The maximum COD and BOD5 recorded values vary between 4519 and 992 mg/L for COD and BOD5, respectively. In contrast, the minimum values reached 4372 and 895 mg/L for the same parameters. These results are close to those reported by Silva et al. [24].

The BOD5/COD ratio makes it possible to determine the age of the leachate (young, intermediate, or stabilized) and its biodegradability. According to Renou et al. [23], BOD5/COD ratios greater than 0.3 characterize young biodegradable leachates, while aged or stabilized leachates have a ratio less than 0.1. This study concludes that the leachate discharges collected from basins B2 and B3 are rich in non-biodegradable materials, while the organic matter of the effluent collected from basins B1 is considered biodegradable.

Furthermore, the presence of suspended matter causes an increase in the turbidity of any liquid. The TSS varies from 0.38 to 0.293 g/L for the three basins, while the turbidity varies from 79 to 73 NTU. These values remain lower than those reported by Mahmud et al. [21].

The absorbance at 254 nm (Abs 254 nm) characterizes conjugated compounds (aromatic rings) and is generally used for the determination of organic matter (Bouaghssas et al. 2022). In the case of the three basins, this absorbance has a maximum value of 0.17 and a minimum value of 0.0.14 (dilution factor 200 times).

In addition, we note that the phenol concentration has a maximum value of 1139.5 mg/L and a minimum. value of 1109 mg/L. These values are higher than those reported by Silva et al. [24], 162 mg/L, and those recorded by Bannama et al. (2010), 37 mg/L. It should be noted that phenol concentration increases over time, given that anaerobic storage promotes the degradation of phenolic derivatives while producing 49 mg/L.

The ammonium ion (NH_4^+) essentially comes from the biodegradation of dead organic matter. The contents recorded during our study vary from 153 to 98 mg/L.

Furthermore, the nitrate levels in the three basins vary from 349 to 283 mg/L. These values remain lower than the maximum value determined by Souabi et al. (2011) for leachate from waste storage centers, being 845 mg/L (Fig. 12.2).

A figure illustrates three leachate samples (raw leachate) R.L, coagulated-flocculated leachate (CFL), and filtered leachate.

***Study* of the design of the proposed treatment sector**

Figure 12.3 illustrates the physicochemical treatment technique proposed for leachate treatment.

Design of treatment basins by nitrification, denitrification, coagulation followed by filtration

Estimated quantity of waste produced for the city of Kénitra

The number of inhabitants in 2014 given by the High Commission for Planning:

The population of Kenitra City in 2014 was 604,069 inhabitants (Po).

Also according to the High Commission for Planning, the urban growth rate for the city of Kenitra is 2.1% (Reference: https://www.hcp.ma/region-kenitra/attachment/1067235/).

In 2040, the population would be 1,036,945 inhabitants.

Fig. 12.2 Comparative study of raw leachate and treated by coagulation-flocculation followed by filtration (basin B1) R.L: Raw leachate, C.F.L: Coagulated-flocculated leachate F.L: Filtered leachate

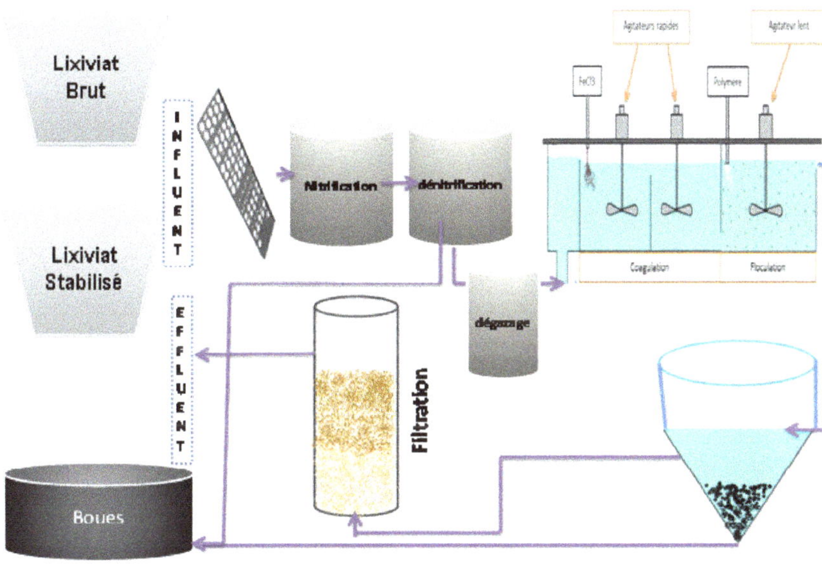

Fig. 12.3 Coagulation flocculation treatment technique followed by filtration

$$P_n = P_0 \times (1 + t)^{n - n_0}$$

Bouyakhssas et al. reference shows that the leachate ratio is 0.26 m^3/tonne of solid waste.

n = 2040 et n = 2014 fdffffffr(

This corresponds to a volume of leachate of Qleachate = 829.6 × 0.26 = 215.7 m^3/day.

We opt for the Qlixiviat design = 216 m^3/day.

The design of a leachate treatment sector (Kénitra landfill) must be carried out based on the leachate produced and the flow of rainwater.

Qc = Qleachate flow + Qrain + 10% Qleachate flow
Qleachate flow = 216 m^3/day
Qrain = 20% leachate flow = 0.2 × 216 = 43.2 m^3/day
Empty volume of the reactor = 10% Qleachate flow = 216 × 0.1 = 21.6 m^3
Q Total = Qdesign = 216 + 43.2 + 21.6 = 281 m^3/day
Q Total = 12 m^3/h Q Total = 0.0033 m^3/s.

12.2.6 Manual Screening

Grid width: st method

The following formula gives the width of the grid:

$$l = \frac{t}{\sin(\phi)}$$

With:

ϕ Angle of inclination (in degrees)
t Draft or Maximum height of water on the grid in m
l width in m

$l = \frac{0.4}{\sin(60)} = 0.46$ m We opt for grid width l = 0.5 m

Calcul de la surface

S inclined = Q/(A × C × Ve)

With A= e/(e + s)

E = 20 mm spacing between bars

s = 10 mm thickness of the bars

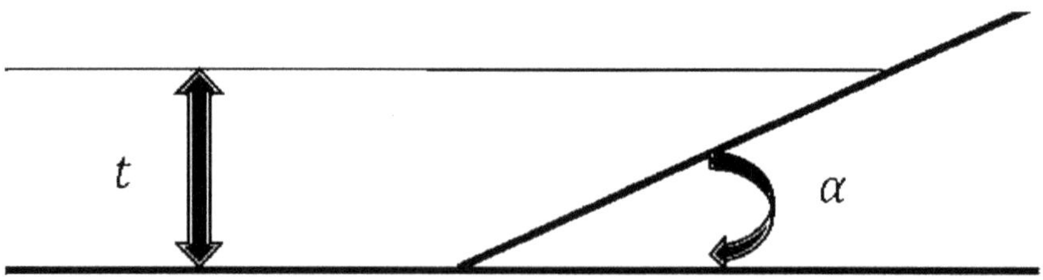

Fig. 12.4 Angle of inclination

A= e/(e + s) = 0.33

C = 0.5

Ve = 0.8 m/s flow velocity

S inclined = 0.0033/(0.8 × 0.33 × 0.5) = 0.025 m^2

Svertical = Sinc × sin 60 = 0.04 × sin 60 = 0.021 m^2

$$0.6\ \text{m/s}\ \ \text{m/s} < \text{Ve} < 1.2$$

Calculation of the width l using the KIRCHMER method

$$l = \frac{S_v.\sin(\alpha)}{h(1-c).\delta}$$

H water height in the water passage channel (0.2–0.3) m
c clogging coefficient 0.45.
δ Coefficient taking into account the screening mode such as δ = 0.5 if the screening is automatic (connected population ≥ 5000 inhabitants).
α angle of inclination of the grid (60°).

Calculating Grid Width

$$l = \frac{0.022 * \sin 60}{0.2 * (1 - 0.45) * 0.5} = 0.35\ \text{m}$$

The first method: l = 0.46 m
The second method: l = 0.55 m
So we opt for l = 0.5 m
Let the length L = 1 m and l = 0.5 m

Sincl = Actual inclined surface = L l = 1 × 0.5 = 0.5 m^2

Verticle area = Sincl × sin 60 = 0.5 × sin 60 = 0.43 m^2

$$H = \frac{Surface\ verticale}{Largeur\ de\ la\ grille} = 0.43/0.5 = 0.87\ \text{m}$$

So we opt for H = 1 m

$$0.7\ \text{m/s} < \text{Ve} < 1.2\ \text{m/s}$$

Calculation of the number of bars in the grid:

The following formula gives the number of spacings in the grid:

Method:

Or

e spacing between bars e = 20 mm
S the thickness of the bars = 10 mm
Ne spacing number
NS Number of bars

l = Ne.e + NS S (L: channel width = grid width = 0.5m)

NS = Ne + 1 So : l = Ne.e + S.(Ne + 1)
l = Ne(e + S) + S and (l − S)/(e + S) = Ne
NS = Ne + 1 = (l + e)/(S + e) = (500 + 20)/(10 + 20) = 17.33 bars

We opt for NS = 18.

Ne = 17 Spacings

12.3 Calculation of Load Loss

The loss calculation method can also be calculated using the KRISCHMER formula.

$$\Delta H = K\frac{Ve^2}{2*g} * \sin\partial = \beta * \left(\frac{s}{e}\right)^{\frac{4}{3}} * \frac{Ve^2}{2*g} * \sin\partial$$

Avec:

$$K = \boldsymbol{\beta} * \left(\frac{s}{e}\right)^{\frac{4}{3}}$$

e spacing between bars; 30 mm
g 9.81 m/s
s bar diameter 8 mm
θ angle of inclination of the grid concerning the horizontal 60°
β coefficient, which takes into account the shape of the bars: 1.79
Ve flow speed upstream of the grid: 0.8 m/s

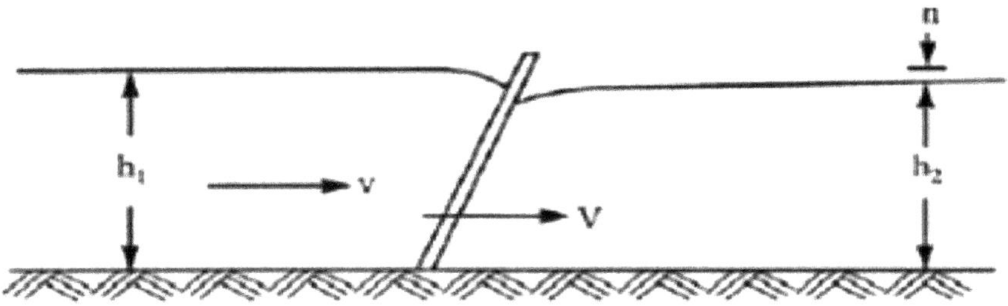

Fig. 12.5 Flow of wastewater through the screen

$$\Delta H = 1.79 * \left(\frac{8}{30}\right)^{\frac{4}{3}} * \frac{0.8^2}{2 * 9.81} * \sin 60$$

$$\Delta H = 0.07 \text{ m} = 70 \text{ mm}$$

Another method for calculating the pressure loss

Source Tchobanoglous G, Burton FL, Stensel HD, 2004; In wastewater engineenng treatment and reuse; METCALF & EDDY, Inc.; the McGraw-Hill Eds., edition international, 316 p.

Source: Lin DL, 2007; In water and wastewater calculations manual; C Clee, Eds; the McGraw-Hill, New York; 582 p.

The figure allows us to calculate the pressure loss in the screen using the Bernoulli equation.

$$h_1 + \frac{\upsilon^2}{2g} = h_2 + \frac{V^2}{2g} = \Delta h$$

$$h = h_1 - h_2 = \frac{V^2 - \upsilon^2}{2gC^2} \quad (12.1)$$

With:

- h_1 the downstream water depth (m).
- h_2 water depth (m).
- h discharge loss (m).
- V Speed through the screen (m/s) = 0.8 m/s.
- υ estimated speed in the screen downstream channel (m/s) = 0.6 m/s.
- g Gravitational constant 9.81 m/s^2.
- C Discharge coefficient whose typical value is $C^2 = 0.7$ for clean screen, 0.6 for mouth screen.

$$\Delta h = \frac{1}{0.7}\left(\frac{0.8 - 0.6^2}{2 * 9.81}\right) = 0.032 \text{ m} \Leftrightarrow 29.4 \text{ mm}$$

Table 12.4 Variation in pollutant loads received by the 3 treatment basins Flow = 281 m^3/d

Parameters	B1 CP kg/d	B2 CP kg/d	B3 CP kg/d
TSS	107	99.3	82.4
COD	1269.9	1240	1228
BDO5	185	251.5	252.7
Phenol	316.6	315.3	311.6
Surfactant	6.5	10.5	13.8

Calculation of air gap and screen dimensions

$L1 = 17 \times 0.02 = 0.34$ and $L2 = 18 \times 0.01 = 0.18$ m

L1: width of the spacing l2 = width of the bars

We chose the bars so that the width is 15 mm and the depth is 38 min.

The width of the channel:

$$\begin{aligned} L3 &= l1 + l2 = \text{Ne} \times \text{e} + \text{Nb} \times \text{s} = 17 \times 0.02 + 18 \times 0.01 \\ &= 0.34 + 0.18 = 0.52\ \text{m} \end{aligned}$$

Screen efficiency:

$$\text{Efficiency}\% = (17/(17 + 18) = /0.34/0.52 \times 100 = 49\%$$

Calculate the average pollutant load in COD, BOD5, TSS, pHenol, surfactants etc.

The table shows the pollutant loads in kg/d for the flow rate of 281 m^3/d.

Table 12.4 illustrates the pollutant load (CP) measured at the three stacking basins (B1, B2 and B3).

These results show that the pollutant loads for the different parameters vary over time. The yield obtained for basins B1 and B2 is not significant.

Calcul du volume de résidu

For urban wastewater, the volume of retained residues (V) expressed in liters per inhabitant per year varies depending on the spacing of the bars.

The volume of leachate produced per inhabitant (lix l/inhabitant)

According to the work carried out by Bouyakhssas et al.

We produce 0.26 m^3/ton of solid waste, i.e. 0.26 l/kg of waste

Solid waste production ratio = 0.7 kg/inhabitant

However, for 0.7 kg DS/inhabitant/day the leachate produced is: $0.26 \times 0.7 = 0.18$ l/inhabitant/day

Leachate produced per year $= 0.18 \times 365 = 65.7$ l/inhabitant/year

Production of leachate by the entire population: Total population $\times$ 65.7

Population 2023 = 508,000 einhabitant

In 2040, the population would be 1,036,945 inhabitants.
Leachate produced = 65.7 × 1,036,945 = 68,127.3 m^3/year

Sizing of the nitrification and denitrification basin

Nitrification

Nitrification is one of the stages of wastewater treatment which aims to transform ammonium (NH_4^+) into nitrate (NO_3^-). Bacteria autotrophs in an aerobic environment carry out this transformation.

The nitrification reaction of ammoniacal nitrogen by autotrophic nitrifying bacteria in the presence of dissolved oxygen can be represented by the following reaction:

$$NH_4^+ + 3/2O_2 \rightarrow NO_2^- + H_2O + 2H^+$$
$$NO_2^- + 1/2O_2 \rightarrow +NO_3^-$$

Denitrification

Different bacteria can carry out the transformation of nitrates into nitrogen according to the following reactions:

$$NO_3^- \rightarrow NO_2^- NO \rightarrow N_2O \rightarrow N_2$$

Calculation of the volume and surface area of the basin to be nitrified (Basin 1 and Basin 3)

We propose to use circular-shaped basins of diameter D.
Residence time for nitrification: 24 h.
Flow rate to nitrify Qc = 281 m^3/d.

Calculation of pelvic volume

$$V = Ts \times Qc = 24 \times 281/24 = 281 \text{ m}^3 \text{ T stay} = 24 \text{ h}$$

Calculation of the surface

Height of the pool = 2 m
V = Sh × H and Sh= V/H = 281/2 = 140.5 m^2 Sh = Horizontal Surface
We adopt Sh = 140 m^2 for a cylindrical basin.
Wc will havc:
Sh = (II D2)/4 and D = (4 Sh/II)1/2 D = 9.5 m We opt for D = 10 m.

Ventilation system design

4.2 g of oxygen is necessary for the oxidation of 1 g of ammoniacal nitrogen.

The daily mass of nitrogen to nitrify [Nnit] is equal to the mass of Kjeldahl nitrogen [TKN inlet] provided by the raw water reduced by:

the mass of refractory particulate organic nitrogen [Nopr] (trapped in the sludge),

The mass of refractory soluble organic nitrogen [Nosr] (rejected),

The mass of nitrogen assimilated essentially by the carbon bacteria [Nass], (incorporated into the sludge),

The mass of ammoniacal nitrogen is released with the purified water [N-NH_4 release].

Case of Basin 1

After 24 h of aeration of 50 L of young leachate (basin 1) the NH_4^+ concentration drops from 221 to 21.5 mg/l. The concentration of TKN in the leachate collected from basin B1 varies around 320 mg/l.

TKN input kg/d = Qc (m^3/d) × TKN (kg/m^3) = 281 × 0.320 = 90 kg/d

NH_4^+ rejection kg/d = 281 × 0.0215= 6

Nnit = TKN entry –N-NH_4 rejection = 90 − 6 = 84 kg/d

The daily oxygen demand corresponding to nitrification (DONit) [expressed in kg O_2/day] is written:

DONit = 4.2 × Nnit = 4.2 × 84 = 353 kg O_2/day (Nnit: quantity of Nitrogen in kg/day)

DONit= 353 kg of O_2/day

QO_2 kg/d) = Volume of air × density of air × proportion of O_2 in air × transfer coefficient

The airflow rate to be blown is determined according to the following expression:

Vair = (O_2kg/d)/(Density x proportion of O_2 in air × O_2 transfer efficiency

Vair = (O_2kg/d)/(1.29 × 0.23 × 0.2) = 353/0.06 = 5884 m^3/d

We give:

- The density of air is 1.29 g/l
- The proportion of O_2 in the air is 23.2% by weight
- Air transfer efficiency is 20%
- The energy required to dissolve O_2

1 kg of O_2 = 0.7 kWh to dissolve the O_2 i.e. Specific intake 1.43 kg of O_2/kWh

The total energy to dissolve O_2 is:

In total (kWh) = O_2 kg/d/1.43 = 353/1.43 = 247

Energy cost = 247 × 0.09 = 22.2 Euro/day with cost of one kWh= 0.09 Euro

The ventilation power to be installed can be determined:

- By dividing the hourly input (kg O_2/h) by the gross specific input of the aerator (kg O_2/kWh) in the case of surface aerators:

P = AH/ASB = (kg O_2/h)/1.42 = 353/24 × 1.42 = 10.4 kWh avec: P = power absorbed at the electric meter (kW) with ASB Gross specific contribution of the aerator: 1.42.

The Table 12.5 summarizes the design parameters of the nitrification basin.

Table 12.5 Nitrification basin design parameters

Parameters	Calculated value
Wastewater flow m^3/j	281
Residence time day	1
V useful m^3	281
Height m	2
Area m^2	140.5
Diameter m	10
O_2 kg d'O_2/day	353
Air volume m^3/d	5884
Pump power kWh	10.4
ETotal useful energy kWh	247
Prix unitaire kWh en Euro	0.09
Energy cost Euro/d	22.2

Case of basin 3

After 24 h of aeration of 50 L of young leachate (basin 1) the NH_4^+ concentration drops from 190.5 to 40 mg/l. The concentration of TKN in the leachate collected from basin B1 varies around 210 mg/l

TKN input kg/d = Qc (m^3/d) × TKN (kg/m^3) = 281 × 0.21 = 59 kg/d

NH_4^+ rejection kg/d = 281 × 0.040 = 11.25

Nnit = TKN entry –N-NH_4 rejection = 59 − 11.25 = 47.75 kg/d

The daily oxygen demand corresponding to nitrification (DONit) [expressed in kg O_2/day] is written:

DONit = 4.2 × Nnit = 4.2 × 47.75 = 200.6 kg O_2/day (Nnit: quantity of Nitrogen in kg/day)

DONit= 200.6 kg of O_2/day

QO_2 (kg/d) = Volume of air × density of air × proportion of O_2 in air × transfer coefficient

The airflow rate to be blown is determined according to the following expression:

Air Volume = (O_2kg/d)/(Density × proportion of O_2 in air × O_2 transfer efficiency

Air Volume = (O_2kg/d)/(1.29 × 0.23 × 0.2) = 200.6/0.06 = 3342.5 m^3/d

*The energy required to dissolve O_2

1 kg of O_2 = 0.7 kWh i.e. Specific intake 1.43 kg of O_2/kWh

The total energy to dissolve O_2 is:

In total (kWh) = O_2 kg/d/1.43= 200.6/ 1.43= 140.3

Energy cost = 140.3 × 0.09 kWh = 12.6 Eur/day with kWh cost = 0.09 Euro

The ventilation power to be installed can be determined:

- By dividing the hourly input (kg O_2/h) by the gross specific input of the aerator (kg O_2/kWh) in the case of surface aerators:

P = AH/ASB = (kg O_2/h)/1.42 = 200.6/24 × 1.42 = 5.9 kWh we adopt P = 6 kWh with

P = power absorbed at the electric meter (kW) with ASB Gross specific contribution of the aerator: 1.42

The design method is the same as that used for Basin 1. Table 12.2 illustrates a summary of the design parameters for nitrification for Basin B3

Vair = (O_2kg/d)/(Density × proportion of O_2 in air × O_2 transfer efficiency)

Vair = (O_2kg/d)/(1.29 × 0.23 × 0.2) = 200.6/0.06= 3342.5 m^3/d

*The energy required to dissolve O2

1 kg of O_2 = 0.7 kWh i.e. Specific intake 1.43 kg of O_2/Kwh

The total energy to dissolve O_2 is:

In total (kWh) = O_2 kg/d/1.43= 200.6/1.43= 140.3

Energy cost = 140.3 × 0.09 kWh = 12.6 Eur/day with kWh cost = 0.09 Euro

The ventilation power to be installed can be determined:

- By dividing the hourly input (kg O_2/h) by the gross specific input of the aerator (kg O_2/kWh) in the case of surface aerators:

P = AH/ASB = (Kg O_2/h)/1.42 = 200.6/24 × 1.42 = 5.9 kWh we adopt P = 6 kWh with

P = power absorbed at the electric meter (kW) with ASB Gross specific contribution of the aerator: 1.42

The design method is the same as that used for Basin 1.

The design parameters are summarized in Table 12.6

Sizing of the anaerobic basin for denitrification

Circular shaped pool

Design flow = 281 m^3/d

Table 12.6 illustrates a summary of the design parameters for nitrification for Basin B3

Parameters	Calculated value
Flow m^3/d	281
Length of stay day	1
V useful m^3	281
Height m	2
Area m^2	140.5
Diameter m	10
O_2 kg of O_2/day	200.6
Air volume m^3/d	3342.5
Pump power kWh	5.9
Total useful energy kWh	140.3
kWh unit price in Euro	0.09
Energy cost Euro/d	12.6

Table 12.7 design parameter of the denitrification basin (Anaerobic basin)

Parameters	Calculated value
Flow m^3/d	281
Length of stay days	3
V useful m^3	843
Sludge produced l/m^3	120
Sludge produced m^3	33.7
Total volume of the basin m^3	876.7
Height m	4
Area m^2	220
Diameter m	17

Residence time: 3 days (80% yield)
Useful volume of the pool $= 281 \times 3 = 843\ m^3$
Sludge produced: 120 ml/l $= 120\ l/\ m^3$
Sludge volume: $0.12 \times 281 = 33.7\ m^3$
Total volume of the basin $= 843 + 33.7 = 876.7\ m^3$
Pool depth: 4 m
Surface area of the pool: Volume/height $= 876.7/4 = 219.2\ m^2$ we take Sh $= 220\ m^2$
Sh $=$ (II D2)/4 and D $=$ (4Sh/II)1/2D $= 16.75$ m we take D $= 17$ m
The results obtained for the design of the denitrification basin (anaerobic basin) are illustrated in Table 12.7.

Degassing basin (circular basin)
Let T be the hydraulic residence time of 10 min.

$$V = \text{Volume of the pool} = Qc \times Tsh = 281 \times 0.0,1666/24 = 2\ m^3 \text{ we opt for } V = 2\ m^3$$

$$\text{Area calculation Horizontal surface} = V/H = 2/1 = 2\ m^2$$

$$\text{Basin diameter} : D = (4Sh/II)1/2 = (4 \times 2/3.14)1/2 = 1.6m \text{ on opte pour } D = 2\ m$$

Table 12.8 illustrates the summary of the design parameters of a degassing basin.
Sizing of the primary decanter
We consider that the basin used is circular (Fig. 12.6).

- Calculation of pelvic volume.

Tdec= 2 h

$$V = Ts \times Qc = 2 \times 281/24 = 23.4\ m^3 \quad \text{We adopt Sh} = 24\ m^3$$

Table 12.8 design parameters of a degassing basin

Parameters	Calculated value
Flow m^3/d	281
Residence time in min	10
V useful m^3	2
Height m	1
Horizontal surface m^2	2
Diameter m	2

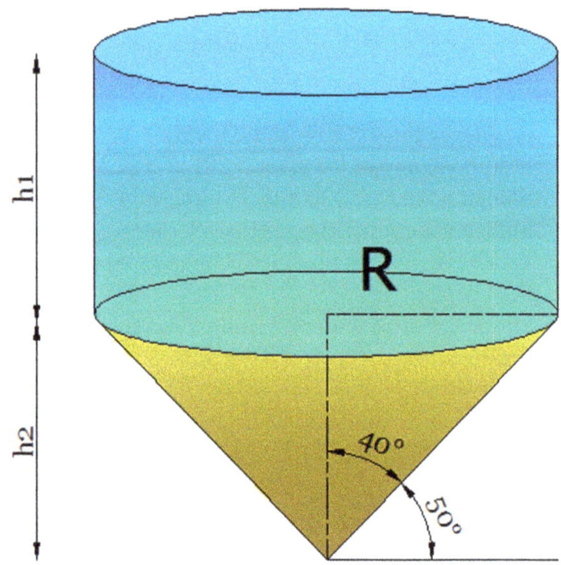

Fig. 12.6 Design of the primary decanter for leachate treatment (voir paragraphe)

$$\text{Pool height} = 4\ \text{m} = \text{H1}$$

- Calculation of the surface.

$$\text{V} = \text{Sh} \times \text{H and Sh} = \text{V/H} = 24/4 = 6\ \text{m}^2 \quad \text{Sh} = \text{ Horizontal Surface}$$

- Calculation of the pool diameter.

$$\text{Sh} = (\text{IID2})/4 \text{ and D} = (4\text{Sh/II})1/2\text{D} = 2.76\ \text{m} \quad \text{we opt for D} = 3\ \text{m}$$

$$\text{V corrected} = \text{Shc} \times \text{H} = 6 \times 4 = 24\text{m}^3$$

it is necessary to take into account the empty volume at the level of the decanter.

empty height = H2 = 0.3 m.

$$Vt = \text{Volume total} = Sh \times (H1 + H2) = 6 \times 4.3 = 25.6\ m^3 \text{ et } \quad \text{we opt } Vt = 26\ m^3$$

Calculation of the cylindrical height: $H = h1 + h2 = 4$ m

We have volume = $Sh1 + S(h2/3)$ with $tg40^\circ = R/h2$□ therefore $h2 = R/(tg40^\circ)$ which implies: volume = $Sh1 + S.R/(3.tg40^\circ)$ (S = Shhorizontal surface).

We therefore obtain the height $h1 = (\text{volume-S } R/(3tg40^\circ))/S = (24\text{-}6\ 1.5/0.84)/6 = 2.2$ m.

$$H = h1 + h2 \text{ and } h2 = H - h1 = 4 - 2.2 = 1.8 \text{ m}$$

Conical volume' = $(h2 \times Sh)/3 = (1.8 \times 6)/3 = 3.6\ m^3$

Cylindrical volume = $24 - 3.6 = 20.4\ m^3$

Length of the chute i.e. treated water outlet (Internal perimeter of the primary decanter):

$L = 2\ \text{II}\ (D/2 - b)$ We take $b = 30$ cm= 0.3 m

Perimeter = $L = 2 \times 3.14\ (1.5 - 0.3) = 7.5$ m

Conical volume = $1.8 \times 6 = 10.8\ m^2$ Cylindrical volume = $h1 \times 6 = 2.2 \times 6 = 13.2\ m^2$

The design parameters are summarized in Table 12.9.

Calculation of reagent costs for coagulation-flocculation

Basin 1

The optimal concentration for treatment: 7 kg/m^3

The optimal concentration for treatment: 7 kg/m^3

Quantity of $FeCl_3$ in kg used per day = $7 \times 281 = 1967$

Quantity of polymer consumed ($0.2\ l/m^3$) = $0.2 \times 281 = 56.2$ l Cost in Euro/l = 2.73

Table 12.9 Design parameters of a primary decanter

Parameters	Calculated value
Flow m^3/d	281
Residence time in hours	2
Conical volume m^3	3.6
Cylindrical volume m^3	20.4
Total Height m	4
Horizontal surface m^2	6
Diameter m	3
Trunking length m	7.5
h1 m	2.2
h2 m	1.8

Table 12.10 Summary of primary decanter design parameter

Parameters	Conception
Quantity of $FeCl_3$ Kg/d Basin 1	1967
Quantity of $FeCl_3$ Kg/d Basin 3	1545.5
Quantity of Flocculant l/d	53.2
Cost of Bassin1 reagents in Euro	684.4
Cost of Basin 3 reagents in Euro	570.6

Cost = 56.2 × 2.73 = 153.3 Euro/d Cost in Euro/Kg polymer = 2.73
(Optimal concentration of $FeCl_3$ 7g/l) = quantity consumed kg × Cost Euro/kg
7 × 281 × 0.27 = 531.1 Euro/day Cost in Euro/liter = 0.27
Total cost = 153.3 + 531.1 = 684.4 Euro/d

Basin 3

The optimal concentration for treatment: 5.5 kg/m^3
Quantity of $FeCl_3$ in Kg used per day = 5.5 × 281 = 1545.5
Quantity of polymer consumed (0.2 l/m^3) = 0.2 × 281 = 56.2 L
Cost = 56.2 × 2.73 = 153.3 Euro/d Cost in Euro/kg polymer = 2.73
(Optimal concentration of $FeCl_3$ 5.5 g/l) = quantity consumed kg × Cost 0.27 Euro/kg
5.5 × 281 × 0.27 = 417.3 Euro/day
Total cost = 153.3 + 417.3 = 570.6 Euro/d

The cost linked to reducing pollution by moving from basin 1 to basin 3 = 684.4–570.6 = 114 Euro/d.

which is the equivalent of 17% reduction in processing cost.

This shows that storage basins can significantly reduce the cost of treatment.

Table 12.10 gives a summary of the primary decanter design parameters.

Sizing of the filtration basin

Calculation of ascent speed, volume and residence time:
For a traditional filling material.
we set a height H = 3 m and D = 3 m.
Either:
Su: unit area
V: volume of the pelvis

$$Su = IID2/4 = (3.14 \times 9)/4 = 7\ m^2$$

$$V = Su \times H = 7 \times 3 = 21\ m^{3*}$$

taking into account the vacuum 10% of the filled volumen.
Total volumen = 21 + 21 × 0.01 = 23.1 m^3 we opt fo Total volume = 23 m^3

- Particle size of materials 100 μm.
- Density of materials: d = 0.95 g/ml

Table 12.11 Summary of filtration basin design parameters (flow calculated by 2030)

Parameters	Conception
Flow (m)	281
H (m)	3
Diameter (m)	7
Mass of filter material in tonnes	20
Particle seze of materials Granulométrie (μm)	100
S unitary (m^2)	7
Empty volume	2.1
Volume (m^3)	21
Total volume (m^3)	23.1
Hydraulic charge (m^3/h)	1.67

- Mass of red bricks: d × V = mass of the material = 0.95 × 21 = 20 tonnes
- Calculation of the Hydraulic Load: CH = flow at the inlet of the bed/Surface of the bed = $Qc/Su = 281/24 \times 7 = 1.67\ m^3/h$.

Table 12.11 summarizes the values of the filtration basin design parameters.

12.4 Conclusion

Leachate discharge is too loaded with polluting materials, which can considerably impact the sanitation network, treatment plants, soil, surface water, groundwater, air, and the marine environment. Indeed, the choice of treatment technique essentially depends on the quality and quantity of the products rejected at the public landfill. The different stages of the process, such as nitrification and denitrification, are justified by the significant presence of ammonium ions and nitrates, which must be reduced before the coagulation-flocculation stage. The latter could be attractive for the substantial reduction in pollution. Filtration on red bricks mixed with sand could finalize the treatment of leachate discharges and subsequently result in treated water meeting the standards to be used for irrigation. The results obtained in the laboratory on a pilot scale showed that the proposed steps are complementary and make it possible to reduce pollution effectively.

However, using the treatment plan to discharge leachate stored in basin 3 is more enjoyable, which presents a reduced pollutant load compared to the quality of raw leachate received by Basin 1. The sector's design based on the determined optimal conditions makes it possible to minimize pollution considerably. Furthermore, calculating the sector's costs remains moderate, which could encourage establishing the industry in developing countries.

References

1. Abouri M, Souabi S, Bahlaoui Fouad, MZ. Baudu M., Moharram,R, Pala,. A. Aerobic treatment of leachate from municipal solid waste in Morocco. In: Proceedings of the institution of civil engineers—waste and resource management, vol 169, pp 92–100
2. Abunama T, Othman F, Alslaibi T, Abualqumboz M (2017) Quantifying the generated and percolated leachate through a landfill's lining system in gaza strip, Palestine. Pol J Environ Stud 26:2455–2461
3. Anouzla A, Malika K, Azoulay, I, Bencheikh, Ghizlane F, Jamal et al (2022) Multi-response optimization of coagulation-flocculation process for stabilized landfill leachate treatment using a coagulant based on an industrial effluent. Desalin Water Treat 254:71–79. https://doi.org/10.5004/dwt.2022.28388
4. Anouzla A, Souabi S, Safi M, Younes, Abrouki Y, Loukili H, Rhbal H (2014) Waste to treat waste of landfill leachates. Int J Environ Prot Policy 2(2):50–. https://doi.org/10.11648/J.IJEPP.20140202.12
5. Arabi M, Sbaa M, Vanclooster M, Darmous A (2020) Impact of the municipal solid waste typology on leachate flow under semi-arid climate—a case study. J Ecol Eng 21:94–101
6. Aziz H, Adlan K, Amilin MN, Mohd Suffian Y, Ramly N, Umar M (2012) Quantification of leachate generation rate from a semi-aerobic landfill in Malaysia. Environ Eng Manag J 11:1581–1585
7. Bakraouy H, Souabi S, Digua K, Dkhissi O, Sabar M, Fadil M (2017) Optimization of the treatment of anaerobic pretreated landfill leachate by acoagulation–flocculation process using experimental design methodology. Process Saf Environ Prot 109:621–630
8. Chiguer H, Khayyat F, El rhaouat O, Rifki R, Bensaid A, Kharrim K, Belghyti D (2016) Evaluation de la charge polluante des lixiviats de la décharge contrôlée de la ville d'Essaouira (MAROC) [Evaluation of the pollution load of leachates from the landfill in the city Essaouira (MOROCCO)]. Int J Innovation Appl Stud 14:863–874 (in French)
9. Croitoru L, Sarraf M (2017) Le coût de la dégradation de l'environnement au Maroc. World Bank
10. Gamar A, Zair T, El Kabriti M, El Hilali F (2018) Study of the impact of the wild dump leachates of the region of El Hajeb (Morocco) on the physicochemical quality of the adjacent water table. Karbala Int J Mod Sci 4 382–392
11. Guaadaoui A, El Yadini, Chiat MK, Jdaini T, Hajjaji SE (2021) Preserving the environment and establishing sustainable development: an overview on the moroccan model. In: E3S Web of conferences, vol 234, p 00065
12. Gupta A, Gaharwar US, Verma A, Rajamani P (2022) Municipal solid waste landfill leachate induced cytotoxicity in root tips of Vicia faba: environmental risk posed by non-engineered landfill. IJBB 59(11):1113–1125
13. Gupta A, Paulraj R (2017) Leachate composition and toxicity assessment: an integrated approach correlating physicochemical parameters and toxicity of leachatesfrom MSW landfill in Delhi. Environ Technol 38(13–14):1599–1605
14. Gupta A, Rajamani P (2015) Toxicity assessment of municipal solid waste landfill leachate collected in different seasons from Okhala landfill site of Delhi. J Biomed Sci Eng 8(06):357
15. Gupta A, Rajamani P (2016) Evaluation of the leachate composition and contamination potential of municipal solid waste landfill sites in Delhi. Int J Environ Waste Manag 18(4):285–302
16. Gupta P, Bhargava A, Kumari R, Lodhi L, Tiwari R, Gupta PK, Bunkar N, Samarth R, Mishra PK (2019) Impairment of mitochondrial-nuclear cross talk in lymphocytes exposed to landfill leachate. Environ Health Insights. 13.1178630219839013
17. Hafidi M (2015) L'impact et la gestion des déchets solides (Région Marrakech-Safi). ISBN: 978-9954-9528-7-0
18. Kastali M, Mouhir L, Chatoui M, Souabi S, Anouzla A (2021) Removal of turbidity and sludge production from industrial process wastewater treatment by a rejection of steel rich in $FeCl_3$ (SIWW). B Res Appl Chem. 11:13359–13376

19. El Kharmouz M, Sbaa M, Abdelhafid C, Saadi S (2013) L'étude de l'impact des lixiviats de l'ancienne décharge publique de la ville d'Oujda (Maroc oriental) sur la qualiré pysicochimique des eaux souterraines et superficielles. Larhyss J 16:105–119
20. Kjeldsen P, Barlaz MA, Rooker AP, Baun A, Ledin A (2002) Present and long-term composition of msw landfill leachate: a review. Crit Rev Solid State Mater Sci 32(4):297–336
21. Mahmud K, Hossain MD, Shams S (2012) Different treatment strategies for highly polluted landfill leachate in developing countries. Waste Manage 32(11):2096–2105
22. Naveen BP, Mahapatra DM, Sitharam TG, Sivapullaiah PV, Ramachandra TV (2017) Physico-chemical and biological characterization of urban municipal landfill leachate. Environ Pollut 220:1–12
23. Renou S, Givaudan JG, Poulain S, Dirassouyan F, Moulin P (2008) Landfill leachate treatment: review and opportunity. J Hazard Mater 150:468–493
24. Silva AC, Dezotti M, Sant'Anna GL (2004) Treatment and detoxification of sanitary landfill leachate. Chemosphere 55:207–214
25. Soudi B, Chrifi H (2007) Options de gestion des déchets solides municipaux adaptées aux contextes des Pays du Sud, Enda Maghreb. ISBN: 9954-444-19-X (In French)

Chapter 13
Leachate Management in Rio De Janeiro: A 2022 Overview

Augusto Ahn Ka, Rafaela Naegele, Pablo Vimercati, Dalton Domingues de Carvalho Neto, Ricardo Soares, and Carlos Eduardo Soares Canejo Pinheiro da Cunha

Abstract While meeting environmental waste disposal requirements, landfills introduce new environmental and public health hazards, including potential operational lapses and conflicts with local communities. 92 municipalities in Rio de Janeiro send waste to 19 sanitary landfills, accounting for 96.94% of the total daily MSW generation in the state. The 19 licensed and operational landfills in Rio de Janeiro are overseen by 16 operators: 10 private entities, two inter-municipal consortia, and four municipal governments. Private operators manage approximately 17,061 tons/day, consortia around 598 tons/day, and municipalities around 259 tons/day. Ciclus Ambiental oversees 8933 tons/day, and Orizon Valorizacao de Residuos manages 6094 tons/day, collectively managing around 15,027 tons/day, representing roughly 80% of the daily waste generation in Rio de Janeiro.

13.1 Introduction

In October of 2010, Statute 12.305 received royal assent, thereby instituting the National Solid Waste Policy (NSWP) within Brazil. Consequently, multifarious governmental echelons embarked upon a meticulous journey of political, technical, and legal orchestration to ameliorate the prevailing paradigm of Municipal Solid Waste (MSW) stewardship within the nation, which had hitherto been marred by the suboptimal culmination of waste in dumps.

Regrettably, notwithstanding concerted endeavours from both public and private sectors, extant dumps persist in every Brazilian state, operating devoid of oversight. This results in deleterious alterations to environmental quality via soil, groundwater, and surface water contamination, whilst simultaneously fostering vector proliferation and precipitating maladies. It is lamentable that vestiges of medieval epochs

A. A. Ka · R. Naegele · P. Vimercati · D. D. de Carvalho Neto · R. Soares (✉) ·
C. E. S. C. P. da Cunha
Solid Waste Treatment, Destination and Final Disposal Research Group at Veiga de Almeida University, Rio de Janeiro, Brazil
e-mail: ricardo.soares@uva.br

A. Anouzla and S. Souabi (eds.), *A Review of Landfill Leachate*, Springer Water,
https://doi.org/10.1007/978-3-031-55513-8_13

still permeate the nation's waste management protocols. Such a culture, predicated upon habitual environmental disregard, obfuscates the requisite ecological foresight imperative for maintaining environmental equilibrium. The imperative metamorphosis in this regard presents novel conundrums for both public and private custodians, as well as the populace at large. The quandary transcends the mere cessation of dumps; it encompasses their rectification and the proposition of fitting disposal stratagems. Landfills, whilst satisfying the prerequisites for environmentally congruous waste disposal, inadvertently epitomise a nascent challenge. The expansion of these disposal vicinities, even when constructed in strict adherence to legal and regulatory edicts, introduces fresh environmental and public health hazards, particularly stemming from potential operational lapses and discord with local communities.

Landfills represent engineered edifices for MSW disposal, meticulously designed and operated to attenuate health and environmental repercussions. As delineated by the NBR 8.419 Brazilian regulation [2], it is a modality for terrestrial solid urban waste disposal, executed without compromising public health and safety, whilst assiduously minimising environmental perturbations. This *modus operandi* harnesses engineering tenets to circumscribe solid waste within the most diminutive feasible expanse, subsequently reducing it to the least permissible volume, and subsequently enveloping it with a stratum of soil post each operational day, or at more frequent junctures if exigent. Tchobanoglous [6] posits that landfills can be analogised to biochemical reactors. The primary inputs encompass the disposed waste and pluvial water; the outputs are the leachate and biogas engendered during the biodegradation of organic constituents and waste leaching. The residuum of this process comprises biodegraded (stable) organic material and inorganic constituents. Landfilling remains the quintessential culmination technique for MSW, even in nations boasting alternative treatment and recuperation methodologies, such as energy-recovery incineration, composting, and recycling.

Vilhena [7] characterises leachate as a fluid of capricious composition that assumes pollutant attributes upon contact with decomposing solid waste. The NBR 8.419 Brazilian regulation [2] elucidates this term as the liquid engendered by the decomposition of substances within solid waste, discernible by its sombre hue, malodorous scent, and elevated biochemical oxygen demand (BOD). The same statute elucidates "leaching" as the liquid-mediated displacement of specific waste-contained substances. It is pertinent to note that leachate is also denominated as wastewater or percolate. Cunha [5] identifies several determinants influencing landfill leachate's genesis and physicochemical attributes, with climate, solid waste, and disposal being paramount.

However, the pivotal role of water in this matrix is incontrovertible. It is via the precipitated water's heightened mobility that solid waste constituents are "rinsed" and solubilised, rendering the resultant leachate's chemical amalgamation exceedingly diverse. Despite assiduous efforts to curtail infiltration, it remains technically implausible to prevent a significant quantum of precipitated water from permeating and potentially engendering leachate. It is salient to acknowledge that even if pluvial water were to remain extraneous to the massif, the routine modus operandi of landfills would inherently produce leachate, attributable to inherent waste humidity, biodecomposition-generated water, and potential subterranean overpressure, inter alia. Moreover, the characteristics undergo temporal evolution, as the landfill's milieu transitions from aerobic to anaerobic. Thus, leachate's composition is inherently mutable, contingent upon the waste's decomposition phase within the landfill. International scholarly literature demarcates this process into three distinct epochs: aerobic, acetogenic, and methanogenic.

Leachate may encompass noxious compounds, including ammonia and heavy metals. Pertaining to nitrogen concentrations, it invariably manifests concentrations of ammoniacal nitrogen. It is noteworthy that in landfills situated within tropical nations, the transition from nascent to mature leachate transpires approximately two years post-initiation, as opposed to the quintet of years observed in temperate climes, attributable to elevated temperatures and intensified precipitation.

Thus, it is cogently inferred that the surveillance of leachate generation and treatment is an indispensable managerial and planning instrument for Brazilian municipalities and states, facilitating the discernment of critical vicinities and operations, in tandem with treatment technology trends. In light of the aforementioned, this scholarly endeavour endeavours to proffer a diagnostic assessment of leachate generation and treatment within the landfills of Rio de Janeiro in 2022, thereby augmenting the dissemination and transparency of pivotal data for the waste sector.

13.1.1 General Objective

A Diagnostic Assessment of Landfill Leachate Generation and Treatment in Rio de Janeiro: The 2022 Landscape.

13.1.2 Specific Objectives

- Ascertain the volume of leachate engendered and the treatment modalities adopted by extant landfills in Rio de Janeiro.
- Chart the diverse leachate treatment technologies employed by MSW disposal locales in Rio de Janeiro.
- Compile an inventory delincating leachate storage within accumulation ponds in Rio de Janeiro for the year 2022.

13.2 Methodology

The inaugural phase of this scholarly endeavour entailed a comprehensive bibliographic review, with the intent of establishing a robust theoretical foundation. This was geared towards discerning both national and international paradigms, legislations, standards, guidelines, and procedures pertinent to waste disposal. The descriptors employed to harmonise synonyms during the data retrieval process included: 'Landfills', 'Controlled Landfills', 'Dumps', 'State of Rio de Janeiro', and 'Sanitary Landfill Leachate)'. Predominant databases consulted encompassed SciELO (Scientific Electronic Library Online), ERIC (Educational Resources Information Center), *Portal Periódicos Capes*, and Google Scholar. The subsequent phase involved assimilating the insights gleaned from the aforementioned review, collating the references, critically dissecting the information, and delving into the environmental licensing procedures for various leachate treatment modalities across the 92 municipalities in Rio de Janeiro with the State Environmental Institute of Rio de Janeiro (INEA). The third phase was dedicated to pinpointing geographical coordinates and charting the identified solutions using the complimentary Google Earth software. The penultimate phase involved ascertaining the areas and storage capacities of the leachate storage ponds via the measurement apparatus within the Google Earth software. The terminal phase culminated in the formulation of a Database to Oversee the Generation and Treatment of Leachate in the State (elaborated in the annex). It is of paramount importance to note that the INEA research encompassed a plethora of public documents contained within the environmental licensing procedures of both public and private entities, either granted or contracted to oversee landfills. Documents scrutinised within each procedure included:

(1) Inspection reports;
(2) Notifications;
(3) Notices of Findings;
(4) Technical Opinions; and
(5) Technical Reports. The Excel software was employed to collate the data and engender the envisaged outcomes.

13.3 Results

During the period under scrutiny, 92 municipalities within Rio de Janeiro dispatched waste to 19 sanitary landfills, licenced to accommodate MSW, either in isolation or in conjunction with Class II Industrial Waste. Collectively, these 76 municipalities produce and dispatch approximately 18,123 tons/day to landfills, accounting for 96.94% of the total daily MSW generation within the state. Of the residual municipalities, six (inclusive Varre-Sai, Santo Antônio de Pádua, Carmo, Duas Barras, Sumidouro, and São José do Vale do Rio Preto) dispatch their generated MSW (98 t/day) to landfill operators in Minas Gerais State, whilst ten others (including Porciúncula, Natividade, Itaperuna, Bom Jesus do Itabapoana, Italva, Cambuci, Cordeiro,

Teresópolis, São Fidélis, and Resende) persist in utilising dumps as the terminal disposal mechanism for their generated MSW (472 t/day).

According to the data collated in this diagnostic assessment, the 19 licenced and operational landfills within Rio de Janeiro are overseen by 16 operators: 10 private entities, 2 inter-municipal consortia, and 4 municipal governments. The private operators manage approximately 17,061 tons/day, the consortia around 598 tons/day, and the municipalities approximately 259 tons/day. It is salient to underscore the contributions of Ciclus Ambiental, overseeing 8933 t/day, and Orizon Valorização de Resíduos, managing 6094 t/day. In tandem, these entities manage approximately 15,027 t/day, representing roughly 80% of ERJ's daily generation.

Of the landfills identified in this study, 13 have internal leachate treatment systems, processing approximately 4262 m^3/day, whilst six dispatch their leachate for external treatment in ETEs and ETDIs authorised by INEA, totalling approximately 282 m^3/day. The findings indicate a predominant trend towards biological systems followed by membrane systems for liquid phase treatment. However, landfills employing geotextile systems, physical–chemical systems followed by membrane systems, and exclusive membrane systems were also identified. It is imperative to note that the primary objective of this study was not to evaluate the efficacy of the various technologies implemented within landfills, but rather to discern the diverse technological pathways.

From this study, 106 stabilisation/aerated lagoons (preliminary treatment) were identified, spanning an area of 266,279 m^2 with a stored volume of approximately 900,626 m^3. The number of lagoons and the leachate storage areas were ascertained using the editing tools within the Google Earth PRO software. According to Jordão and Pessoa (Jordão and Pessoa 2011), stabilisation ponds typically possess depths ranging between 1.0 and 2.0 m, whilst aerated ponds have depths between 3.0 and 4.0 m. Thus, an average depth of 3.0 m was employed as the technical criterion for calculating the reported volume. The table in Annex I consolidates the data pertaining to each assessed landfill.

13.4 Conclusions

Post this scholarly endeavour, it was discerned that amongst the 19 operational landfills in Rio de Janeiro, 13 possess internal leachate treatment systems, whilst six dispatch their leachate for external treatment in ETEs and ETDIs authorised by the pertinent environmental bodies. Biological systems, succeeded by membrane systems, emerge as the predominant trend in liquid-phase treatment. However, it is of paramount importance to underscore that, as of the current scenario (October 2022), there exist 106 stabilisation/aerated lagoons (preliminary treatment) within the state's landfills, spanning an area of 266,279 m^2 and storing a volume of approximately 900,626 m^3. Given the aforementioned, the prevailing scenario is deemed disconcerting due to the diminished technological rigour in lagoon construction compared to base waterproofing systems, coupled with the proliferation of lagoons devoid of oversight by the competent environmental bodies. The excess of lagoons indicates a technical deficiency in the treatments implemented, and there exists a palpable risk of exacerbating the situation during periods of intense precipitation (spanning November to March).

Acknowledgements Our profound gratitude is extended to the State Environment Institute of Rio de Janeiro (INEA).

Annex: Leachate Storage Volume Calculations for ERJ Landfills

Final disposal	Amount of leachate storage ponds	Total area of leachate storage at the landfill (m^2)	Estimated volume of leachate stored in the landfill (m^3) (Calculate $h = 3.0$ m)
Seropédica landfill (CTR RIO)	24	124,720.31	374,160.93
Nova Iguaçu landfill	6	6542.09	19,626.27
São Gonçalo landfill	11	45,655.92	136,967.76
Itaboraí landfill	14	31,331.00	93,993.00
São Pedro da Aldeia landfill	7	13,936.00	41,808.00
Campos dos Goytacazes Landfill	2	3300.00	9900.00
Barra Mansa landfill	9	22,655.00	169,754.00
Paracambi landfill	7	2524.00	7572.00
Três Rios landfill	6	716.10	2148.30
Macaé landfill (Zadar)	6	9269.00	27,807.00
Vassouras landfill	2	1583.00	4749.00
Angra dos Reis Landfill	2	105.00	315.00
Nova Friburgo landfill	2	932.00	2796.00
Santa Maria Madalena landfill	2	605.00	1815.00
Macaé landfill (Essencis)	2	1263.00	3789.00
Sapucaia landfill	1	332.00	996.00
Macuco landfill	2	88.90	266.70

(continued)

(continued)

Final disposal	Amount of leachate storage ponds	Total area of leachate storage at the landfill (m^2)	Estimated volume of leachate stored in the landfill (m^3) (Calculate h = 3.0 m)
Niterói landfill	0	0.00	0.00
Saquarema landfill	1	721.00	2163.00
Total	106	266,279.32	900,626.96

References

1. ASSOCIAÇÃO BRASILEIRA DE NORMAS TÉCNICAS—ABNT (1985) NBR 8.849 Apresentação de Projetos de Aterros Controlados de Resíduos Sólidos Urbanos: classificação. Rio de Janeiro
2. ASSOCIAÇÃO BRASILEIRA DE NORMAS TÉCNICAS—ABNT (1992) NBR 8.419: Apresentação de projetos de aterros sanitários de resíduos sólidos urbanos: procedimento. Rio de Janeiro
3. Brasil. Lei 12.305 (2010) Institui a Política Nacional de Resíduos Sólidos de 02 de agosto de 2010; decreto n°7.404, de 23 de dezembro de 2010. Brasília
4. Cunha CESCP (2022) Gestão Integrada de Resíduos Sólidos: Múltiplas perspectivas para um gerenciamento sustentável e circular. Editora Freitas Bastos, 123p
5. Cunha CESCP (2009) Gestão de resíduos perigosos em refinarias de petróleo. 128 p. Dissertação (Mestrado em Engenharia Sanitária e Ambiental)Programa de Pós-Graduação em Engenharia Ambiental (PEAMB), Universidade do Estado do Rio de Janeiro (UERJ), Rio de Janeiro
6. Tchobanoglous G, Kreith F (2002) Handbook of solid waste management. 2nd ed. McGRAW-HILL Companies, Califórnia, 834 p
7. Vilhena A (2010) Lixo Municipal: Manual de gerenciamento integrado. 3rd ed. CEMPRE, São Paulo, 350 p

GPSR Compliance
The European Union's (EU) General Product Safety Regulation (GPSR) is a set of rules that requires consumer products to be safe and our obligations to ensure this.

If you have any concerns about our products, you can contact us on

ProductSafety@springernature.com

In case Publisher is established outside the EU, the EU authorized representative is:

Springer Nature Customer Service Center GmbH
Europaplatz 3
69115 Heidelberg, Germany

www.ingramcontent.com/pod-product-compliance
Ingram Content Group UK Ltd.
Pitfield, Milton Keynes, MK11 3LW, UK
UKHW022212051026
15624UKWH00001B/127
9783031555121